Essential Calculus-based

PHYSICS

Study Guide Workbook

Volume 1: The Laws of Motion

$$y_{cm} = \frac{1}{m}\int y\,dm$$

$$m = \int dm = \int \sigma\,dA = \frac{\sigma\pi R^2}{2}$$

$$y_{cm} = \frac{\sigma}{m}\int_{r=0}^{R}\int_{\theta=0}^{\pi}(r\sin\theta)\,r\,dr\,d\theta$$

$$dm = \sigma dA$$

$$y_{cm} = \frac{\sigma}{m}\int_{r=0}^{R} r^2[-\cos\pi - (-\cos 0)]\,dr$$

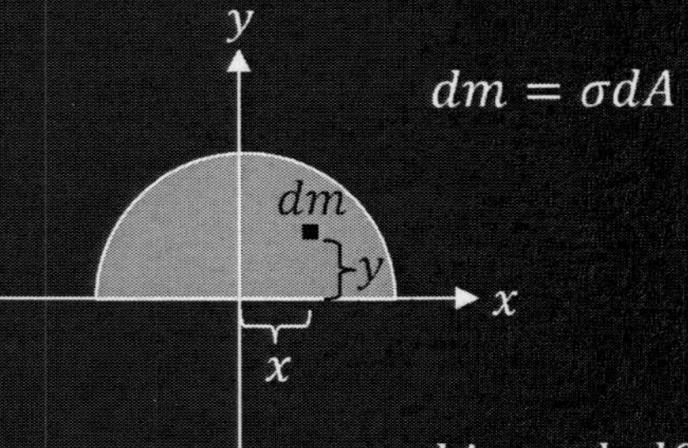

$$y_{cm} = \frac{2\sigma}{m}\int_{r=0}^{R} r^2\,dr = \frac{2\sigma R^3}{3m}$$

$$dA = rdrd\theta$$

$$y_{cm} = \frac{2\sigma R^3}{3}\frac{2}{\sigma\pi R^2} = \frac{4R}{3\pi}$$

$$y = r\sin\theta$$

Chris McMullen, Ph.D.

Essential Calculus-based Physics Study Guide Workbook
Volume 1: The Laws of Motion
Learn Physics with Calculus Step-by-Step

Chris McMullen, Ph.D.
Physics Instructor
Northwestern State University of Louisiana

Updated edition: February, 2017

www.monkeyphysicsblog.wordpress.com
www.improveyourmathfluency.com
www.chrismcmullen.wordpress.com

Zishka Publishing

ISBN: 978-1-941691-15-1

Textbooks > Science > Physics
Study Guides > Workbooks> Science

CONTENTS

Introduction 5
Chapter 1 – Algebra Essentials 7
Chapter 2 – Calculus Essentials 21
Chapter 3 – One-dimensional Uniform Acceleration 35
Chapter 4 – One-dimensional Motion with Calculus 47
Chapter 5 – Geometry Essentials 53
Chapter 6 – Motion Graphs 61
Chapter 7 – Two Objects in Motion 69
Chapter 8 – Net and Average Values 75
Chapter 9 – Trigonometry Essentials 83
Chapter 10 – Vector Addition 95
Chapter 11 – Projectile Motion 109
Chapter 12 – Two-dimensional Motion with Calculus 117
Chapter 13 – Newton's Laws of Motion 123
Chapter 14 – Applications of Newton's Second Law 129
Chapter 15 – Hooke's Law 143
Chapter 16 – Uniform Circular Motion 149
Chapter 17 – Uniform Circular Motion with Newton's Second Law 157
Chapter 18 – Newton's Law of Gravity 165
Chapter 19 – Satellite Motion 175
Chapter 20 – The Scalar Product 181
Chapter 21 – Work and Power 185
Chapter 22 – Conservation of Energy 205
Chapter 23 – One-dimensional Collisions 229
Chapter 24 – Two-dimensional Collisions 243
Chapter 25 – Rocket Propulsion 251
Chapter 26 – Techniques of Integration and Coordinate Systems 259
Chapter 27 – Center of Mass 279
Chapter 28 – Uniform Angular Acceleration 299
Chapter 29 – The Vector Product 305
Chapter 30 – Torque 311
Chapter 31 – Static Equilibrium 319
Chapter 32 – Moment of Inertia 331
Chapter 33 – A Pulley Rotating without Slipping 355
Chapter 34 – Rolling without Slipping 361
Chapter 35 – Conservation of Angular Momentum 369
Hints, Intermediate Answers, and Explanations 377

EXPECTATIONS

Prerequisites:

- The student should have learned basic calculus skills, including how to find the derivative of a polynomial or how to integrate over a polynomial. This book reviews these calculus skills in Chapter 2. **Additional calculus skills are taught as needed.**
- The student should know some basic algebra skills, such as how to solve for x by combining like terms and isolating the unknown. This book reviews the quadratic equation, the method of substitution, and other algebra skills in Chapter 1.
- The student should have prior exposure to trigonometry. This book reviews trig skills which are essential to physics in Chapter 9.
- The student should be familiar with how to find a few geometric measures, like the area of a circle or the perimeter of a rectangle. This book reviews useful geometric formulas in Chapter 5.
- **No prior physics knowledge is needed to use this book.**

Use:

- This book is intended to serve as a supplement for students who are attending physics lectures, reading a physics textbook, or reviewing physics fundamentals.
- The goal is to help students quickly find the most essential material.

Concepts:

- Each chapter reviews relevant definitions, concepts, laws, or equations needed to understand how to solve the problems.
- This book does not provide a comprehensive review of every concept from physics, but does cover most physics concepts that are involved in solving problems.

Strategies:

- Each chapter describes the problem-solving strategy needed to solve the problems at the end of the chapter.
- This book covers the kinds of fundamental problems which are commonly found in standard physics textbooks.

Help:

- Every chapter includes representative examples with step-by-step solutions and explanations. These examples should serve as a guide to help students solve similar problems at the end of each chapter.
- Each problem includes the main answer(s) on the same page as the question. At the **back of the book**, you can find hints, intermediate answers, directions to help walk you through the steps of each solution, and explanations regarding common issues that students encounter when solving the problems. It's very much like having your own **physics tutor** at the back of the book to help you solve each problem.

INTRODUCTION

The goal of this study guide workbook is to provide practice and help carrying out essential problem-solving strategies that are standard in first-semester physics. The aim here is not to overwhelm the student with comprehensive coverage of every type of problem, but to focus on the main strategies and techniques with which most physics students struggle.

This workbook is not intended to serve as a substitute for lectures or for a textbook, but is rather intended to serve as a valuable supplement. Each chapter includes a concise review of the essential information, a handy outline of the problem-solving strategies, and examples which show step-by-step how to carry out the procedure. This is not intended to teach the material, but is designed to serve as a time-saving review for students who have already been exposed to the material in class or in a textbook. Students who would like more examples or a more thorough introduction to the material should review their lecture notes or read their textbooks.

Every exercise in this study guide workbook applies the same strategy which is solved step-by-step in at least one example within the chapter. Study the examples and then follow them closely in order to complete the exercises. Many of the exercises are broken down into parts to help guide the student through the exercises. Each exercise tabulates the corresponding answers on the same page. Students can find additional help in the hints section at the back of the book, which provides hints, answers to intermediate steps, directions to walk students through every solution, and explanations regarding issues that students commonly ask about.

Every problem in this book can be solved without the aid of a calculator. You may use a calculator if you wish, though it is a valuable skill to be able to perform basic math without relying on a calculator.

The mathematics is so beautiful that the universe had to exist in order to represent it.

— Chris McMullen, Ph.D.

1 ALGEBRA ESSENTIALS

Students who struggle with calculus often aren't fluent in fundamental algebra skills.

Relevant Terminology

In the expression x^n, the quantity n can be referred to either as the **power** or as the **exponent** of the quantity x, while the quantity x is called the **base**.

Essential Concepts

The following rules from algebra are used frequently in calculus.

- When multiplying two different powers of the same base, add the powers:
$$x^m x^n = x^{m+n}$$
- When dividing two different powers of the same base, subtract the powers:
$$\frac{x^m}{x^n} = x^{m-n}$$
- A negative power means to find the reciprocal (or multiplicative inverse).
$$x^{-1} = \frac{1}{x} \quad , \quad x^{-m} = \frac{1}{x^m}$$
- A negative power in the denominator moves the quantity to the numerator.
$$\frac{1}{x^{-1}} = x \quad , \quad \frac{1}{x^{-m}} = x^m$$
- A power of one is implied when no power is shown.
$$x^1 = x$$
- Anything (except zero) raised to the power of zero equals one.
$$x^0 = 1 \qquad (x \neq 0)$$
- When an expression in parentheses is raised to a power, everything inside the parentheses is affected.
$$(ax)^m = a^m x^m \quad , \quad (x^m)^n = x^{mn} \quad , \quad (a^m x^n)^p = a^{mp} x^{np}$$
$$\left(\frac{a}{x}\right)^m = \frac{a^m}{x^m} \quad , \quad \left(\frac{1}{x^m}\right)^n = \frac{1}{x^{mn}} = x^{-mn} \quad , \quad \left(\frac{a^m}{x^n}\right)^p = \frac{a^{mp}}{x^{np}} = a^{mp} x^{-np}$$
- An exponent of one-half makes a squareroot.
$$x^{1/2} = \sqrt{x} \quad , \quad x^{-1/2} = \frac{1}{x^{1/2}} = \frac{1}{\sqrt{x}} \quad , \quad (ax)^{1/2} = \sqrt{ax} = \sqrt{a}\sqrt{x} \quad , \quad (ax)^{-1/2} = \frac{1}{\sqrt{ax}}$$
- Fractional exponents involve roots.
$$x^{m/n} = \sqrt[n]{x^m} = \left(\sqrt[n]{x}\right)^m$$
- A squareroot has multiple solutions. **<u>Don't</u>** disregard negative roots in physics.
$$\sqrt{x^2} = \pm x$$

- When a squareroot multiplies itself, the squareroot is effectively removed:
$$\sqrt{x}\sqrt{x} = x$$
This is because $\sqrt{x}\sqrt{x} = x^{1/2}x^{1/2} = x^{1/2+1/2} = x^1 = x$.
- To add or subtract fractions, first find a **common denominator**.
$$\frac{1}{x} + \frac{1}{y} = \frac{1}{x}\frac{y}{y} + \frac{1}{y}\frac{x}{x} = \frac{y}{xy} + \frac{x}{xy} = \frac{y+x}{xy}$$
Following is an example:
$$\frac{1}{2} + \frac{1}{3} = \frac{1}{2}\frac{3}{3} + \frac{1}{3}\frac{2}{2} = \frac{3}{6} + \frac{2}{6} = \frac{3+2}{6} = \frac{5}{6}$$
- To find the **reciprocal** of a fraction, swap the numerator and the denominator.
$$\left(\frac{x}{y}\right)^{-1} = \frac{y}{x}$$
Following are a few examples of finding the reciprocal (or inverse) of a fraction:
$$\left(\frac{3}{4}\right)^{-1} = \frac{4}{3} \quad , \quad \left(\frac{1}{5}\right)^{-1} = \frac{5}{1} = 5 \quad , \quad 6^{-1} = \frac{1}{6}$$
- To reduce a fraction, divide both the numerator and denominator by the greatest common factor. In the following example, the greatest common factor is 4 (meaning that 4 is the largest integer that evenly divides into both the numerator and the denominator):
$$\frac{16}{20} = \frac{16 \div 4}{20 \div 4} = \frac{4}{5}$$
- To multiply fractions, simply multiply their numerators and their denominators.
$$\frac{a}{b} \times \frac{c}{d} = \frac{ac}{bd}$$
Following is an example:
$$\frac{2}{3} \times \frac{6}{7} = \frac{2 \times 6}{3 \times 7} = \frac{12}{21} = \frac{12 \div 3}{21 \div 3} = \frac{4}{7}$$
The fraction $\frac{12}{21}$ was reduced by dividing the numerator and denominator both by 3.
- To divide fractions, **multiply by the reciprocal**.
$$\frac{a}{b} \div \frac{c}{d} = \frac{a}{b} \times \frac{d}{c} = \frac{ad}{bc}$$
Following is an example:
$$\frac{3}{4} \div \frac{1}{2} = \frac{3}{4} \times \frac{2}{1} = \frac{3 \times 2}{4 \times 1} = \frac{6}{4} = \frac{6 \div 2}{4 \div 2} = \frac{3}{2}$$
The fraction $\frac{6}{4}$ was reduced by dividing the numerator and denominator both by 2.
- In order to express your answer in standard form, rationalize the denominator. In the example below, multiply both the numerator and denominator by $\sqrt{x}$ and use the rule $\sqrt{x}\sqrt{x} = x$.
$$\frac{1}{\sqrt{x}} = \frac{1}{\sqrt{x}}\frac{\sqrt{x}}{\sqrt{x}} = \frac{\sqrt{x}}{x}$$

Example: Simplify the expression $x^4 x$.
Apply the rule $x^m x^n = x^{m+n}$ with $m = 4$ and $n = 1$ (since $x^1 = x$).

$$x^4 x = x^4 x^1 = x^{4+1} = x^5$$

Example: Evaluate the expression $\frac{10^{-10}10^{24}}{(10^6)^2}$ without using a calculator.

Apply the rule $x^m x^n = x^{m+n}$, the rule $\frac{x^m}{x^n} = x^{m-n}$, and the rule $(x^m)^n = x^{mn}$.

$$\frac{10^{-10}10^{24}}{(10^6)^2} = \frac{10^{-10+24}}{10^{6\times 2}} = \frac{10^{14}}{10^{12}} = 10^{14-12} = 10^2 = 10 \times 10 = 100$$

Example: Simplify the expression $\frac{x}{\sqrt{x}}$.

There is more than one way to solve this problem. One way is to recognize that $x = x^1$ and $\sqrt{x} = x^{1/2}$ and apply the rule $\frac{x^m}{x^n} = x^{m-n}$.

$$\frac{x}{\sqrt{x}} = \frac{x^1}{x^{1/2}} = x^{1-1/2} = x^{1/2} = \sqrt{x}$$

A simpler way is to note that $\sqrt{x}\sqrt{x} = x$ and therefore $x \div \sqrt{x} = \sqrt{x}$. (Note that $\frac{x}{\sqrt{x}}$ is the same as $x \div \sqrt{x}$).

Example: Express $(3x)^{-1/2}$ in standard form.

Apply the rule $(ax)^{-1/2} = \frac{1}{\sqrt{ax}}$.

$$(3x)^{-1/2} = \frac{1}{\sqrt{3x}}$$

To put this in standard form, we must rationalize the denominator. Multiply the numerator and denominator both by $\sqrt{3x}$ and note that $\sqrt{3x}\sqrt{3x} = 3x$.

$$(3x)^{-1/2} = \frac{1}{\sqrt{3x}}\frac{\sqrt{3x}}{\sqrt{3x}} = \frac{\sqrt{3x}}{3x}$$

Example: Add the following fractions: $\frac{1}{\sqrt{x}} + \frac{1}{x}$.

To add fractions, first find a common denominator.

$$\frac{1}{\sqrt{x}} + \frac{1}{x} = \frac{1}{\sqrt{x}}\frac{\sqrt{x}}{\sqrt{x}} + \frac{1}{x} = \frac{\sqrt{x}}{x} + \frac{1}{x} = \frac{\sqrt{x}+1}{x}$$

Example: What is the reciprocal of $\sqrt{x}$?

Apply the rule $x^{1/2} = \sqrt{x}$. Negate the power to find the reciprocal. The reciprocal of $\sqrt{x}$ is $x^{-1/2}$, which equals $\frac{1}{\sqrt{x}}$. Rationalize the denominator: $x^{-1/2} = \frac{1}{\sqrt{x}} = \frac{1}{\sqrt{x}}\frac{\sqrt{x}}{\sqrt{x}} = \frac{\sqrt{x}}{x}$. The reciprocal of $\sqrt{x}$ is $\frac{\sqrt{x}}{x}$.

1. Simplify the following expression.

$x^2x^3x^4 =$

Answer: x^9

2. Simplify the following expression.

$\frac{x^4x^5}{x^6} =$

Answer: x^3

3. Simplify the following expression.

$\frac{x^{-7}}{x^{-8}} =$

Answer: x

4. Simplify the following expression.

$\frac{x^3x^{-6}}{x^{-2}x^5} =$

Answer: x^{-6}

5. Simplify the following expression.

$x\sqrt{x} =$

Want help? Check the hints section at the back of the book.

Answer: $x^{3/2}$

6. Simplify the following expression.

$$\frac{x^{3/2}}{\sqrt{x}} =$$

Answer: x

7. Express $(4x)^{-1/2}$ in standard form.

Answer: $\frac{\sqrt{x}}{2x}$

8. Subtract the following fractions:

$$\frac{2}{x^2} - \frac{3}{x^3} =$$

Answer: $\frac{2x-3}{x^3}$

9. Add the following fractions:

$$\frac{1}{\sqrt{x}} + \sqrt{x} =$$

Want help? Check the hints section at the back of the book.

Answer: $\frac{x\sqrt{x}+\sqrt{x}}{x}$

Distributive Property and Factoring

Recall the distributive property from algebra:

$$x(y+z) = xy + xz$$

Following are a few examples of the **distributive** property.

$$2(x+1) = 2x + (2)(1) = 2x + 2$$
$$-(x+1) = -x - 1$$
$$x(x+3) = x^2 + 3x$$
$$-(2-x) = -2 - (-x) = -2 + x$$
$$5x(2x^2+4) = 5x(2x^2) + 5x(4) = 10x^3 + 20x$$

A particularly useful application of the distributive property is the **foil** method:

$$(w+x)(y+z) = wy + wz + xy + xz$$

Following is an example of the foil method.

$$(x+2)(x-3) = x^2 - 3x + 2x - 6 = x^2 - x - 6$$

The opposite (or inverse) of the distributive property is called **factoring**. It's often helpful to look for common factors in an expression and factor them out, as in these examples:

$$x^2 + 2x = x(x+2)$$
$$3x^3 - 6x^2 = 3x^2(x-2)$$
$$\frac{1}{x^2} + \frac{1}{x} = \frac{1}{x}\left(\frac{1}{x} + 1\right) = \frac{1}{x}\left(\frac{1}{x} + \frac{x}{x}\right) = \frac{1}{x}\left(\frac{1+x}{x}\right) = \frac{1+x}{x^2} = \frac{x+1}{x^2}$$

With squareroots, we factor out perfect squares in order to express an answer in standard form. Following is an example.

$$\sqrt{12} = \sqrt{(4)(3)} = \sqrt{4}\sqrt{3} = 2\sqrt{3}$$

Example: Apply the distributive property to $2x(3x^4 - 8)$.

$$2x(3x^4 - 8) = 2x(3x^4) + 2x(-8) = 6x^5 - 16x$$

Example: Apply the foil method to $(2x-3)(4x+5)$.

$$(2x-3)(4x+5) = 2x(4x) + 2x(5) - 3(4x) - 3(5) = 8x^2 + 10x - 12x - 15$$
$$(2x-3)(4x+5) = 8x^2 - 2x - 15$$

Example: Factor the expression $10x^6 + 15x^4$.

The greatest common factor to both terms is $5x^4$:

$$10x^6 + 15x^4 = 5x^4(2x^2 + 3)$$

As a check, $5x^4(2x^2) = 10x^6$ and $5x^4(3) = 15x^4$.

Example: Express $\sqrt{50}$ in standard form.

Factor out any perfect squares. We can factor 50 as $(25)(2)$, where 25 is a perfect square because $5^2 = 25$.

$$\sqrt{50} = \sqrt{(25)(2)} = \sqrt{25}\sqrt{2} = 5\sqrt{2}$$

10. Apply the distributive property to $6x(x + 9)$.

Answer: $6x^2 + 54x$

11. Apply the distributive property to $-3x^2(5x^6 - 2x^4)$.

Answer: $-15x^8 + 6x^6$

12. Apply the distributive property to $\sqrt{x}(x + \sqrt{x})$.

Answer: $x^{3/2} + x$

13. Apply the foil method to $(3x + 2)(x + 5)$.

Answer: $3x^2 + 17x + 10$

14. Apply the foil method to $(6x - 4)^2$.

Want help? Check the hints section at the back of the book.

Answer: $36x^2 - 48x + 16$

15. Factor the expression $x^3 - 4x$.

Answer: $x(x^2 - 4)$

16. Factor the expression $8x^9 + 12x^6$.

Answer: $4x^6(2x^3 + 3)$

17. Factor the expression $45x^7 - 18x^5 + 27x^3$.

Answer: $9x^3(5x^4 - 2x^2 + 3)$

18. Express $\sqrt{18}$ in standard form.

Answer: $3\sqrt{2}$

19. Express $\sqrt{108}$ in standard form.

Want help? Check the hints section at the back of the book.

Answer: $6\sqrt{3}$

Quadratic Equation Strategy

A quadratic equation consists of a term with the variable squared (x^2), a term with the variable not raised to a power (x), and a constant term. The equation below shows one example of a quadratic equation.

$$3x^2 + 2x - 5 = 0$$

The equation below offers another example. The like terms ($6x$ and $2x$) can be combined together and the terms can be reordered so that it has the same form as the above example.

$$6x + 1 = 8x^2 + 2x$$

To solve a quadratic equation, follow these steps:

1. If there are more than three terms, first combine like terms together.
2. Bring all three terms to the same side of the equation. Write the squared term first, then the linear term, followed by the constant term: $ax^2 + bx + c = 0$.
3. By comparison, identify the coefficients a, b, and c of these three terms.
4. Plug the values of a, b, and c into the **quadratic formula**:

$$x = \frac{-b \pm \sqrt{b^2 - 4ac}}{2a}$$

Example: Solve the equation below.

$$5x^2 + 4x = 3x^2 + 6x + 12$$

Combine like terms: $5x^2 - 3x^2$ reduces to $2x^2$, while $4x - 6x$ reduces to $-2x$.
Bring every term to the same side of the equation, in the order $ax^2 + bx + c = 0$.

$$2x^2 - 2x - 12 = 0$$

Identify the constants a, b, and c by comparing $2x^2 - 2x - 12 = 0$ with $ax^2 + bx + c = 0$.

$$a = 2 \quad , \quad b = -2 \quad , \quad c = -12$$

Plug the values of a, b, and c into the quadratic formula:

$$x = \frac{-b \pm \sqrt{b^2 - 4ac}}{2a} = \frac{-(-2) \pm \sqrt{(-2)^2 - 4(2)(-12)}}{2(2)}$$

$$x = \frac{2 \pm \sqrt{4 + 96}}{4} = \frac{2 \pm \sqrt{100}}{4} = \frac{2 \pm 10}{4}$$

$$x = \frac{2 + 10}{4} = \frac{12}{4} = 3 \quad \text{or} \quad x = \frac{2 - 10}{4} = \frac{-8}{4} = -2$$

$$x = 3 \text{ or } x = -2$$

Note: There are generally two solutions to the quadratic equation.

20. Use the quadratic formula to solve for x.

$$2x^2 - 2x - 40 = 0$$

Answers: −4, 5

21. Use the quadratic formula to solve for y.

$$3y - 27 + 2y^2 = 0$$

Want help? Check the hints section at the back of the book.

Answers: $-\frac{9}{2}, 3$

22. Use the quadratic formula to solve for t.

$$6t = 8 - 2t^2$$

Answers: $-4, 1$

23. Use the quadratic formula to solve for x.

$$1 + 25x - 5x^2 = 8x - 3x^2 + 9$$

Want help? Check the hints section at the back of the book.

Answers: $\frac{1}{2}, 8$

Method of Substitution Strategy

To solve a system of equations with the method of substitution, follow these steps:

1. If there are two equations in two unknowns (like x and y):
    - Isolate y in one equation. That is, solve for y in terms of x.
    - Substitute this expression in parentheses for y in the unused equation.
    - Distribute the coefficient in order to remove the parentheses. Solve for x.
    - Plug the value of x into the equation from the first step where y was isolated.
2. If there are three equations in three unknowns (like x, y, and z):
    - Solve for z in terms of x and y in one equation. Substitute this expression in parenthesis for z in both of the unused equations.
    - Now you will have two equations in two unknowns (x and y). Solve this system using the method from Step 1.

Example: Solve the system of equations below.

$$5x + 2y = 16$$
$$8x - 3y = 7$$

Isolate y in the first equation. Bring $5x$ to the right and then divide both sides by 2:

$$2y = 16 - 5x$$
$$y = 8 - \frac{5x}{2}$$

Substitute this expression in for y in the unused equation:

$$8x - 3\left(8 - \frac{5x}{2}\right) = 7$$

Distribute the -3. Remember that two minuses make a plus: $(-3)(-5) = +15$.

$$8x - 24 + \frac{15x}{2} = 7$$

Combine like terms to isolate x. First add 24 to both sides to bring the constants together.

$$8x + \frac{15x}{2} = 31$$

Now make a common denominator to add $8x$ to $\frac{15x}{2}$. Write $8x$ as $\frac{16x}{2}$.

$$\frac{16x}{2} + \frac{15x}{2} = 31$$
$$\frac{31x}{2} = 31$$
$$x = 2$$

Plug this value for x into the previous equation where y had been isolated:

$$y = 8 - \frac{5x}{2} = 8 - \frac{5(2)}{2} = 8 - 5 = 3$$

The solution is $x = 2$ and $y = 3$.

24. Use the method of substitution to solve this system of equations for each unknown.

$$3x + 2y = 18$$
$$8x - 5y = 17$$

Answers: 4, 3

25. Use the method of substitution to solve this system of equations for each unknown.

$$4y + 3z = 10$$
$$5y - 2z = -22$$

Want help? Check the hints section at the back of the book.

Answers: –2, 6

26. Use the method of substitution to solve this system of equations for each unknown.

$$3x - 4y + 2z = 44$$
$$5y + 6z = 29$$
$$2x + z = 13$$

Want help? Check the hints section at the back of the book.

Answers: 2, –5, 9

2 CALCULUS ESSENTIALS

Essential Concepts

A polynomial expression consists of terms of the form ax^b where a and b are constants. The constant a is called a coefficient while the constant b is a power of x. The following expression is an example of a polynomial: $2x^4 + 5x^3 - 4x^2 - 8$.

The notation $\frac{dy}{dx}$ is used to represent a derivative of the function y with respect to x.

Following is the formula to find the **derivative** of a polynomial term of the form ax^b.

$$\frac{d}{dx}(ax^b) = bax^{b-1}$$

The exponent b comes down to multiply the coefficient a and the exponent is reduced by 1 (from b to $b - 1$). This technique is illustrated in the examples that follow.

Strategy for Finding the First Derivative of a Polynomial Expression

To find the first derivative of a polynomial expression, follow these steps:

1. The derivative of an expression of the form ax^b (where a and b are constants) is:
$$\frac{d}{dx}(ax^b) = bax^{b-1}$$
Note the following special cases:
    - If a variable is inside of a squareroot, use the substitution $\sqrt{x} = x^{1/2}$.
    - If a variable and constant appear together inside of a squareroot, use the substitution $\sqrt{cx} = (cx)^{1/2} = c^{1/2}x^{1/2}$.
    - If a variable appears in a denominator, use the substitution $\frac{1}{x^n} = x^{-n}$.
2. If there are two or more terms, find the derivative of each term and then add the results together:
$$\frac{d}{dx}(y_1 + y_2 + \cdots + y_N) = \frac{dy_1}{dx} + \frac{dy_2}{dx} + \cdots + \frac{dy_N}{dx}$$
The notation $\cdots$ means "and so on" and the symbol N represents the total number of terms. Following is an example of what this means when there are $N = 3$ terms:
$$\frac{d}{dx}(6x^3 - 3x^2 - 12x) = \frac{d}{dx}(6x^3) + \frac{d}{dx}(-3x^2) + \frac{d}{dx}(-12x)$$
In this example, $y_1 = 6x^3$, $y_2 = -3x^2$, and $y_3 = -12x$.

Example: Evaluate the following derivative with respect to x.

$$\frac{d}{dx}(3x^4)$$

Compare $3x^4$ with ax^b to see that $a = 3$ and $b = 4$. Plug these values into the formula $\frac{d}{dx}(ax^b) = bax^{b-1}$:

$$\frac{d}{dx}(3x^4) = (4)(3)x^{4-1} = 12x^3$$

Example: Evaluate the following derivative with respect to t.

$$\frac{d}{dt}(t^3)$$

Since this derivative is with respect to t (instead of the usual x), the variable t is playing the role of x. Therefore, we will rewrite the formula $\frac{d}{dx}(ax^b) = bax^{b-1}$ as $\frac{d}{dt}(at^b) = bat^{b-1}$ (where we simply changed x to t). Compare t^3 with at^b to see that $a = 1$ and $b = 3$. (Recall from algebra that a coefficient of 1 is implied when you don't see a coefficient.) Plug these values into the formula $\frac{d}{dt}(at^b) = bat^{b-1}$:

$$\frac{d}{dt}(t^3) = (3)(1)t^{3-1} = 3t^2$$

Example: Evaluate the following derivative with respect to x.

$$\frac{d}{dx}(4x^{-2})$$

Compare $4x^{-2}$ with ax^b to see that $a = 4$ and $b = -2$. Plug these values into the formula $\frac{d}{dx}(ax^b) = bax^{b-1}$:

$$\frac{d}{dx}(4x^{-2}) = (-2)(4)x^{-2-1} = -8x^{-3}$$

Example: Evaluate the following derivative with respect to t.

$$\frac{d}{dt}\left(\frac{1}{t}\right)$$

Since this derivative is with respect to t (instead of x), the variable t is playing the role of x. Therefore, we will rewrite the formula $\frac{d}{dx}(ax^b) = bax^{b-1}$ as $\frac{d}{dt}(at^b) = bat^{b-1}$. Since t appears in the denominator of $\frac{1}{t}$, we use the rule $\frac{1}{t^n} = t^{-n}$ to write $\frac{1}{t}$ as t^{-1}. Compare t^{-1} with at^b to see that $a = 1$ and $b = -1$. (Recall from algebra that a coefficient of 1 is implied when you don't see a coefficient.) Plug these values into the formula $\frac{d}{dt}(at^b) = bat^{b-1}$. In the last step of the line below, we again apply the rule $t^{-n} = \frac{1}{t^n}$.

$$\frac{d}{dt}\left(\frac{1}{t}\right) = \frac{d}{dt}(t^{-1}) = (-1)(1)t^{-1-1} = -t^{-2} = -\frac{1}{t^2}$$

Example: Evaluate the following derivative with respect to x.

$$\frac{d}{dx}(\sqrt{2x})$$

Recall from algebra that $\sqrt{x} = x^{1/2}$. The trick to this problem is to write $\sqrt{2x}$ as $2^{1/2}x^{1/2}$. Compare $2^{1/2}x^{1/2}$ with ax^b to see that $a = 2^{1/2}$ and $b = \frac{1}{2}$. Plug these values into the formula $\frac{d}{dx}(ax^b) = bax^{b-1}$:

$$\frac{d}{dx}(\sqrt{2x}) = \frac{d}{dx}(2^{1/2}x^{1/2}) = \left(\frac{1}{2}\right)(2^{1/2})x^{1/2-1} = \left(\frac{2^{1/2}}{2}\right)x^{-1/2}$$

Recall the rule from algebra that $\frac{x^m}{x^n} = x^{m-n}$. We can use this to replace $\frac{2^{1/2}}{2}$ with $2^{1/2-1}$.

$$\frac{d}{dx}(\sqrt{2x}) = 2^{1/2-1}x^{-1/2} = 2^{-1/2}x^{-1/2}$$

Recalling from algebra the rule that $x^{-1} = \frac{1}{x}$, we can write $2^{-1/2}x^{-1/2}$ as $\frac{1}{2^{1/2}x^{1/2}}$ which is the same as $\frac{1}{\sqrt{2x}}$.

$$\frac{d}{dx}(\sqrt{2x}) = \frac{1}{2^{1/2}x^{1/2}} = \frac{1}{\sqrt{2x}}$$

Rationalize the denominator by multiplying both the numerator and denominator by $\sqrt{2x}$ (because $\sqrt{2x}\sqrt{2x} = 2x$). **If you're struggling with the algebra here, review Chapter 1.**

$$\frac{d}{dx}(\sqrt{2x}) = \frac{1}{\sqrt{2x}}\frac{\sqrt{2x}}{\sqrt{2x}} = \frac{\sqrt{2x}}{2x}$$

Example: Evaluate the following derivative with respect to x.

$$\frac{d}{dx}\left(\frac{4x^{1/2}}{5}\right)$$

Compare $\frac{4x^{1/2}}{5}$ with ax^b to see that $a = \frac{4}{5}$ and $b = \frac{1}{2}$. Plug these values into the formula $\frac{d}{dx}(ax^b) = bax^{b-1}$:

$$\frac{d}{dx}\left(\frac{4x^{1/2}}{5}\right) = \left(\frac{1}{2}\right)\left(\frac{4}{5}\right)x^{1/2-1} = \frac{4}{10}x^{-1/2} = \frac{2}{5}x^{-1/2} = \frac{2}{5}\frac{1}{\sqrt{x}}$$

Note that $\frac{4}{10}$ reduces to $\frac{2}{5}$ if you divide both the numerator and denominator by 2. In the last step, we combined the rules from algebra that $x^{-1} = \frac{1}{x}$ and $x^{1/2} = \sqrt{x}$ to write $x^{-1/2} = \frac{1}{\sqrt{x}}$. In a math course, we would rationalize the denominator by multiplying both the numerator and denominator by $\sqrt{x}$ (applying the rule $\sqrt{x}\sqrt{x} = x$). **If you're struggling to follow the algebra in these examples, you should review Chapter 1.**

$$\frac{d}{dx}\left(\frac{4x^{1/2}}{5}\right) = \frac{2}{5}\frac{1}{\sqrt{x}}\frac{\sqrt{x}}{\sqrt{x}} = \frac{2\sqrt{x}}{5x}$$

Example: Evaluate the following derivative with respect to u.

$$\frac{d}{du}(2u^3 - 5u^2 + 6)$$

Since this derivative is with respect to u (instead of x), the variable u is playing the role of x. Therefore, we will rewrite the formula $\frac{d}{dx}(ax^b) = bax^{b-1}$ as $\frac{d}{du}(au^b) = bau^{b-1}$. According to the strategy, we should first find the derivative of each term:

- The first term is $2u^3$. Compare $2u^3$ with au^b to see that $a = 2$ and $b = 3$.
$$\frac{d}{du}(2u^3) = (3)(2)u^{3-1} = 6u^2$$
- The second term is $-5u^2$. Compare $-5u^2$ with au^b to see that $a = -5$ and $b = 2$.
$$\frac{d}{du}(-5u^2) = (2)(-5)u^{2-1} = -10u^1 = -10u$$
Recall the rule from algebra that $x^1 = x$ (or in this case $u^1 = u$).
- The third term is 6. Compare 6 with au^b to see that $a = 6$ and $b = 0$. (Recall the rule from algebra that $x^0 = 1$, or in this case $u^0 = 1$.) Alternatively, you could just remember that the derivative of a constant equals zero. For example, $\frac{d}{dx}c = 0$.
$$\frac{d}{du}(6) = (0)(6)u^{0-1} = 0$$

Add these three derivatives together:

$$\frac{d}{du}(2u^3 - 5u^2 + 6) = \frac{d}{du}(2u^3) + \frac{d}{du}(-5u^2) + \frac{d}{du}(6)$$
$$\frac{d}{du}(2u^3 - 5u^2 + 6) = 6u^2 - 10u + 0 = 6u^2 - 10u$$

27. Evaluate the following derivative with respect to x.

$$\frac{d}{dx}(6x^5) =$$

Answer: $30x^4$

28. Evaluate the following derivative with respect to t.

$$\frac{d}{dt}(t) =$$

Want help? Check the hints section at the back of the book.

Answer: 1

29. Evaluate the following derivative with respect to x.

$$\frac{d}{dx}(2x^{-3}) =$$

Answer: $-6x^{-4}$

30. Evaluate the following derivative with respect to t.

$$\frac{d}{dt}\left(\frac{3}{t^6}\right) =$$

Answer: $-18t^{-7}$

31. Evaluate the following derivative with respect to x.

$$\frac{d}{dx}(3x^8 - 6x^5 + 9x^2 - 4) =$$

Answer: $24x^7 - 30x^4 + 18x$

32. Evaluate the following derivative with respect to u.

$$\frac{d}{du}\left(\frac{1}{\sqrt{u}}\right) =$$

Want help? Check the hints section at the back of the book.

Answer: $-\frac{\sqrt{u}}{2u^2}$

Essential Concepts

The following notation is used to represent an **integral** of the function y over x:

$$\int y\,dx$$

What the above integral means is this: What function could you take the derivative of with respect to x and obtain y as a result? The answer is called the anti-derivative. A way to perform an integral is to find the anti-derivative of the function in the integrand (in the above example, y, which is a function of x, is the integrand). Here is an example:

- $\int 3x^2\,dx$ means, the derivate of what function would result in $3x^2$ as the answer?
- The anti-derivative of $3x^2$ equals x^3. Why? Because $\frac{d}{dx}(x^3) = 3x^2$.
- Therefore one answer to $\int 3x^2\,dx$ is x^3. However, that's **not** the *only* answer.
- Another answer to $\int 3x^2\,dx$ is $x^3 + 9$. Why? Because $\frac{d}{dx}(x^3 + 9) = 3x^2$.
- The general answer to $\int 3x^2\,dx$ is $x^3 + c$, where c is a constant. If you take a derivative of a constant, you get zero. Therefore, a constant c shows up in *all* anti-derivatives.

The general formula for an **anti-derivative** of an expression of the form ax^b can be obtained by inverting the formula to find a derivative. In the derivative, the exponent comes down to multiply the coefficient and the exponent is reduced by 1. The inversion of this is as follows: The exponent is raised by 1 and we divide the coefficient by the new exponent (the old exponent plus 1). We also obtain a constant of integration as discussed previously.

$$\int ax^b\,dx = \frac{ax^{b+1}}{b+1} + c \qquad (b \neq -1)$$

For those who are curious, when $b = -1$ you get a natural logarithm instead. In this book, we will focus on polynomial expressions for which $b \neq -1$.

$$\int x^{-1}\,dx = \int \frac{dx}{x} = \ln(x) + c$$

In physics, we generally perform definite integrals rather than work with a constant of integration. In a **definite integral**, the anti-derivative is evaluated at the upper and lower limits of the integral and the two results are subtracted as follows:

$$\int_{x=x_i}^{x_f} ax^b\,dx = \left[\frac{ax^{b+1}}{b+1}\right]_{x=x_i}^{x_f} = \frac{ax_f^{b+1}}{b+1} - \frac{ax_i^{b+1}}{b+1}$$

In a definite integral, the constant (c) **cancels** out during the subtraction. The technique of performing a definite integral is illustrated in the examples that follow.

Strategy for Integrating Over a Polynomial Expression

To perform a definite integral over a polynomial expression, follow these steps:

1. First find the anti-derivative of the polynomial expression in the integrand. The anti-derivative of an expression of the form ax^b (where a and b are constants, and where $b \neq -1$) is:
$$\int ax^b\,dx = \frac{ax^{b+1}}{b+1} + c \qquad (b \neq -1)$$
We will ignore the constant c because it will **cancel** out in a **definite** integral. Note the following special cases:
   - If a variable is inside of a squareroot, use the substitution $\sqrt{x} = x^{1/2}$.
   - If a variable and constant appear together inside a squareroot, use the substitution $\sqrt{cx} = (cx)^{1/2} = c^{1/2}x^{1/2}$.
   - If a variable appears in a denominator, use the substitution $\frac{1}{x^n} = x^{-n}$.
2. If there are two or more terms, find the anti-derivative of each term and then add the results together:
$$\int (y_1 + y_2 + \cdots + y_N)\,dx = \int y_1\,dx + \int y_2\,dx + \cdots + \int y_N\,dx$$
The notation $\cdots$ means "and so on" and the symbol N represents the total number of terms. Following is an example where there are $N = 3$ terms:
$$\int (4x^5 - 2x^3 + 8x)\,dx = \int 4x^5\,dx + \int (-2x^3)\,dx + \int 8x\,dx$$
3. Evaluate the anti-derivative at the upper limit and at the lower limit and subtract the two results. For example, for the case where there is just one polynomial term in the integrand, the definite integral will look like this:
$$\int_{x=x_i}^{x_f} ax^b\,dx = \left[\frac{ax^{b+1}}{b+1}\right]_{x=x_i}^{x_f} = \frac{ax_f^{b+1}}{b+1} - \frac{ax_i^{b+1}}{b+1}$$

Example: Perform the following definite integral: $\int_{x=1}^{2} 6x^2\,dx$.

Compare $6x^2$ with ax^b to see that $a = 6$ and $b = 2$. The anti-derivative is $\frac{ax^{b+1}}{b+1}$.

$$\int_{x=1}^{2} 6x^2\,dx = \left[\frac{6x^{2+1}}{2+1}\right]_{x=1}^{2} = \left[\frac{6x^3}{3}\right]_{x=1}^{2} = [2x^3]_{x=1}^{2}$$

The notation $[2x^3]_{x=1}^{2}$ means to evaluate the function $2x^3$ when $x = 2$, then evaluate the function $2x^3$ when $x = 1$, and finally to subtract the two results: $2(2)^3 - 2(1)^3$.

$$\int_{x=1}^{2} 6x^2\,dx = [2x^3]_{x=1}^{2} = 2(2)^3 - 2(1)^3 = 2(8) - 2(1) = 16 - 2 = 14$$

Example: Perform the following definite integral: $\int_{u=2}^{4} u\,du$.

Compare u with au^b to see that $a = 1$ and $b = 1$ since $1u^1 = u$. The anti-derivative is $\frac{au^{b+1}}{b+1}$.

$$\int_{u=2}^{4} u\,du = \left[\frac{u^{1+1}}{1+1}\right]_{u=2}^{4} = \left[\frac{u^2}{2}\right]_{u=2}^{4}$$

The notation $\left[\frac{u^2}{2}\right]_{u=2}^{4}$ means to evaluate the function $\frac{u^2}{2}$ when $u = 4$, then evaluate the function $\frac{u^2}{2}$ when $u = 2$, and finally to subtract the two results: $\frac{(4)^2}{2} - \frac{(2)^2}{2}$.

$$\int_{u=2}^{4} u\,du = \left[\frac{u^2}{2}\right]_{u=2}^{4} = \frac{(4)^2}{2} - \frac{(2)^2}{2} = \frac{16}{2} - \frac{4}{2} = 8 - 2 = 6$$

Example: Perform the following definite integral: $\int_{x=-3}^{-1} 12x^{-2}\,dx$.

Compare $12x^{-2}$ with ax^b to see that $a = 12$ and $b = -2$. The anti-derivative is $\frac{ax^{b+1}}{b+1}$.

$$\int_{x=-3}^{-1} 12x^{-2}\,dx = \left[\frac{12x^{-2+1}}{-2+1}\right]_{x=-3}^{-1} = \left[\frac{12x^{-1}}{-1}\right]_{x=-3}^{-1} = [-12x^{-1}]_{x=-3}^{-1} = \left[-\frac{12}{x}\right]_{x=-3}^{-1}$$

In the last step, we used the rule that $x^{-1} = \frac{1}{x}$. The notation $\left[-\frac{12}{x}\right]_{x=-3}^{-1}$ means to evaluate the function $-\frac{12}{x}$ when $x = -1$, then evaluate the function $-\frac{12}{x}$ when $x = -3$, and finally to subtract the two results: $-\frac{12}{-1} - \left(-\frac{12}{-3}\right)$. Be careful with the signs: There is one minus sign between the terms from the subtraction, each term has its own minus sign from $-\frac{12}{x}$, and both endpoints are negative ($x_i = -3$ and $x_f = -1$). The result is that the first term has two minus signs $-\frac{12}{-1}$ and it is followed by three minus signs $-\left(-\frac{12}{-3}\right)$.

$$\int_{x=-3}^{-1} 12x^{-2}\,dx = \left[-\frac{12}{x}\right]_{x=-3}^{-1} = -\frac{12}{-1} - \left(-\frac{12}{-3}\right)$$

Treat these minus signs carefully. Two minus signs make a plus sign: $-\frac{12}{-1} = +12$. Three minus signs make a minus sign $-\left(-\frac{12}{-3}\right) = -(4) = -4$.

$$\int_{x=-3}^{-1} 12x^{-2}\,dx = \left[-\frac{12}{x}\right]_{x=-3}^{-1} = (+12) - (+4) = 12 - 4 = 8$$

Example: Perform the following definite integral: $\int_{t=1}^{2} \frac{16\,dt}{t^3}$.

Since t appears in the denominator of $\frac{16}{t^3}$, we use the rule $\frac{1}{t^n} = t^{-n}$ to write $\frac{16}{t^3}$ as $16t^{-3}$. Compare $16t^{-3}$ with at^b to see that $a = 16$ and $b = -3$. The anti-derivative is $\frac{at^{b+1}}{b+1}$.

$$\int_{t=1}^{2} 16\,t^{-3}\,dt = \left[\frac{16t^{-3+1}}{-3+1}\right]_{t=1}^{2} = \left[\frac{16t^{-2}}{-2}\right]_{t=1}^{2} = [-8t^{-2}]_{t=1}^{2} = \left[-\frac{8}{t^2}\right]_{t=1}^{2}$$

In the last step, we used the rule that $t^{-2} = \frac{1}{t^2}$. The notation $\left[-\frac{8}{t^2}\right]_{t=1}^{2}$ means to evaluate the function $-\frac{8}{t^2}$ when $t = 2$, then evaluate the function $-\frac{8}{t^2}$ when $t = 1$, and finally to subtract the two results: $-\frac{8}{2^2} - \left(-\frac{8}{1^2}\right)$.

$$\int_{t=1}^{2} \frac{16\,dt}{t^3} = \left[-\frac{8}{t^2}\right]_{t=1}^{2} = -\frac{8}{2^2} - \left(-\frac{8}{1^2}\right) = -\frac{8}{4} + \frac{8}{1} = -2 + 8 = 6$$

Example: Perform the following definite integral: $\int_{x=9}^{36} \sqrt{x}\,dx$.

The trick to this problem is to recall the rule from algebra that $\sqrt{x} = x^{1/2}$. Compare $x^{1/2}$ with ax^b to see that $a = 1$ and $b = \frac{1}{2}$ (since a coefficient of 1 is implied when you don't see a coefficient). The anti-derivative is $\frac{ax^{b+1}}{b+1}$.

$$\int_{x=9}^{36} x^{1/2}\,dx = \left[\frac{x^{1/2+1}}{\frac{1}{2}+1}\right]_{x=9}^{36}$$

Add $\frac{1}{2} + 1$ by making a common denominator: $\frac{1}{2} + 1 = \frac{1}{2} + \frac{2}{2} = \frac{1+2}{2} = \frac{3}{2}$.

$$\int_{x=9}^{36} x^{1/2}\,dx = \left[\frac{x^{3/2}}{3/2}\right]_{x=9}^{36}$$

To divide by $\frac{3}{2}$, multiply by its reciprocal $\left(\frac{2}{3}\right)$.

$$\int_{x=9}^{36} x^{1/2}\,dx = \left[\frac{2x^{3/2}}{3}\right]_{x=9}^{36}$$

The notation $\left[\frac{2x^{3/2}}{3}\right]_{x=9}^{36}$ means to evaluate the function $\frac{2x^{3/2}}{3}$ when $x = 36$, then evaluate the function $\frac{2x^{3/2}}{3}$ when $x = 9$, and finally to subtract the two results: $\frac{2(36)^{3/2}}{3} - \frac{2(9)^{3/2}}{3}$.

$$\int_{x=9}^{36} \sqrt{x}\,dx = \frac{2(36)^{3/2}}{3} - \frac{2(9)^{3/2}}{3} = \frac{2(\sqrt{36})^3}{3} - \frac{2(\sqrt{9})^3}{3} = \frac{2(6)^3}{3} - \frac{2(3)^3}{3} = 144 - 18 = 126$$

Example: Perform the following definite integral: $\int_{t=1}^{8} \frac{4t^{1/3}}{3} dt$.

Compare $\frac{4t^{1/3}}{3}$ with at^b to see that $a = \frac{4}{3}$ and $b = \frac{1}{3}$. The anti-derivative is $\frac{at^{b+1}}{b+1}$.

$$\int_{t=1}^{8} \frac{4t^{1/3}}{3} dt = \left[\left(\frac{4}{3}\right)\frac{t^{1/3+1}}{\frac{1}{3}+1}\right]_{t=1}^{8}$$

Add $\frac{1}{3} + 1$ by making a common denominator: $\frac{1}{3} + 1 = \frac{1}{3} + \frac{3}{3} = \frac{1+3}{3} = \frac{4}{3}$.

$$\int_{t=1}^{8} \frac{4t^{1/3}}{3} dt = \left[\left(\frac{4}{3}\right)\frac{t^{4/3}}{4/3}\right]_{t=1}^{8}$$

To divide by $\frac{4}{3}$, multiply by its reciprocal $\left(\frac{3}{4}\right)$.

$$\int_{t=1}^{8} \frac{4t^{1/3}}{3} dt = \left[\frac{3}{4}\left(\frac{4}{3}\right)t^{4/3}\right]_{t=1}^{8} = \left[t^{4/3}\right]_{t=1}^{8}$$

The notation $\left[t^{4/3}\right]_{t=1}^{8}$ means to evaluate the function $t^{4/3}$ when $t = 8$, then evaluate the function $t^{4/3}$ when $t = 1$, and finally to subtract the two results: $8^{4/3} - 1^{4/3}$.

$$\int_{t=1}^{8} \frac{4t^{1/3}}{3} dt = \left[t^{4/3}\right]_{t=1}^{8} = 8^{4/3} - 1^{4/3} = \left(\sqrt[3]{8}\right)^4 - \left(\sqrt[3]{1}\right)^4 = 2^4 - 1^4 = 16 - 1 = 15$$

Example: Perform the following definite integral: $\int_{x=0}^{1} (15x^4 - 6x^2 + 3)\, dx$.

According to the strategy, we should first find the anti-derivative of each term:

- The first term is $15x^4$. Compare $15x^4$ with ax^b to see that $a = 15$ and $b = 4$.

$$\int 15x^4\, dx = \frac{15x^{4+1}}{4+1} = \frac{15x^5}{5} = 3x^5$$

- The second term is $-6x^2$. Compare $-6x^2$ with ax^b to see that $a = -6$ and $b = 2$.

$$\int -6x^2\, dx = \frac{-6x^{2+1}}{2+1} = -\frac{6x^3}{3} = -2x^3$$

- The third term is 3. Compare 3 with ax^b to see that $a = 3$ and $b = 0$. (Recall the rule from algebra that $x^0 = 1$.)

$$\int 3\, dx = \frac{3x^{0+1}}{0+1} = \frac{3x^1}{1} = 3x$$

Add these three anti-derivatives together and evaluate the definite integral over the limits:

$$\int_{x=0}^{1} (15x^4 - 6x^2 + 3)\, dx = [3x^5 - 2x^3 + 3x]_{x=0}^{1}$$

$$= [3(1)^5 - 2(1)^3 + 3(1)] - [3(0)^5 - 2(0)^3 + 3(0)]$$

$$= 3 - 2 + 3 - 0 + 0 - 0 = 4$$

33. Perform the following definite integral.

$$\int_{x=0}^{2} 8x^3 \, dx =$$

Answer: 32

34. Perform the following definite integral.

$$\int_{t=-5}^{5} t^4 \, dt =$$

Answer: 1250

35. Perform the following definite integral.

$$\int_{x=3}^{5} 900 \, x^{-3} \, dx =$$

Want help? Check the hints section at the back of the book.

Answer: 32

36. Perform the following definite integral.

$$\int_{u=1}^{3} \frac{81\,du}{u^4} =$$

Answer: 26

37. Perform the following definite integral.

$$\int_{x=4}^{9} \frac{3\,dx}{\sqrt{x}} =$$

Want help? Check the hints section at the back of the book.

Answer: 6

38. Perform the following definite integral.

$$\int_{t=1}^{16} 14\, t^{3/4}\, dt =$$

Answer: 1016

39. Perform the following definite integral.

$$\int_{x=2}^{4} (12x^3 - 9x^2)\, dx =$$

Want help? Check the hints section at the back of the book.

Answer: 552

40. Perform the following definite integral.

$$\int_{x=-1}^{2} (12x^5 + 24x^3 - 48x)\,dx =$$

Want help? Check the hints section at the back of the book.

Answer: 144

3 ONE-DIMENSIONAL UNIFORM ACCELERATION

Relevant Terminology

Velocity – a combination of speed and direction.
Acceleration – the instantaneous rate at which velocity is changing.
Uniform acceleration – the acceleration is constant; velocity changes at a constant rate.
Net displacement – a straight line from the initial position to the final position.

Important Distinctions

Velocity tells you both how fast an object is moving and which way the object is headed. Speed only tells you how fast an object is moving, with no information about the direction it is headed.

An object must be **changing velocity** in order to have **acceleration**.

Net displacement does not depend on how the object reaches the final position. Unlike the total distance traveled, the net displacement only depends on where the object starts and where the object finishes.

Equations of Uniform Acceleration

$$\Delta x = v_{x0}t + \frac{1}{2}a_x t^2$$
$$v_x = v_{x0} + a_x t$$
$$v_x^2 = v_{x0}^2 + 2a_x \Delta x$$

Symbols and SI Units

Symbol	Name	SI Units
Δx	net displacement	m
v_{x0}	initial velocity	m/s
v_x	final velocity	m/s
a_x	acceleration	m/s^2
t	time	s

One-dimensional Uniform Acceleration Strategy

An object experiences uniform acceleration if its acceleration remains constant. To solve a problem with uniform acceleration in one dimension, follow these steps:

1. Draw a diagram of the path. Label the initial position (i), final position (f), and the positive x-direction.
2. Identify the unknown symbol and three known symbols. You must know three of the following symbols: $\Delta x, v_{x0}, v_x, a_x,$ and t. See the table of symbols on the previous page. The units can help: For example, only a_x can be expressed in m/s^2.
3. Use the following equations to solve for the unknown. Think about which symbol you're solving for and which symbols you know to help you choose the right equation.

$$\Delta x = v_{x0}t + \frac{1}{2}a_x t^2$$
$$v_x = v_{x0} + a_x t$$
$$v_x^2 = v_{x0}^2 + 2a_x \Delta x$$

Example: A monkey drives a bananamobile with uniform acceleration. If the bananamobile travels 200 m in a time of 4.0 s with an acceleration of 6.0 m/s^2, what was the initial velocity of the bananamobile?

Begin with a labeled diagram.

i • ——————→ f

$+x$

The unknown we are looking for is v_{x0}. List the three knowns.

$$v_{x0} = ? \quad , \quad \Delta x = 200 \text{ m} \quad , \quad t = 4.0 \text{ s} \quad , \quad a_x = 6.0 \text{ m/s}^2$$

Based on this, it would be simplest to use the first equation of uniform acceleration:

$$\Delta x = v_{x0}t + \frac{1}{2}a_x t^2$$

Plug the knowns into this equation. To avoid clutter, suppress the units until the end.

$$200 = v_{x0}(4) + \frac{1}{2}(6)(4)^2$$

This simplifies to:

$$200 = 4v_{x0} + 48$$

Subtract 48 from both sides to isolate the unknown term:

$$152 = 4v_{x0}$$

Divide both sides by 4 to solve for the unknown:

$$v_{x0} = 38 \text{ m/s}$$

More examples: You can find more examples of uniform acceleration on pages 42-43.

41. A monkey drives a bananamobile with uniform acceleration. Starting from rest, the bananamobile travels 90 m in a time of 6.0 s.

Draw/label a diagram with the path, initial (i), final (f), and an arrow showing $+x$.

Based on the question below, list the three symbols that you know (based on your labeled diagram) along with their values and SI units. **Tip**: Consult the chart on page 35.

________________ ________________ ________________

What is the acceleration of the bananamobile?

Write the relevant equation. Choose from the three equations of uniform acceleration.

Rewrite the equation with numbers plugged in.

Solve for the unknown.

Want help? Check the hints section at the back of the book.

Answer: 5.0 m/s^2

42. A mechanical monkey toy has an initial speed of 15 m/s, has uniform acceleration of -4.0 m/s^2, and travels for 6.0 seconds.

Draw/label a diagram with the path, initial (i), final (f), and an arrow showing $+x$.

Based on the question below, list the three symbols that you know (based on your labeled diagram) along with their values and SI units. **Tip**: Consult the chart on page 35.

________________ ________________ ________________

What is the final velocity of the mechanical monkey?

Write the relevant equation. Choose from the three equations of uniform acceleration.

Rewrite the equation with numbers plugged in.

Solve for the unknown.

Want help? Check the hints section at the back of the book.

Answer: –9.0 m/s

43. A monkey drives a bananamobile with uniform acceleration, beginning with a speed of 10 m/s and ending with a speed of 30 m/s. The acceleration is 8.0 m/s^2.

Draw/label a diagram with the path, initial (i), final (f), and an arrow showing $+x$.

Based on the question below, list the three symbols that you know (based on your labeled diagram) along with their values and SI units. **Tip**: Consult the chart on page 35.

__________________ __________________ __________________

How far does the monkey travel during this time?

Write the relevant equation. Choose from the three equations of uniform acceleration.

Rewrite the equation with numbers plugged in.

Solve for the unknown.

Want help? Check the hints section at the back of the book.

Answer: 50 m

Relevant Terminology

Free fall – the motion of an object (whether falling or rising) when the only force acting on it is the force of gravity.
Gravitational acceleration – the acceleration of an object in free fall.
Vacuum – a region of space that does not contain any matter of any kind (not even air).

Common Assumptions

In most physics courses and textbooks, the following assumptions are implied in all problems except when a problem explicitly states otherwise:

- Neglect any effects of air resistance. Assume that all objects fall in vacuum.
- Assume that all objects are near the surface of the earth.
- Assume that the change in altitude is small enough that gravitational acceleration is approximately uniform throughout the motion.

Important Values

Memorize the values of gravitational acceleration near the surface of the earth and moon:

- Near the surface of the earth, use $a_y = -9.81 \text{ m/s}^2$ for free fall problems.
- Near the surface of the moon, use $a_y = -1.62 \text{ m/s}^2$ for free fall problems.

Rounding note: In this workbook, we will round these values of gravitational acceleration so that every exercise may be solved without the need of a calculator:

- We will use $a_y \approx -10 \text{ m/s}^2$ near the surface of the earth.
- We will use $a_y \approx -\frac{8}{5} \text{ m/s}^2$ near the surface of the moon.

Essential Concepts

An object that is freely falling experiences approximately uniform acceleration:

- Near the surface of the earth, a freely falling object loses approximately 10 m/s of speed each second when it is rising upward.
- Near the surface of the earth, a freely falling object gains approximately 10 m/s of speed each second when it is falling downward.

Freely falling objects experience the same acceleration regardless of how much they weigh. The motion of a freely falling object does **<u>not</u>** depend on how heavy or light the object is. **Test this out**: Drop an eraser and book from rest from the same height. You should observe that both objects reach the floor at approximately the same time.

Free Fall Strategy

To solve a problem with an object in free fall, follow these steps:

1. Draw a diagram of the path. Label the initial position (i), final position (f), and the positive y-direction. Choose $+y$ to be upward regardless of the motion of the object.
2. The acceleration will equal $a_y \approx -10 \text{ m/s}^2$ unless the problem specifically states that it is not falling near earth's surface. Near the moon, use $a_y \approx -\frac{8}{5} \text{ m/s}^2$ instead.
3. Identify the unknown symbol and three known symbols. You must know three of the following symbols: $\Delta y, v_{y0}, v_y, a_y,$ and t. See the previous note regarding a_y. See the table of symbols below. The units can help: For example, only t can be expressed in seconds (s).
4. Use the following equations to solve for the unknown. Think about which symbol you're solving for and which symbols you know to help you choose the right equation.

$$\Delta y = v_{y0} t + \frac{1}{2} a_y t^2$$

$$v_y = v_{y0} + a_y t$$

$$v_y^2 = v_{y0}^2 + 2 a_y \Delta y$$

Symbols and SI Units

Symbol	Name	SI Units
Δy	net displacement	m
v_{y0}	initial velocity	m/s
v_y	final velocity	m/s
a_y	acceleration	m/s^2
t	time	s

Getting the Signs Right

Use the following sign conventions:

- a_y is negative for all free fall problems. (Draw $+y$ upward.)
- Δy is negative if the final position (f) is **below** the initial position (i).
- v_{y0} is negative if the object is moving downward in the initial position.
- v_y is negative if the object is moving downward in the final position.

Important Notes

In free fall, we use y instead of x. That's because y is **vertical** whereas x is **horizontal**.

If an object makes an impact in the final position, the final velocity means just before impact (not after it lands). In this case, the final velocity will **not** be zero. (Final velocity only equals zero if the final position is at the very top of the trajectory.)

Similarly, if the initial position is where an object is launched or thrown, the initial velocity means just after it is released (not before it is launched). In this case, the initial velocity will **not** be zero. (Initial velocity only equals zero if the object is dropped from rest.)

Example: A monkey leans over the edge of a building and drops a banana from rest. The banana strikes the ground 3.0 s later. Approximately, how tall is the building?

Begin with a labeled diagram. Choose the $+y$-direction to be upward even though the banana falls downward. This choice makes it easier to reason out the signs correctly:

- Δy is negative because the final position (f) is below the initial position (i).
- v_y is negative because the banana is moving downward in the final position (f).
- a_y is negative because the force of gravity is downward. For **all** free fall problems, regardless of the motion of the object, a_y is negative (if you choose the $+y$-direction to be upward).

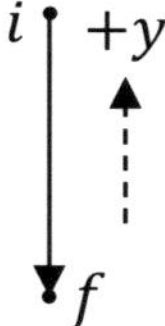

The unknown we are looking for is Δy. List the three knowns.

$$\Delta y = ? \quad , \quad t = 3.0 \text{ s} \quad , \quad v_{y0} = 0 \quad , \quad a_y = -9.81 \text{ m/s}^2 \approx -10 \text{ m/s}^2$$

We rounded 9.81 m/s² to 10 m/s² so that we can solve the problem without a calculator. Based on this, it would be simplest to use the first equation of uniform acceleration:

$$\Delta y = v_{y0}t + \frac{1}{2}a_y t^2$$

Plug the knowns into this equation. To avoid clutter, suppress the units until the end.

$$\Delta y = 0(3) + \frac{1}{2}(-9.81)(3)^2 \approx \frac{1}{2}(-10)(3)^2$$

The approximately equal sign ($\approx$) reflects that $9.81 \approx 10$. The above equation simplifies to:

$$\Delta y \approx -45 \text{ m}$$

If we neglect the height of the monkey, the building is approximately 45 m tall. The minus sign indicates that the final position (f) is below the initial position (i).

Example: A monkey leans over the edge of a cliff and throws a banana straight upward with a speed of 30 m/s. The banana lands on the ground, 35 m below its starting position. How fast is the banana moving just before impact?

Begin with a labeled diagram.

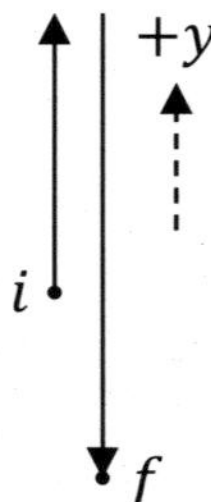

The unknown we are looking for is v_y. List the three knowns.

$$v_y = ? \quad , \quad v_{y0} = 30 \text{ m/s} \quad , \quad \Delta y = -35 \text{ m} \quad , \quad a_y = -9.81 \text{ m/s}^2 \approx -10 \text{ m/s}^2$$

We rounded 9.81 m/s² to 10 m/s² so that we can solve the problem without a calculator. The reason that Δy is negative is because the final position (f) is below the initial position (i). Based on the list above, it would be simplest to use the third equation of uniform acceleration:

$$v_y^2 = v_{y0}^2 + 2a_y\Delta y$$

Plug the knowns into this equation. To avoid clutter, suppress the units until the end.

$$v_y^2 = (30)^2 + 2(-9.81)(-35) \approx (30)^2 + 2(-10)(-35)$$

This simplifies to:

$$v_y^2 \approx 900 + 700 = 1600$$

Squareroot both side to solve for the unknown:

$$v_y \approx \sqrt{1600}$$

Note that there are two answers to $\sqrt{1600}$ since $(-40)^2$ and 40^2 both equal 1600. Thus, we must consider both -40 and 40 as possible solutions to $\sqrt{1600}$. In this example, the negative root (-40) is correct because v_y is negative. The reason for this is that the banana is moving downward in the final position.

$$v_y \approx -40 \text{ m/s}$$

Therefore, just before impact, the banana is moving approximately 40 m/s. (The final velocity is negative because the banana is moving downward, but the final speed is positive because speed does not include direction.)

Note: It was **<u>not</u>** necessary to split the trip up into two parts (one for the trip up and another for the trip back down). You can work with the complete trip and solve the problem correctly in one step rather than two. (Try solving this example with two separate trips to convince yourself that you can get the same answer either way.)

44. On Planet Fyzx, a chimpanzee astronaut drops a 500-g banana from rest from a height of 36 m above the ground and the banana strikes the ground 4.0 s later.

Draw/label a diagram with the path, initial (i), final (f), and an arrow showing $+y$.

Based on the question below, list the three symbols that you know (based on your labeled diagram) along with their values and SI units. **Hint**: One of these numbers is **negative**.

__________________ __________________ __________________

What is the acceleration of the banana?

Write the relevant equation. Choose from the three equations of uniform acceleration.

Rewrite the equation with numbers plugged in.

Solve for the unknown.

Want help? Check the hints section at the back of the book.

Answer: $-\frac{9}{2}\text{m/s}^2$

45. A monkey leans over the edge of a cliff and throws a banana straight upward with a speed of 20 m/s. The banana lands on the ground, 60 m below its starting position.

Draw/label a diagram with the path, initial (i), final (f), and an arrow showing $+y$

Based on the question below, list the three symbols that you know (based on your labeled diagram) along with their values and SI units. **Hint**: Two of these numbers are **negative**.

__________________ __________________ __________________

For how much time is the banana in the air?

Write the relevant equation. Choose from the three equations of uniform acceleration.

Rewrite the equation with numbers plugged in.

Solve for the unknown.

Want help? Check the hints section at the back of the book.

Answer: 6.0 s

46. Monk Jordan leaps straight upward. He spends exactly 1.0 second in the air.

Draw/label a diagram with the path, initial (i), final (f), and an arrow showing $+y$. **Tip**: Read the question below before you decide where to draw initial (i) and final (f).

Based on the question below, list the three symbols that you know (based on your labeled diagram) along with their values and SI units. **Hints**: One number is **negative**. The time is **not** 1.0 second.

__________________ __________________ __________________

With what initial speed does Monk Jordan jump?

Write the relevant equation. Choose from the three equations of uniform acceleration.

Rewrite the equation with numbers plugged in.

Solve for the unknown.

Want help? Check the hints section at the back of the book.

Answer: 5.0 m/s

4 ONE-DIMENSIONAL MOTION WITH CALCULUS

Calculus-based Motion Equations

Velocity (v_x) is a first-derivative of position (x) with respect to time (t).

$$v_x = \frac{dx}{dt}$$

Acceleration (a_x) is a first-derivative of velocity (v_x) with respect to time (t). Acceleration (a_x) is also a second-derivative of position (x) with respect to time (t).

$$a_x = \frac{dv_x}{dt} = \frac{d^2x}{dt^2}$$

Net displacement (Δx) is the definite integral of velocity (v_x) over time (t). Recall that net displacement is a straight line from the initial position to the final position (see Chapter 3).

$$\Delta x = \int_{t=t_0}^{t} v_x \, dt$$

The **change in velocity** ($v_x - v_{x0}$) is the definite integral of acceleration (a_x) over time (t). Therefore, final velocity (v_x) equals initial velocity (v_{x0}) plus the definite integral of acceleration (a_x) over time (t).

$$v_x = v_{x0} + \int_{t=t_0}^{t} a_x \, dt$$

Symbols and SI Units

Symbol	Name	SI Units
x	position coordinate	m
Δx	net displacement	m
v_{x0}	initial velocity	m/s
v_x	velocity	m/s
a_x	acceleration	m/s^2
t_0	initial time	s
t	time	s

One-dimensional Calculus-based Motion Strategy

If a one-dimensional motion problem gives you position (x), velocity (v_x), or acceleration (a_x) as a function of time, follow this strategy. How you solve the problem depends on what you're given and what you're looking for, as described below.

If you're given **position** (x) as a function of time (t):

1. To find net displacement, simply plug the final time into the position function to find $x(t)$, plug the initial time into the position function to find $x(t_0)$, and subtract the two results. Notation: $x(t)$ means "x evaluated at t." It is **not** x times t.
$$\Delta x = x(t) - x(t_0)$$
2. To find velocity, first take a derivative of the position function with respect to time and then plug the desired time into the velocity function.
$$v_x = \frac{dx}{dt}$$
3. To find acceleration, take two successive derivatives of the position function with respect to time and then plug the desired time into the acceleration function.
$$a_x = \frac{d^2x}{dt^2}$$

If you're given **velocity** (v_x) as a function of time (t):

4. To find net displacement, find the definite integral of the velocity function over the desired time interval.
$$\Delta x = \int_{t=t_0}^{t} v_x \, dt$$
5. To find velocity, simply plug the specified time into the velocity function.
6. To find acceleration, first take a derivative of the velocity function with respect to time and then plug the desired time into the acceleration function.
$$a_x = \frac{dv_x}{dt}$$

If you're given **acceleration** (a_x) as a function of time (t):

7. To find net displacement, first do Step 8 below and then do Step 4 above. Be sure to use the velocity function in Step 4 (and not a numerical value for velocity).
8. To find velocity, first find the definite integral of the acceleration function over the desired time interval and then add this result to the initial velocity.
$$v_x = v_{x0} + \int_{t=t_0}^{t} a_x \, dt$$
9. To find acceleration, simply plug the specified time into the acceleration function.

Example: The position of a monkey is given according to the following equation, where SI units have been suppressed: $x = 3t^4$. Determine the velocity of the monkey at $t = 3.0$ s.

Velocity is the first derivative of position with respect to time.

$$v_x = \frac{dx}{dt} = \frac{d}{dt}(3t^4) = (4)(3)t^{4-1} = 12t^3$$

Evaluate the velocity function at $t = 3.0$ s:

$$v_x(\text{at } t = 3 \text{ s}) = 12(3)^3 = 12(27) = 324 \text{ m/s}$$

Example: The velocity of a gorilla is given according to the following equation, where SI units have been suppressed: $v_x = 9t^2 - 16t$. (A) Determine the net displacement of the gorilla from $t = 1.0$ s until $t = 4.0$ s. (B) Find the acceleration of the gorilla at $t = 4.0$ s.

(A) Net displacement is the definite integral of velocity over time.

$$\Delta x = \int_{t=t_0}^{t} v_x \, dt = \int_{t=1}^{4} (9t^2 - 16t) \, dt$$

Find the anti-derivative of each term and then evaluate the definite integral over the limits following the strategy from Chapter 2. If you don't feel fluent with your integration, you should review the integration section of Chapter 2.

$$\Delta x = \left[\frac{9t^{2+1}}{2+1} - \frac{16t^{1+1}}{1+1}\right]_{t=1}^{4} = \left[\frac{9t^3}{3} - \frac{16t^2}{2}\right]_{t=1}^{4} = [3t^3 - 8t^2]_{t=1}^{4}$$

$$\Delta x = [3(4)^3 - 8(4)^2] - [3(1)^3 - 8(1)^2] = (192 - 128) - (3 - 8) = 64 - (-5) = 69 \text{ m}$$

(B) Acceleration is the first derivative of velocity with respect to time.

$$a_x = \frac{dv_x}{dt} = \frac{d}{dt}(9t^2 - 16t) = (2)(9)t^{2-1} - (1)(16)t = 18t - 16$$

Evaluate the acceleration function at $t = 4.0$ s:

$$a_x(\text{at } t = 4 \text{ s}) = 18(4) - 16 = 72 - 16 = 56 \text{ m/s}^2$$

Example: The acceleration of a lemur is given according to the following equation, where SI units have been suppressed: $a_x = 12t$. The initial velocity of the lemur is 18 m/s at $t = 0$. Determine the velocity of the lemur at $t = 3.0$ s.

Velocity equals initial velocity plus the definite integral of acceleration over time.

$$v_x = v_{x0} + \int_{t=t_0}^{t} a_x \, dt = 18 + \int_{t=0}^{3} 12t \, dt$$

Find the anti-derivative and then evaluate the definite integral over the limits following the strategy from Chapter 2.

$$v_x = 18 + \left[\frac{12t^{1+1}}{1+1}\right]_{t=0}^{3} = 18 + \left[\frac{12t^2}{2}\right]_{t=0}^{3} = 18 + [6t^2]_{t=0}^{3}$$

$$v_x = 18 + 6(3)^2 - 6(0)^2 = 18 + 54 = 72 \text{ m/s}$$

47. The position of a bananamobile is given according to the following equation, where SI units have been suppressed.

$$x = 4t^3 - 8t^2 + 64$$

Write an equation in symbols to determine the velocity of the bananamobile.

Determine the velocity of the bananamobile at $t = 5.0$ s.

Write an equation in symbols to determine the acceleration of the bananamobile.

Determine the acceleration of the bananamobile at $t = 7.0$ s.

Want help? Check the hints section at the back of the book.

Answers: 220 m/s, 152 m/s^2

48. As you begin your physics exam, a monkey grabs your pencil and runs according to the following equation, where SI units have been suppressed.

$$v_x = \sqrt{2t}$$

Write an equation in symbols to determine the net displacement of the monkey.

Determine the net displacement of the monkey from $t = 2.0$ s until $t = 8.0$ s.

Write an equation in symbols to determine the acceleration of the monkey.

Determine the acceleration of the monkey at $t = 8.0$ s.

Want help? Check the hints section at the back of the book.

Answers: $\frac{56}{3}$ m, $\frac{1}{4}$ m/s^2

49. The acceleration of a chimpanzee is given according to the following equation, where SI units have been suppressed. The initial velocity of the chimpanzee is 18 m/s at $t = 0$.

$$a_x = 24t^2$$

Write an equation in symbols to determine the velocity of the chimpanzee.

Determine the velocity of the chimpanzee at $t = 3.0$ s.

Write an equation in symbols to determine the net displacement of the chimpanzee.

Determine the net displacement of the chimpanzee from $t = 0$ until $t = 3.0$ s.

Want help? Check the hints section at the back of the book.

Answers: 234 m/s, 216 m

5 GEOMETRY ESSENTIALS

Handy Formulas

Shape	Quantity	Formula
Square (L, L)	Area	$A = L^2$
Rectangle (W, L)	Perimeter	$P = 2L + 2W$
	Area	$A = LW$
Triangle (h, b)	Area	$A = \frac{1}{2}bh$
Right triangle (c, a, b)	Pythagorean theorem	$a^2 + b^2 = c^2$
Circle (D, R)	Diameter	$D = 2R$
	Area	$A = \pi R^2$
	Circumference	$C = \pi D$ $C = 2\pi R$
Circular Arc (R, θ, s, R)	Arc length	$s = R\theta$ Note: θ must be in radians.

Relevant Terminology

Perimeter – the total distance around the edges of a polygon.
Hypotenuse – the longest side of a right triangle. See side c in the diagram below.
Legs – the two shortest sides of a right triangle. See sides a and b in the diagram below.
Pythagorean theorem – the sum of the squares of the legs of any right triangle equals the square of the hypotenuse.

$$a^2 + b^2 = c^2$$

Circumference – the total distance around the edge of a circle.
Arc length – the distance partway around the edge of a circle.

Symbols and SI Units

Symbol	Name	SI Units
P	perimeter	m
A	area	m^2
L	length	m
W	width	m
b, h	base and height of a triangle	m
a, b	legs of a right triangle	m
c	hypotenuse	m
C	circumference	m
D	diameter	m
R	radius	m
s	arc length	m
θ	angle	radians

Note: The symbol for angle (θ) is the lowercase Greek letter theta.

Geometric Formula Strategy

To solve a problem involving a geometric formula, follow these steps:

1. Find the relevant geometric formula from the table on page 53.
2. Plug in the known values and solve for the unknown.

Example: Find the perimeter and area of the rectangle illustrated below.

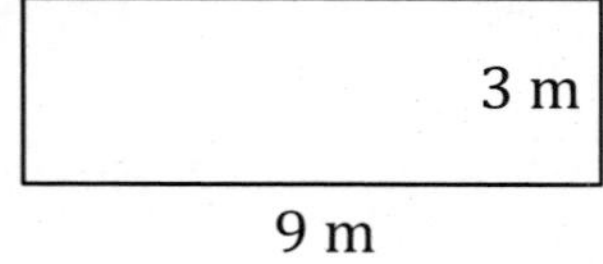

Identify the width and length from the figure:

$$L = 9 \text{ m} \quad , \quad W = 3 \text{ m}$$

Plug these values into the formula for perimeter:

$$P = 2L + 2W = 2(9) + 2(3) = 18 + 6 = 24 \text{ m}$$

Now plug the length and width into the formula for area:

$$A = LW = (9)(3) = 27 \text{ m}^2$$

Example: Find the area of the triangle illustrated below.

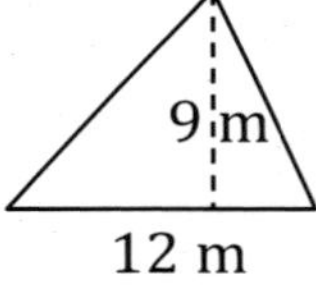

Identify the base and height from the figure:

$$b = 12 \text{ m} \quad , \quad h = 9 \text{ m}$$

Plug these values into the formula for area:

$$A = \frac{1}{2}bh = \frac{1}{2}(12)(9) = 54 \text{ m}^2$$

Example: Determine the hypotenuse of the right triangle illustrated below.

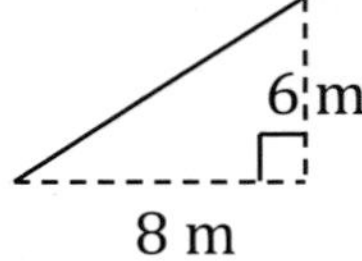

Use the Pythagorean theorem:

$$a^2 + b^2 = c^2$$
$$6^2 + 8^2 = c^2$$
$$36 + 64 = 100 = c^2$$

Squareroot both sides to solve for the unknown:

$$c = \sqrt{100} = 10 \text{ m}$$

Example: Determine the unknown side of the right triangle illustrated below.

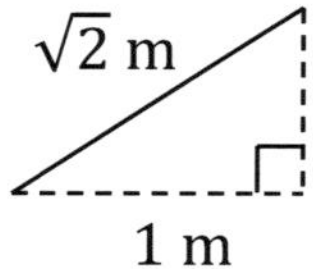

Use the Pythagorean theorem:

$$a^2 + b^2 = c^2$$

$$a^2 + 1^2 = \left(\sqrt{2}\right)^2$$

Recall from algebra that $\left(\sqrt{x}\right)^2 = \sqrt{x}\sqrt{x} = x$. Therefore, $\left(\sqrt{2}\right)^2 = 2$:

$$a^2 + 1 = 2$$

Subtract 1 from both sides to isolate the unknown term:

$$a^2 = 2 - 1 = 1$$

Squareroot both sides to solve for the unknown:

$$a = \sqrt{1} = 1 \text{ m}$$

Example: Find the diameter, circumference, and area of the circle illustrated below.

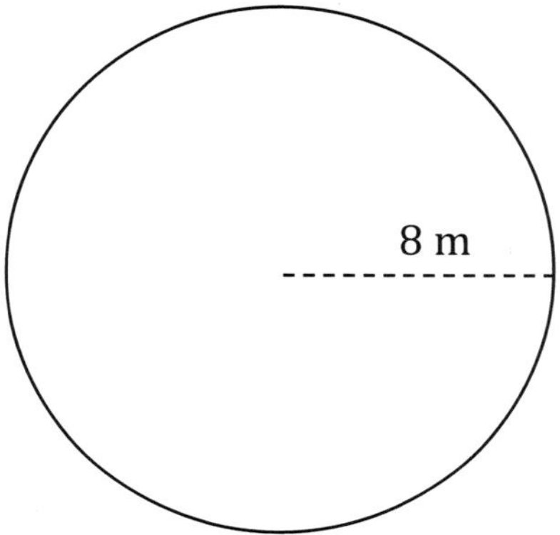

The indicated radius is

$$R = 8 \text{ m}$$

Plug the radius into the formula for diameter:

$$D = 2R = 2(8) = 16 \text{ m}$$

Next plug the radius into the formula for circumference:

$$C = 2\pi R = 2\pi(8) = 16\pi \text{ m}$$

Finally, plug the radius into the formula for area:

$$A = \pi R^2 = \pi(8)^2 = 64\pi \text{ m}^2$$

50. Find the perimeter and area of the rectangle illustrated below.

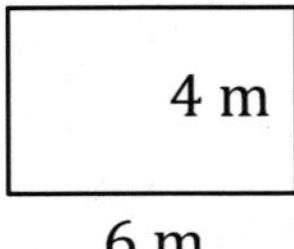

Answers: 20 m, 24 m^2

51. Find the perimeter and area of the right triangle illustrated below.

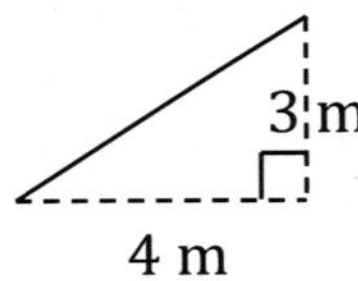

Want help? Check the hints section at the back of the book.

Answers: 12 m, 6 m^2

52. If the area of a square is 36 m^2, what is its perimeter?

Answer: 24 m

53. Determine the hypotenuse of the right triangle illustrated below.

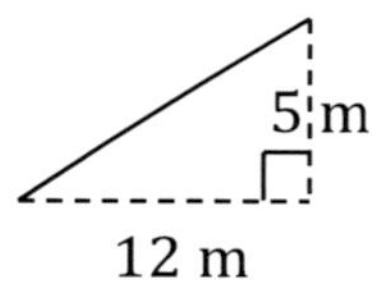

Want help? Check the hints section at the back of the book.

Answer: 13 m

54. Determine the unknown side of the right triangle illustrated below.

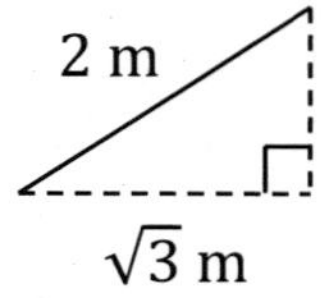

Answer: 1 m

55. Determine the length of the diagonal of the rectangle illustrated below.

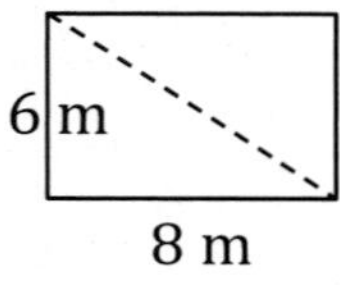

Want help? Check the hints section at the back of the book.

Answer: 10 m

56. Find the radius, circumference, and area of the circle illustrated below.

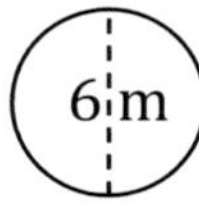

Answers: 3 m, 6π m, 9π m^2

57. If the area of a circle is 16π m^2, what is its circumference?

Want help? Check the hints section at the back of the book.

Answer: 8π m

6 MOTION GRAPHS

Relevant Terminology

Slope – the rise over run of a graph. For a curve, the slope refers to the tangent line.
Area – the region between the curve and the horizontal axis.

Finding Slope

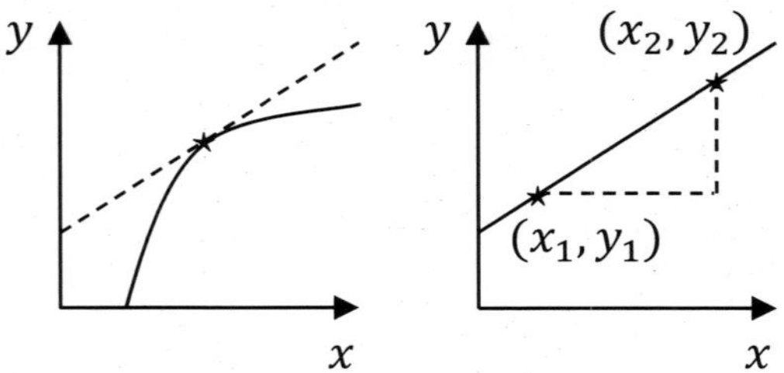

Use this method to find the slope of a graph:

1. If the graph is a curve, draw a tangent line that matches the slope of the curve at the desired point. See the left figure above.
2. Choose two points on the line (**<u>not</u>** the curve) which are far apart (this reduces relative interpolation error). See the right figure above.
3. Read the coordinates of each point, (x_1, y_1) and (x_2, y_2).
4. Plug these coordinates into the slope equation:

$$slope = \frac{y_2 - y_1}{x_2 - x_1}$$

Finding Area

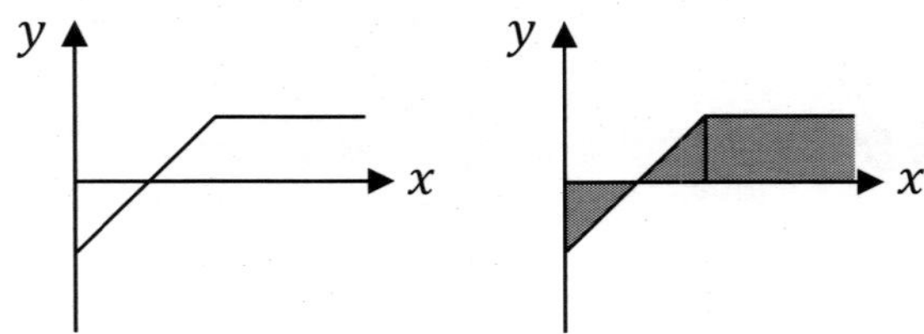

Use this method to find the area between a curve and the horizontal axis:

1. Divide the region between the curve and the horizontal axis into triangles and rectangles as best you can. See the example above.
2. Find the area of each rectangle and triangle.
3. Add these areas together.

Motion Graph Strategy

To solve a problem with a motion graph, first check to see whether position (x), velocity (v_x), or acceleration (a_x) is plotted on the vertical axis.

For a graph of **position** (x) as a function of time (t):

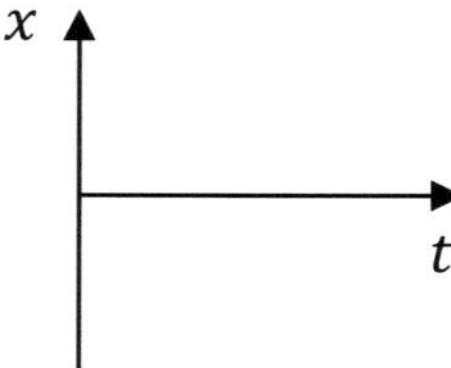

1. Just read the graph directly to find **net displacement** or **total distance traveled**. For the total distance traveled, add up each distance traveled forward or backward in absolute values (as in the first example that follows). For net displacement, use the formula $ND = x_f - x_i$, where x_f is the final position and x_i is the initial position.
2. Find the slope of the tangent line to find **velocity**.

For a graph of **velocity** (v_x) as a function of time (t):

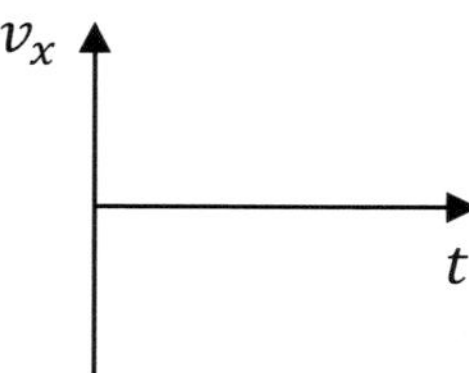

3. Find the area between the curve and the horizontal axis to determine the **net displacement** or the **total distance traveled**. For net displacement, any area below the horizontal axis is negative. For total distance traveled, all areas are positive.
4. Just read the graph directly to find **velocity**.
5. Find the slope of the tangent line to find **acceleration**.

For a graph of **acceleration** (a_x) as a function of time (t):

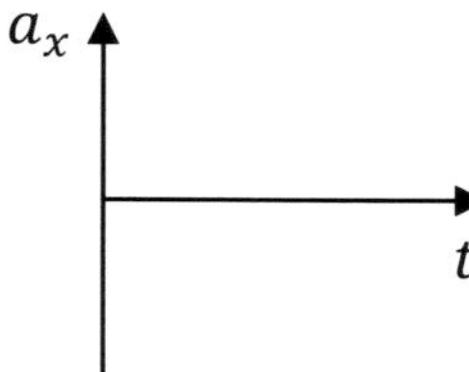

6. Find the area between the curve and the horizontal axis to determine the **change in velocity**. Any area below the horizontal axis is negative. Then use this equation:
$$v_x = v_{x0} + area$$
7. Just read the graph directly to find **acceleration**.

Example: The graph below shows the position of a bananamobile as a function of time.

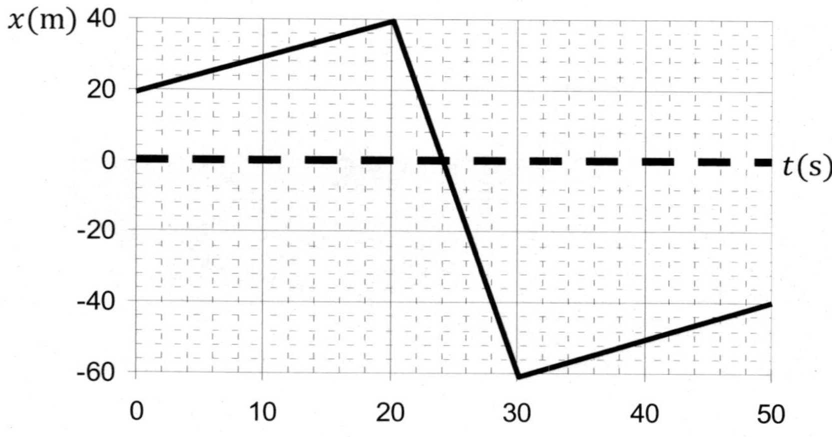

(A) Determine the net displacement.

To find net displacement from a position graph, read the initial and final values of position:

$$x_i = 20 \text{ m}$$
$$x_f = -40 \text{ m}$$

Use the formula for net displacement:

$$ND = x_f - x_i = -40 - 20 = -60 \text{ m}$$

(B) Determine the total distance traveled.

To find total distance traveled from a position graph, read the position values to determine how far the object moves forward or backward in each segment of the trip, and then add these increments in absolute values:

- For the first 20 s, the object moves forward 20 m (from $x = 20$ m to $x = 40$ m).
- For the next 10 s, the object moves backward 100 m (from $x = 40$ m to $x = -60$ m).
- For the last 20 s, the object moves forward 20 m (from $x = -60$ m to $x = -40$ m).

Add these increments up in absolute values:

$$TDT = |d_1| + |d_2| + |d_3| = |20| + |-100| + |20| = 20 + 100 + 20 = 140 \text{ m}$$

(C) Determine the velocity at $t = 25$ s.

To find velocity from a position graph, find the slope. We want the slope of the line where $t = 25$ s. Let's use the endpoints of this line. Read off the coordinates of the endpoints:

$$(t_1, x_1) = (20 \text{ s}, 40 \text{ m})$$
$$(t_2, x_2) = (30 \text{ s}, -60 \text{ m})$$

Plug these values into the equation for slope:

$$v_x = \frac{x_2 - x_1}{t_2 - t_1} = \frac{-60 - 40}{30 - 20} = \frac{-100}{10} = -10 \text{ m/s}$$

The velocity is negative because the slope is negative: The object is moving backward.

Note that time (t) is on the horizontal axis, while position (x) is on the vertical axis.

Example: The graph below shows the velocity of a bananamobile as a function of time.

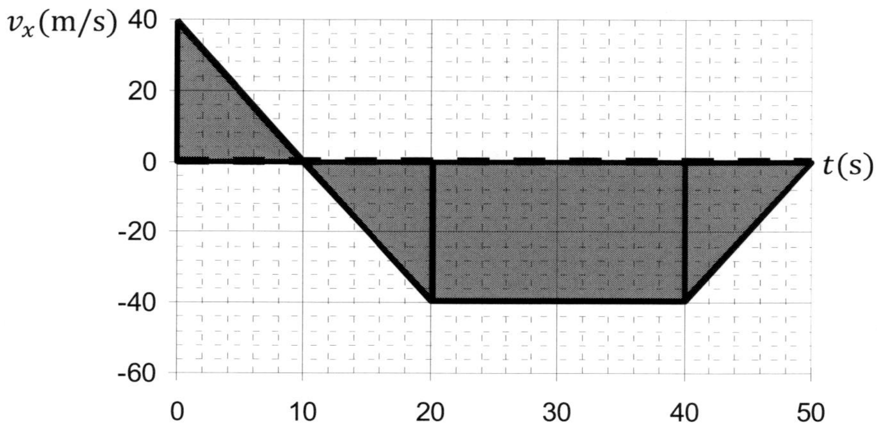

(A) Determine the net displacement.

To find net displacement from a velocity graph, divide the region between the curve (in this case the "curve" is made up of straight lines) and the horizontal axis into triangles and rectangles. See the three triangles and rectangle in the diagram above. Find these areas:

$$A_1 = \frac{1}{2}b_1h_1 = \frac{1}{2}(10)(40) = 200 \text{ m}$$
$$A_2 = \frac{1}{2}b_2h_2 = \frac{1}{2}(10)(-40) = -200 \text{ m}$$
$$A_3 = L_3W_3 = (20)(-40) = -800 \text{ m}$$
$$A_4 = \frac{1}{2}b_4h_4 = \frac{1}{2}(10)(-40) = -200 \text{ m}$$

Note: The last three areas are negative because they lie below the t-axis. (However, if we had been finding total distance traveled, we would instead make all of the areas positive.) Add these four areas together:

$$ND = A_1 + A_2 + A_3 + A_4 = 200 - 200 - 800 - 200 = -1000 \text{ m}$$

(B) Determine the speed at $t = 30$ s.

To find velocity from a velocity graph, simply read the graph. At $t = 30$ s, the velocity is –40 m/s. This means that the object is moving 40 m/s backward. The speed is 40 m/s.

(C) Determine the acceleration at $t = 45$ s.

To find acceleration from a velocity graph, find the slope. We want the slope of the line where $t = 45$ s. Read off the coordinates of the endpoints:

$$(t_1, v_{1x}) = (40 \text{ s}, -40 \text{ m/s})$$
$$(t_2, v_{2x}) = (50 \text{ s}, 0)$$

Plug these values into the equation for slope:

$$a_x = \frac{v_{2x} - v_{1x}}{t_2 - t_1} = \frac{0 - (-40)}{50 - 40} = \frac{40}{10} = 4.0 \text{ m/s}^2$$

Note that time (t) is on the horizontal axis, while velocity (v_x) is on the vertical axis.

Example: The graph below shows the acceleration of a bananaplane as a function of time. The initial velocity of the bananaplane is 50 m/s.

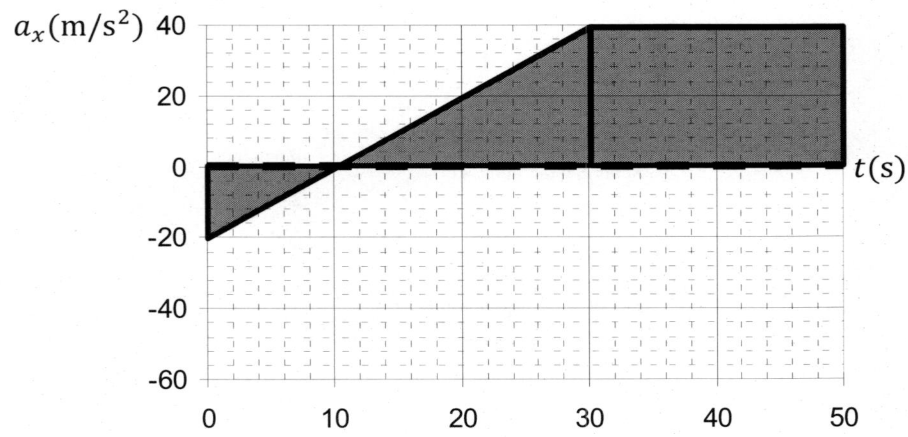

(A) Determine the final velocity of the bananaplane.
To find velocity from an acceleration graph, divide the region between the curve (in this case the "curve" is made up of straight lines) and the horizontal axis into triangles and rectangles. See the two triangles and rectangle in the diagram above. Find these areas:

$$A_1 = \frac{1}{2}b_1h_1 = \frac{1}{2}(10)(-20) = -100 \text{ m}$$

$$A_2 = \frac{1}{2}b_2h_2 = \frac{1}{2}(20)(40) = 400 \text{ m}$$

$$A_3 = L_3W_3 = (20)(40) = 800 \text{ m}$$

Note: The first area is negative because it lies below the t-axis.
Add these three areas together to determine the total area:

$$area = A_1 + A_2 + A_3 = -100 + 400 + 800 = 1100 \text{ m}$$

Plug this area into the equation for final velocity:

$$v_x = v_{x0} + area = 50 + 1100 = 1150 \text{ m/s}$$

(B) Determine the acceleration at $t = 20$ s.
To find acceleration from an acceleration graph, simply read the graph. At $t = 20$ s, the acceleration is 20 m/s^2.

WHY FIND SLOPE AND AREA?

The answer has to do with calculus:

- Velocity is a derivative of position with respect to time, and acceleration is a derivative of velocity with respect to time. In calculus, a derivative represents the slope of the tangent line.
- Net displacement is the integral of velocity, and change in velocity is the integral of acceleration. In calculus, a definite integral represents the area under the curve.

58. A monkey drives a jeep. The position as a function of time is illustrated below.

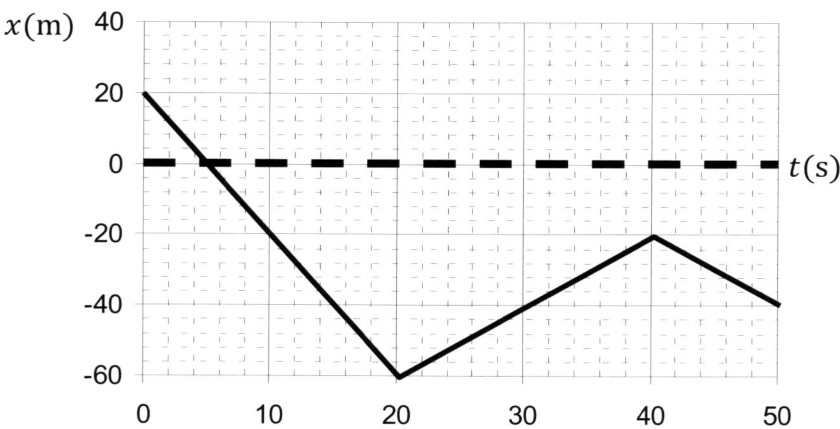

(A) Find the total distance traveled and net displacement for the trip.
First circle the appropriate word:

read slope area

Now show your work:

(B) Find the velocity of the jeep at $t =$ 10 s.
First circle the appropriate word:

read slope area

Now show your work:

Want help? Check the hints section at the back of the book.

Answers: 140 m, –60 m, –4.0 m/s

59. A monkey rides a bicycle. The velocity as a function of time is illustrated below.

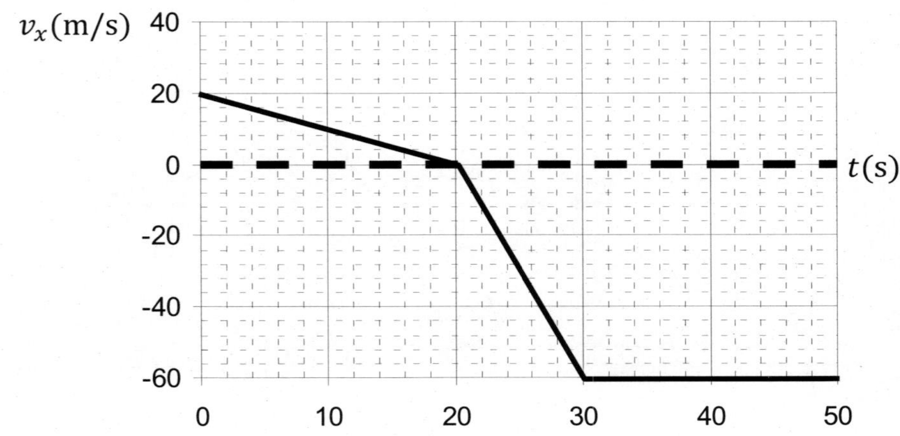

(A) What is the bicycle's acceleration at $t = 25$ seconds?
First circle the appropriate word:

read slope area

Now show your work:

(B) What is the bicycle's net displacement for the whole trip?
First circle the appropriate word:

read slope area

Now show your work:

Want help? Check the hints section at the back of the book.

Answers: -6.0 m/s^2, -1.3 km

60. A monkey flies a bananaplane. The acceleration as a function of time is illustrated below. The initial velocity of the bananaplane is 150 m/s.

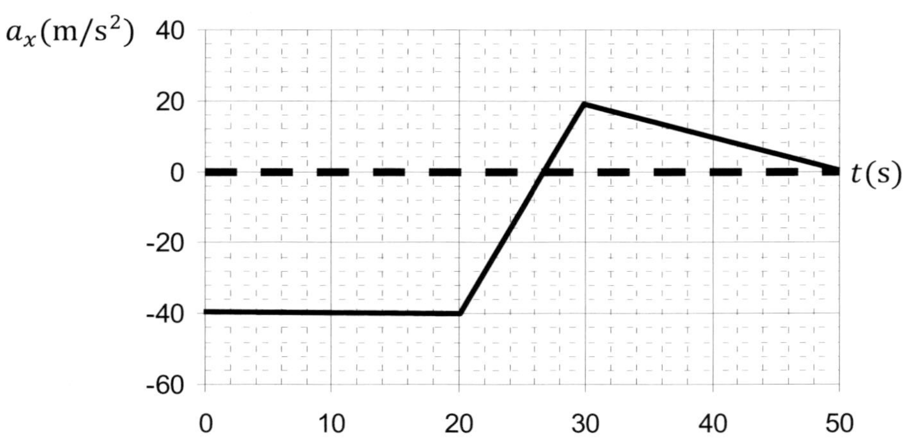

(A) What is the final velocity of the bananaplane?
First circle the appropriate word:

read slope area

Now show your work:

(B) When is the acceleration of the bananaplane equal to zero?
First circle the appropriate word:

read slope area

Now show your work:

Want help? Check the hints section at the back of the book.

Answers: approximately –545 m/s, 26.7 s and 50.0 s

7 TWO OBJECTS IN MOTION

Two Objects in Motion Strategy

For a motion problem where one object is chasing another, or where two objects head toward one another, follow these steps:

1. Draw a diagram showing the two paths. Label the initial positions (i_1 and i_2), final positions (f_1 and f_2), and the coordinate (x or y, as appropriate).
2. Identify the unknown symbol and three known symbols. Use subscripts for any quantities that may be different between objects 1 and 2. For example, if the objects have different net displacements, distinguish between Δx_1 and Δx_2, but if they both start and finish in the same positions, call them both Δx. Similarly, if one object has a headstart, distinguish between t_1 and t_2, but if they both start and finish at the same time, call both times t.
3. Write an equation for the net displacement of each object using subscripts as appropriate (see the previous note).
$$\Delta x_1 = v_{10} t_1 + \frac{1}{2} a_1 t_1^2$$
$$\Delta x_2 = v_{20} t_2 + \frac{1}{2} a_2 t_2^2$$
For a vertical problem, use y in place of x. **Special case**: If either object travels with constant velocity, its acceleration will equal zero (since acceleration is the instantaneous rate at which velocity changes).
4. Write an equation of constraint. If the problem gives you information relating to distance, your **equation of constraint** may look something like
$$\Delta x_1 + \Delta x_2 = d \quad \text{or} \quad \Delta x_1 - \Delta x_2 = d$$
However, if the problem gives you information relating to time (like a headstart), your equation of constraint may look more like
$$t_1 = t_2 + \Delta t \quad \text{or} \quad t_2 = t_1 + \Delta t$$
The examples that follow will show you how to reason out two common types of constraints.
5. Use the method of substitution (described Chapter 1), as illustrated in the following examples.

Example: Two monkeys, initially 600 m apart, begin running directly toward one another at the same time. One monkey uniformly accelerates from rest at $\frac{1}{2}$ m/s^2, while the other monkey uniformly accelerates from rest at $\frac{1}{4}$ m/s^2. Where do the two monkeys meet?

Begin with a labeled diagram. They begin in different positions, but meet up at the end.

i_1 •————▶ f ◀———• i_2

$+x$ ⟶

List the knowns, using subscripts where appropriate. The acceleration of the second monkey is negative because he is heading in the opposite direction.

$$d = 600 \text{ m} \quad , \quad v_{10} = 0 \quad , \quad a_1 = \frac{1}{2} \text{ m/s}^2 \quad , \quad v_{20} = 0 \quad , \quad a_2 = -\frac{1}{4} \text{ m/s}^2$$

Note that neither Δx_1 nor Δx_2 equals 600 m, as discussed below. Write an equation for the net displacement of each monkey, using subscripts for any quantities that may be different. Since they spend the same amount of time running, use t for time instead of t_1 and t_2.

$$\Delta x_1 = v_{10}t + \frac{1}{2}a_1 t^2$$

$$\Delta x_2 = v_{20}t + \frac{1}{2}a_2 t^2$$

Plug the knowns into this equation. To avoid clutter, suppress the units until the end.

$$\Delta x_1 = 0 + \frac{1}{2}\left(\frac{1}{2}\right)t^2 = \frac{1}{4}t^2$$

$$\Delta x_2 = 0 + \frac{1}{2}\left(-\frac{1}{4}\right)t^2 = -\frac{1}{8}t^2$$

Write an equation of constraint. Together, the two monkeys travel a total distance equal to 600 m. Since the second monkey's net displacement is negative, we must include a minus sign to make them "add" up to 600 m (since two minuses make a plus):

$$\Delta x_1 - \Delta x_2 = d$$

$$\Delta x_1 - \Delta x_2 = 600$$

Substitute the net displacement equations into the equation of constraint:

$$\frac{1}{4}t^2 - \left(-\frac{1}{8}t^2\right) = 600$$

$$\frac{1}{4}t^2 + \frac{1}{8}t^2 = \left(\frac{1}{4} + \frac{1}{8}\right)t^2 = \left(\frac{2}{8} + \frac{1}{8}\right)t^2 = \frac{3}{8}t^2 = 600$$

$$t = \sqrt{\frac{8}{3}(600)} = \sqrt{\frac{4800}{3}} = \sqrt{1600} = 40 \text{ s}$$

Plug the time into either equation for net displacement:

$$\Delta x_1 = \frac{1}{4}t^2 = \frac{1}{4}(40)^2 = \frac{1}{4}1600 = 400 \text{ m}$$

Therefore, they meet 400 m from where the first monkey started.

Example: A gorilla and chimpanzee are initially parked side by side on a street. The chimpanzee uniformly accelerates from rest at 4.0 m/s^2. After waiting 6.0 s, the gorilla uniformly accelerates from rest at 16.0 m/s^2. When does the gorilla catch the chimpanzee?

Begin with a labeled diagram. They both begin and finish in the same positions.

$$i \bullet \longrightarrow f \quad +x$$

List the knowns, using subscripts where appropriate.

$$\Delta t = 6.0 \text{ s} \quad , \quad v_{10} = 0 \quad , \quad a_1 = 4.0 \text{ m/s}^2 \quad , \quad v_{20} = 0 \quad , \quad a_2 = 16.0 \text{ m/s}^2$$

Note that 6.0 s is the **difference** between the two times (see below). Write an equation for the net displacement of each object, using subscripts for any quantities that may be different. Since one has a headstart, use t_1 and t_2, but since they both start and finish in the same position, use Δx instead of Δx_1 and Δx_2.

$$\Delta x = v_{10} t_1 + \frac{1}{2} a_1 t_1^2$$

$$\Delta x = v_{20} t_2 + \frac{1}{2} a_2 t_2^2$$

Plug the knowns into this equation. To avoid clutter, suppress the units until the end.

$$\Delta x = 0 + \frac{1}{2} 4 t_1^2 = 2 t_1^2$$

$$\Delta x = 0 + \frac{1}{2} 16 t_2^2 = 8 t_2^2$$

Write an equation of constraint. Since the chimpanzee starts 6.0 s sooner than the gorilla, the chimpanzee spends more time running. Therefore, the chimpanzee's time (t_1) spent driving must be 6.0 s greater than the gorilla's time (t_2) spent driving:

$$t_1 = t_2 + \Delta t$$

$$t_1 = t_2 + 6$$

Substitute this time constraint into the first net displacement equation:

$$\Delta x = 2 t_1^2 = 2(t_2 + 6)^2 = 2(t_2^2 + 12 t_2 + 36) = 2 t_2^2 + 24 t_2 + 72$$

Now set the two net displacement equations equal to each other (since both equal Δx):

$$2 t_2^2 + 24 t_2 + 72 = 8 t_2^2$$

Combine like terms and reorder terms to put this quadratic equation in standard form:

$$-6 t_2^2 + 24 t_2 + 72 = 0$$

Use the quadratic formula (see Chapter 1):

$$t_2 = \frac{-b \pm \sqrt{b^2 - 4ac}}{2a} = \frac{-24 \pm \sqrt{24^2 - 4(-6)(72)}}{2(-6)} = \frac{-24 \pm \sqrt{2304}}{-12} = \frac{-24 \pm 48}{-12}$$

$$t_2 = -2.0 \text{ s or } 6.0 \text{ s}$$

They meet after the gorilla has driven for 6.0 s, which means that they meet after the chimpanzee has driven for 12.0 s (since $t_1 = t_2 + 6 = 6 + 6 = 12.0$ s).

61. Two monkeys, initially 1600 m apart, begin running directly toward one another at the same time. One monkey uniformly accelerates from rest at $\frac{1}{8}$ m/s^2, while the other monkey runs with a constant speed of 15 m/s.

Draw/label a diagram with the paths, initial positions, final position, and a coordinate axis.

Based on the question below, list the symbols that you know (based on your labeled diagram) along with their values and SI units.

What is the net displacement of each monkey when they meet?

Write two equations – one for the net displacement of each monkey – using subscripts.

Write the equation of constraint.

Solve for the unknowns.

Want help? Check the hints section at the back of the book.

Answer: 400 m, –1200 m

62. A monkey steals his uncle's banana and runs away with a constant speed of 9.0 m/s, while his uncle uniformly accelerates from rest at 4.0 m/s^2. The thief has a 2.0-s headstart.

Draw/label a diagram with the paths, initial position, final position, and a coordinate axis.

Based on the question below, list the symbols that you know (based on your labeled diagram) along with their values and SI units.

What is the net displacement of each monkey when the thief is caught?

Write two equations – one for the net displacement of each monkey – using subscripts.

Write the equation of constraint.

Solve for the unknown.

Want help? Check the hints section at the back of the book.

Answer: 72 m

63. A monkey at the top of a 90-m tall cliff parachutes downward with a constant speed of 5.0 m/s at the same time as a monkey at the bottom of the cliff throws a banana straight upward with an initial speed of 40 m/s.

Draw/label a diagram with the paths, initial positions, final position, and a coordinate axis.

Based on the question below, list the symbols that you know (based on your labeled diagram) along with their values and SI units.

Where is the banana when it reaches the parachuting monkey?

Write two equations – one for the net displacement of each monkey – using subscripts.

Write the equation of constraint.

Solve for the unknown.

Want help? Check the hints section at the back of the book.

Answer: 75 m above the ground

8 NET AND AVERAGE VALUES

Relevant Terminology

Total distance traveled – the sum total of all the distances traveled.
Net displacement – a straight line from the initial position to the final position.

Important Distinction

Net displacement is the shortest distance between the initial position (i) and final position (f). Net displacement does not depend on the path taken. Total distance traveled is the sum of all the distances. Total distance traveled does depend on the path taken.

Strategy to Find Net or Total Values

If a problem asks you to find the total distance traveled or the net displacement of an object, follow these steps:

1. Draw a diagram of the path. Label the initial position (i) and final position (f).
2. Determine the quantity needed:
   - For the total distance traveled (TDT), add up all of the distances for each part of the trip.
   - For the net displacement (ND), find the shortest distance between the initial position (i) and final position (f).

Symbols and SI Units

Symbol	Name	SI Units
TDT	total distance traveled	m
ND	net displacement	m

Net Displacement

Δx and Δy are the components of the net displacement (ND) vector. If an object travels along two the sides of a right triangle, use the Pythagorean theorem $ND = \sqrt{\Delta x^2 + \Delta y^2}$ to find the magnitude of the net displacement. If instead an object travels back and forth along a straight line, use the formula $ND = \Delta x_1 + \Delta x_2 + \cdots + \Delta x_N$ to determine the net displacement, noting that one or more of the Δx's may be negative.

Example: A monkey travels 400 m east and then 300 m north. Find the total distance traveled and the magnitude of the net displacement for the whole trip.

Begin with a labeled diagram.

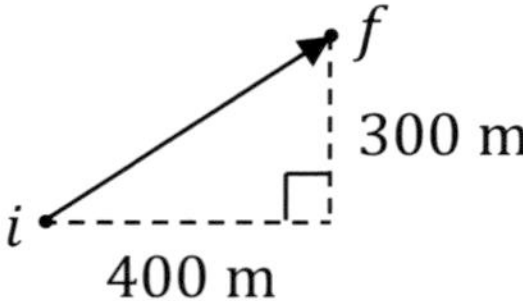

The total distance traveled is the sum of the two distances:

$$TDT = d_1 + d_2 = 400 + 300 = 700 \text{ m}$$

The net displacement is the hypotenuse of the right triangle:

$$ND = \sqrt{d_1^2 + d_2^2} = \sqrt{300^2 + 400^2}$$

If you're not using a calculator, it's convenient to factor out 100^2.

$$ND = \sqrt{100^2(3^2 + 4^2)}$$

This works since $100^2(3^2 + 4^2) = (100 \times 3)^2 + (100 \times 4)^2 = 300^2 + 400^2$.

$$ND = \sqrt{100^2}\sqrt{3^2 + 4^2} = 100\sqrt{9 + 16} = 100\sqrt{25} = 500 \text{ m}$$

Example: A monkey walks 60 m east, 60 m north, 60 m west, and then 60 m south. Find the total distance traveled and the magnitude of the net displacement for the whole trip.

Begin with a labeled diagram.

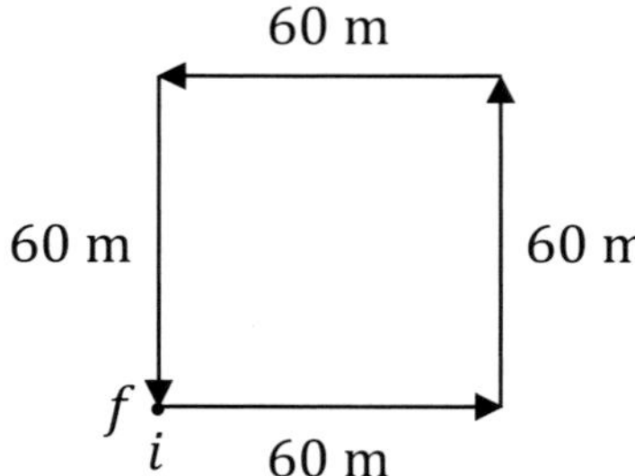

The total distance traveled is the sum of the four distances:

$$TDT = d_1 + d_2 + d_3 + d_4 = 60 + 60 + 60 + 60 = 240 \text{ m}$$

The net displacement is zero since the final position is the same as the initial position for the total trip. Note that the four segments of the trip form a square.

64. A monkey travels 15 m east, then 70 m west, and finally 35 m east.

Draw/label your diagram here, including initial (i), final (f), and the path.

Write an equation in symbols to find the total distance traveled.

Rewrite the equation with numbers plugged in. Complete the calculation.

Write an equation in symbols to find the net displacement.

Rewrite the equation with numbers plugged in. Complete the calculation.

Want help? Check the hints section at the back of the book.

Answers: 120 m, 20 m west

Relevant Terminology

Average speed – the total distance traveled divided by the elapsed time.
Average velocity – the net displacement divided by the elapsed time.
Average acceleration – the change in velocity divided by the elapsed time.

Strategy to Find Average Values

If a problem asks you to find average speed, average velocity, or average acceleration, follow these steps:

1. Draw a diagram of the path. Label the initial position (i) and final position (f).
2. Determine the total time of the trip (TT).
3. Determine the intermediate quantity needed:
   - For **average speed**, find the total distance traveled (TDT).
   - For **average velocity**, find the net displacement (ND).
   - For **average acceleration**, find the initial velocity (v_i) and final velocity (v_f).
4. Plug the values from Steps 2 and 3 into the appropriate equation below.

$$\begin{matrix}\text{ave.}\\\text{spd.}\end{matrix} = \frac{TDT}{TT}$$

$$\begin{matrix}\text{ave.}\\\text{vel.}\end{matrix} = \frac{ND}{TT}$$

$$\begin{matrix}\text{ave.}\\\text{accel.}\end{matrix} = \frac{v_f - v_i}{TT}$$

Symbols and SI Units

Symbol	Name	SI Units
TT	total time	s
TDT	total distance traveled	m
ND	net displacement	m
ave. spd.	average speed	m/s
ave. vel.	average velocity	m/s
ave. accel.	average acceleration	m/s^2

Example: A monkey drives 40 m/s to the east for 120 m, and then drives 60 m/s to the west for 120 m. What are the monkey's average speed and average velocity?

Begin with a labeled diagram.

i_1, f_2 ——— f_1, i_2 ; $+x$

The average speed is **not** the average of 40 m/s and 60 m/s. It is **not** 50 m/s. The reason is that the monkey spends more time driving 40 m/s and less time driving 60 m/s. It's really a weighted average, but you can find the answer simply using the equation for average speed. Let's first find the total time. The time for each part of the trip equals the distance divided the by speed for that part of the trip:

$$t_1 = \frac{d}{v_1} = \frac{120}{40} = 3.0 \text{ s} \quad , \quad t_2 = \frac{d}{v_2} = \frac{120}{60} = 2.0 \text{ s}$$

The total time for the whole trip is therefore:

$$TT = t_1 + t_2 = 3 + 2 = 5.0 \text{ s}$$

The total distance traveled is:

$$TDT = d + d = 120 + 120 = 240 \text{ m}$$

Divide the total distance traveled by the total time to find the average speed:

$$\begin{matrix}\text{ave.}\\ \text{spd.}\end{matrix} = \frac{TDT}{TT} = \frac{240}{5} = 48 \text{ m/s}$$

As predicted, the average speed is less than 50 m/s because the monkey spent more time traveling at the slower speed. The average speed is 48 m/s. In contrast, the average velocity is zero because the net displacement is zero (since the initial position and final position are identical for the total trip).

Example: A bananamobile accelerates from 18 m/s to 50 m/s in 4.0 s. What is the average acceleration?

Begin with a labeled diagram.

i ——— f ; $+x$

List the three knowns.

$$v_i = 18 \text{ m/s} \quad , \quad v_f = 50 \text{ m/s} \quad , \quad TT = 4.0 \text{ s}$$

Plug these values into the equation for average acceleration:

$$\begin{matrix}\text{ave.}\\ \text{accel.}\end{matrix} = \frac{v_f - v_i}{TT} = \frac{50 - 18}{4} = \frac{32}{4} = 8.0 \text{ m/s}^2$$

65. A gorilla skates 120 m to the west for 3.0 s, then skates 180 m to the east for 7.0 s.

Draw/label your diagram here, including initial (i), final (f), and the path.

Write an equation in symbols to find the total time.

Rewrite the equation with numbers plugged in. Complete the calculation.

Write an equation in symbols to find the total distance traveled.

Rewrite the equation with numbers plugged in. Complete the calculation.

Write an equation in symbols to find the net displacement.

Rewrite the equation with numbers plugged in. Complete the calculation.

Write an equation in symbols to find the average speed.

Rewrite the equation with numbers plugged in. Complete the calculation.

Write an equation in symbols to find the average velocity.

Rewrite the equation with numbers plugged in. Complete the calculation.

Want help? Check the hints section at the back of the book.

Answers: 10.0 s, 300 m, 60 m east, 30 m/s, 6.0 m/s east

66. A chimpanzee travels 60 m to the east for 4.0 s, then skates 80 m to the south for 16.0 s.

Draw/label your diagram here, including initial (i), final (f), and the path.

Write an equation in symbols to find the total time.

Rewrite the equation with numbers plugged in. Complete the calculation.

Write an equation in symbols to find the total distance traveled.

Rewrite the equation with numbers plugged in. Complete the calculation.

Write an equation in symbols to find the magnitude of the net displacement.

Rewrite the equation with numbers plugged in. Complete the calculation.

Write an equation in symbols to find the average speed.

Rewrite the equation with numbers plugged in. Complete the calculation.

Write an equation in symbols to find the magnitude of the average velocity.

Rewrite the equation with numbers plugged in. Complete the calculation.

Want help? Check the hints section at the back of the book.

Answers: 20 s, 140 m, 100 m, 7.0 m/s, 5.0 m/s

67. A gorilla initially traveling 5.0 m/s to the east uniformly accelerates for 4.0 s, by which time the gorilla has a velocity of 25.0 m/s to the east. The gorilla then maintains constant velocity for 10.0 s. Next, the gorilla uniformly decelerates for 6.0 s until coming to rest.

Draw/label your diagram here, including initial (i), final (f), and the path.

Write an equation in symbols to find the total time.

Rewrite the equation with numbers plugged in. Complete the calculation.

Write an equation in symbols to find the average acceleration.

Rewrite the equation with numbers plugged in. Complete the calculation.

Want help? Check the hints section at the back of the book.

Answer: -0.25 m/s^2 (the – sign means to the west)

9 TRIGONOMETRY ESSENTIALS

Right Triangle Trig Strategy

In a right triangle, each trig function expresses a ratio of two particular sides. The three basic trig functions are the sine ($\sin\theta$), cosine ($\cos\theta$), and tangent ($\tan\theta$).

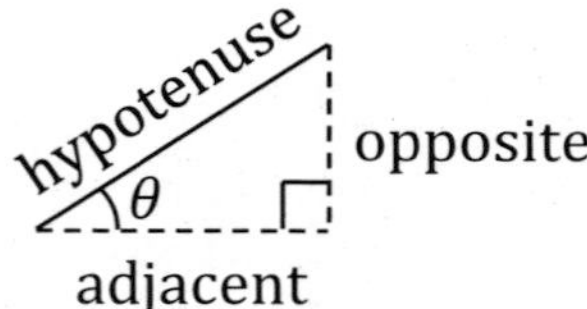

To apply trigonometry to a right triangle, follow these steps:

1. Look at the right triangle and identify the sides which are opposite and adjacent to the desired angle. Also identify the hypotenuse. See the illustration above.
2. If needed, use the **Pythagorean theorem** (Chapter 5) to solve for an unknown side.
3. Plug the relevant sides into the formula for the trig function:

$$\sin\theta = \frac{\text{opposite}}{\text{hypotenuse}}$$
$$\cos\theta = \frac{\text{adjacent}}{\text{hypotenuse}}$$
$$\tan\theta = \frac{\text{opposite}}{\text{adjacent}}$$

Example: Find the sine, cosine, and tangent of θ in the diagram below.

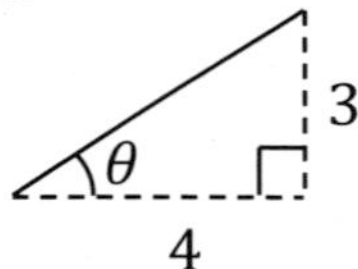

First use the Pythagorean theorem (Chapter 5) to solve for the unknown side:

$$c^2 = a^2 + b^2 = 3^2 + 4^2 = 9 + 16 = 25$$
$$c = \sqrt{25} = 5$$

Identify the sides opposite and adjacent to θ. Also identify the hypotenuse.

opposite	3
adjacent	4
hypotenuse	5

Plug these values into the formulas for the trig functions:

$$\sin\theta = \frac{\text{opp.}}{\text{hyp.}} = \frac{3}{5} \quad , \quad \cos\theta = \frac{\text{adj.}}{\text{hyp.}} = \frac{4}{5} \quad , \quad \tan\theta = \frac{\text{opp.}}{\text{adj.}} = \frac{3}{4}$$

68. Find the sine, cosine, and tangent of θ in the diagram below.

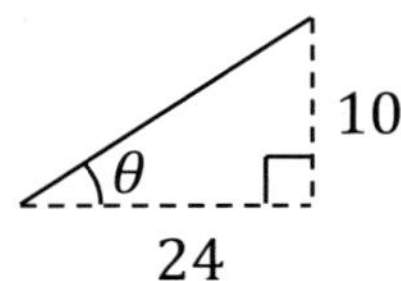

Answers: $\frac{5}{13}, \frac{12}{13}, \frac{5}{12}$

69. Find the sine, cosine, and tangent of θ in the diagram below.

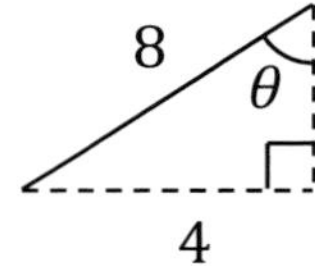

Want help? Check the hints section at the back of the book.

Answers: $\frac{1}{2}, \frac{\sqrt{3}}{2}, \frac{\sqrt{3}}{3}$

Standard Trig Values

θ	0°	30°	45°	60°	90°
$\sin\theta$	0	$\frac{1}{2}$	$\frac{\sqrt{2}}{2}$	$\frac{\sqrt{3}}{2}$	1
$\cos\theta$	1	$\frac{\sqrt{3}}{2}$	$\frac{\sqrt{2}}{2}$	$\frac{1}{2}$	0
$\tan\theta$	0	$\frac{\sqrt{3}}{3}$	1	$\sqrt{3}$	undef.

Note: The tangent of 90° is undefined.

Memorization Tip

There is a simple method to reproduce the above chart:

1. Write the values 0 thru 4 in order: 0, 1, 2, 3, 4.
2. Squareroot each number: $0, 1, \sqrt{2}, \sqrt{3}, 2$. (Note that $\sqrt{4} = 2$.)
3. Divide each number by two: $0, \frac{1}{2}, \frac{\sqrt{2}}{2}, \frac{\sqrt{3}}{2}, 1$. These are the values of $\sin\theta$.
4. Write the numbers backwards: $1, \frac{\sqrt{3}}{2}, \frac{\sqrt{2}}{2}, \frac{1}{2}, 0$. These are the values of $\cos\theta$.
5. Divide the previous two rows: $0, \frac{\sqrt{3}}{3}, 1, \sqrt{3}$, undef. These are the values of $\tan\theta$. Note that $\frac{1}{\sqrt{3}} = \frac{\sqrt{3}}{3}$. (Multiply the numerator and denominator by $\sqrt{3}$.)

Standard Trig Values Strategy

To find the sine, cosine, or tangent of a standard angle in Quadrant I, follow these steps:

1. Find the angle on the top row of the chart. (If the angle you need isn't on the chart, use a calculator instead of the chart.)
2. Find the row for the desired trig function.
3. Read off the trig value.

Example: Find the sine, cosine, and tangent of 30°.
According to the chart:

$$\sin 30° = \frac{1}{2} \quad , \quad \cos 30° = \frac{\sqrt{3}}{2} \quad , \quad \tan 30° = \frac{\sqrt{3}}{3}$$

70. Practice evaluating trig functions at standard angles.

(A) $\sin 60° =$

(B) $\cos 45° =$

(C) $\tan 30° =$

(D) $\sin 45° =$

(E) $\cos 30° =$

(F) $\tan 60° =$

(G) $\sin 90° =$

(H) $\cos 90° =$

(I) $\tan 45° =$

(J) $\sin 30° =$

(K) $\cos 60° =$

(L) $\tan 90° =$

(M) $\sin 0° =$

(N) $\cos 0° =$

(O) $\tan 0° =$

(P) $\sin 60° =$

(Q) $\cos 30° =$

(R) $\tan 45° =$

Answers: (A) $\frac{\sqrt{3}}{2}$ (B) $\frac{\sqrt{2}}{2}$ (C) $\frac{\sqrt{3}}{3}$ (D) $\frac{\sqrt{2}}{2}$ (E) $\frac{\sqrt{3}}{2}$ (F) $\sqrt{3}$ (G) 1 (H) 0 (I) 1 (J) $\frac{1}{2}$
(K) $\frac{1}{2}$ (L) undef. (M) 0 (N) 1 (O) 0 (P) $\frac{\sqrt{3}}{2}$ (Q) $\frac{\sqrt{3}}{2}$ (R) 1

Relevant Terminology

Quadrant – one of four regions of the two-dimensional Cartesian coordinate system defined by x and y. The four Quadrants are labeled I, II, III, and IV in a counterclockwise sense, as illustrated below.
Reference angle – the smallest angle with either the positive or negative x-axis (whereas the argument of a trig function is instead counter-clockwise from the positive x-axis).

Trig Values Beyond Quadrant I

Each trig function yields negative values in specific quadrants, as indicated below.

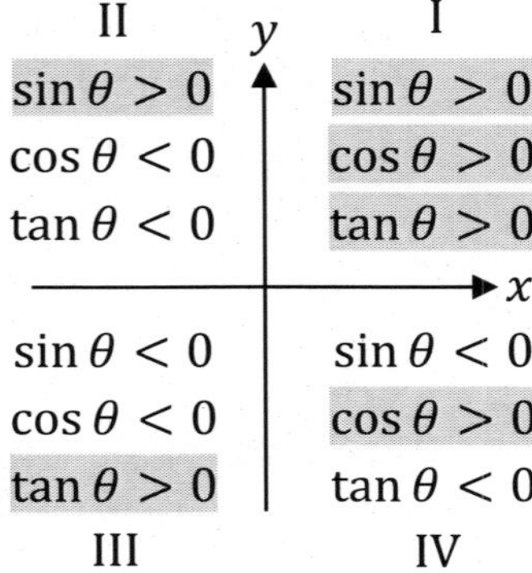

Memorization Tip

One way to remember which trig functions are negative in which quadrants is to memorize the sentence, "Apes study trig calculations." The first letter of each word will help you remember that "all" are positive in Quadrant I, "sine" is positive in Quadrant II, "tangent" is positive in Quadrant III, and "cosine" is positive in Quadrant IV. Everything else is negative.

Reference Angle

When not using a calculator, the way to deal with trig functions outside of Quadrant I is to determine the reference angle. The reference angle is the smallest angle with either the positive or negative x-axis, whereas the angle that appears in the argument of the function is counterclockwise from the positive x-axis. These angles are shown visually below.

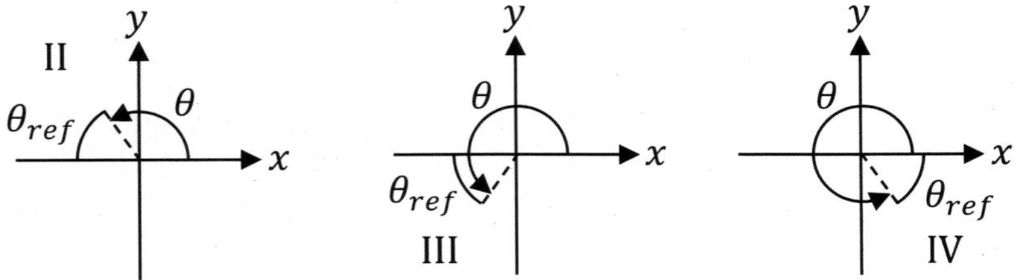

Strategy for Trig Values Beyond Quadrant I

To evaluate a trig function at a standard angle in Quadrants II-IV, follow these steps:

1. Is the angle a multiple of 30° or 45°? If so, continue onto Step 2. If the angle is **not** a multiple of 30° or 45°, use a calculator instead.
2. First determine which Quadrant the given angle lies in:
   - If $90° < \theta < 180°, \theta$ lies in Quadrant II.
   - If $180° < \theta < 270°, \theta$ lies in Quadrant III.
   - If $270° < \theta < 360°, \theta$ lies in Quadrant IV.
3. Next, decide if the trig function is positive or negative:
   - In Quadrant II, sine is positive, while cosine and tangent are negative.
   - In Quadrant III, tangent is positive, while sine and cosine are negative.
   - In Quadrant IV, cosine is positive, while sine and tangent are negative.
4. Now determine the reference angle (θ_{ref}) from the given angle (θ):
   - In Quadrant II, $\theta_{ref} = 180° - \theta$.
   - In Quadrant III, $\theta_{ref} = \theta - 180°$.
   - In Quadrant IV, $\theta_{ref} = 360° - \theta$.
5. Evaluate the trig function at the reference angle using the chart on page 85.
6. Combine your answers from Step 3 and Step 5, as in the following examples.
7. Note the following special cases:

$\sin 180° = 0, \cos 180° = -1, \tan 180° = 0, \sin 270° = -1, \cos 270° = 0, \tan 270° = \text{undef.}$

Example: Evaluate $\sin 210°$.
210° lies in Quadrant III. Sine is negative in Quadrant III. The reference angle is:

$$\theta_{ref} = \theta - 180° = 210° - 180° = 30°$$

Therefore,

$$\sin 210° = -\sin 30° = -\frac{1}{2}$$

Example: Evaluate $\cos 315°$.
315° lies in Quadrant IV. Cosine is positive in Quadrant IV. The reference angle is:

$$\theta_{ref} = 360° - \theta = 360° - 315° = 45°$$

Therefore,

$$\cos 315° = +\cos 45° = \frac{\sqrt{2}}{2}$$

Example: Evaluate $\tan 120°$.
120° lies in Quadrant II. Tangent is negative in Quadrant II. The reference angle is:

$$\theta_{ref} = 180° - \theta = 180° - 120° = 60°$$

Therefore,

$$\tan 120° = -\tan 60° = -\sqrt{3}$$

71. Evaluate each trig function at the indicated angle.

(A) $\sin 150° =$

(B) $\cos 240° =$

(C) $\tan 300° =$

(D) $\sin 315° =$

(E) $\cos 135° =$

(F) $\tan 210° =$

(G) $\sin 240° =$

(H) $\cos 180° =$

(I) $\tan 225° =$

(J) $\sin 180° =$

(K) $\cos 270° =$

(L) $\tan 300° =$

(M) $\sin 300° =$

(N) $\cos 330° =$

(O) $\tan 180° =$

(P) $\sin 270° =$

(Q) $\cos 225° =$

(R) $\tan 150° =$

Answers: (A) $\frac{1}{2}$ (B) $-\frac{1}{2}$ (C) $-\sqrt{3}$ (D) $-\frac{\sqrt{2}}{2}$ (E) $-\frac{\sqrt{2}}{2}$ (F) $\frac{\sqrt{3}}{3}$ (G) $-\frac{\sqrt{3}}{2}$ (H) -1 (I) 1 (J) 0
(K) 0 (L) $-\sqrt{3}$ (M) $-\frac{\sqrt{3}}{2}$ (N) $\frac{\sqrt{3}}{2}$ (O) 0 (P) -1 (Q) $-\frac{\sqrt{2}}{2}$ (R) $-\frac{\sqrt{3}}{3}$

Inverse Trig Functions

An inverse trig function asks the question, "At which angles could this trig function be evaluated and obtain its argument as the answer?"

- $\sin\theta = x$ uses an angle (θ) in its argument and returns a ratio (x) as its answer.
- $\sin^{-1} x = \theta$ uses a ratio (x) in its argument and returns an angle (θ) as its answer.

Here is an example:

- $\sin 30°$ equals $\frac{1}{2}$.
- $\sin^{-1}\left(\frac{1}{2}\right)$ equals 30° or 150° because $\sin 30°$ and $\sin 150°$ both equal $\frac{1}{2}$.
- $\sin^{-1}\left(\frac{1}{2}\right)$ means: For which angle(s) θ does $\sin\theta$ equal $\frac{1}{2}$?

Strategy to Evaluate Inverse Trig Functions

To evaluate an inverse trig function, follow these steps:

1. First ignore the sign of the argument. Can you find the argument on the chart on page 85? If so, read the reference angle from the chart. If not, is the argument one of the special values listed in Step 4 below? If so, use the special value listed below. Otherwise, use a calculator to determine one of the two answers. (If using a calculator, you'll still need to complete the steps below to determine the second answer. First use the formula from page 88 to find the reference angle based on the calculator's answer, but beware that if the calculator's angle is negative, you must first add 360° before using the formula to find the reference angle.)
2. Now look at the sign of the argument. Determine the two Quadrants in which the specified trig function has that sign.
   - In Quadrant I, all trig functions are positive.
   - In Quadrant II, sine is positive, while cosine and tangent are negative.
   - In Quadrant III, tangent is positive, while sine and cosine are negative.
   - In Quadrant IV, cosine is positive, while sine and tangent are negative.
3. Determine θ from the reference angle (θ_{ref}) in each Quadrant from Step 2 using the formulas below. These two angles are the two answers to the inverse trig function.
   - In Quadrant I, $\theta = \theta_{ref}$.
   - In Quadrant II, $\theta = 180° - \theta_{ref}$.
   - In Quadrant III, $\theta = 180° + \theta_{ref}$.
   - In Quadrant IV, $\theta = 360° - \theta_{ref}$.
4. Note the following special cases:

$$\sin^{-1} 0 = 0° \text{ or } 180°, \sin^{-1}(-1) = 270°, \cos^{-1} 0 = 90° \text{ or } 270°$$

$$\cos^{-1}(-1) = 180°, \tan^{-1} 0 = 0° \text{ or } 180°, \tan^{-1}(\text{undef.}) = 90° \text{ or } 270°$$

Note: There are usually two answers to an inverse trig function.

Example: Evaluate $\sin^{-1}\left(-\frac{\sqrt{2}}{2}\right)$.

First, ignore the sign of the argument. That leaves $\frac{\sqrt{2}}{2}$. According to the chart on page 85:

$$\sin 45° = \frac{\sqrt{2}}{2}$$

This means that the reference angle is 45°. Now look at the sign of the argument: In this problem, the argument is negative. The sine function is negative in Quadrants III and IV. Use the formulas to find the Quadrant III and IV angles from the reference angle:

$$\theta = 180° + \theta_{ref} = 180° + 45° = 225°$$

$$\theta = 360° - \theta_{ref} = 360° - 45° = 315°$$

The two answers are 225° and 315°.

Example: Evaluate $\cos^{-1}\left(-\frac{1}{2}\right)$.

First, ignore the sign of the argument. That leaves $\frac{1}{2}$. According to the chart on page 85:

$$\cos 60° = \frac{1}{2}$$

This means that the reference angle is 60°. Now look at the sign of the argument: In this problem, the argument is negative. The cosine function is negative in Quadrants II and III. Use the formulas to find the Quadrant II and III angles from the reference angle:

$$\theta = 180° - \theta_{ref} = 180° - 60° = 120°$$

$$\theta = 180° + \theta_{ref} = 180° + 60° = 240°$$

The two answers are 120° and 240°.

Example: Evaluate $\tan^{-1}\left(\frac{\sqrt{3}}{3}\right)$.

First find the reference angle. According to the chart on page 85:

$$\tan 30° = \frac{\sqrt{3}}{3}$$

This means that the reference angle is 30°. Now look at the sign of the argument: In this problem, the argument is positive. The tangent function is positive in Quadrants I and III. Use the formulas to find the Quadrant I and III angles from the reference angle:

$$\theta = \theta_{ref} = 30°$$

$$\theta = 180° + \theta_{ref} = 180° + 30° = 210°$$

The two answers are 30° and 210°.

Example: Evaluate $\sin^{-1}(-1)$.

This is one of the special values from Step 7 on page 88. The answer is 270°.

72. Evaluate each inverse trig function.

(A) $\sin^{-1}\left(-\frac{\sqrt{3}}{2}\right) =$

(B) $\cos^{-1}\left(\frac{\sqrt{2}}{2}\right) =$

(C) $\tan^{-1}(-1) =$

(D) $\sin^{-1}\left(\frac{1}{2}\right) =$

(E) $\cos^{-1}\left(-\frac{\sqrt{3}}{2}\right) =$

(F) $\tan^{-1}(-\sqrt{3}) =$

(G) $\sin^{-1}\left(\frac{\sqrt{2}}{2}\right) =$

(H) $\cos^{-1}\left(\frac{1}{2}\right) =$

(I) $\tan^{-1}(\sqrt{3}) =$

(J) $\sin^{-1}(1) =$

(K) $\cos^{-1}\left(\frac{\sqrt{3}}{2}\right) =$

(L) $\tan^{-1}(1) =$

(M) $\sin^{-1}\left(\frac{\sqrt{3}}{2}\right) =$

(N) $\cos^{-1}(-1) =$

(O) $\tan^{-1}(0) =$

(P) $\sin^{-1}(0) =$

Answers: (A) 240°, 300° (B) 45°, 315° (C) 135°, 315° (D) 30°, 150° (E) 150°, 210° (F) 120°, 300° (G) 45°, 135° (H) 60°, 300° (I) 60°, 240° (J) 90° (K) 30°, 330° (L) 45°, 225° (M) 60°, 120° (N) 180° (O) 0°, 180° (P) 0°, 180°

Secant, Cosecant, and Cotangent

The secant ($\sec\theta$), cosecant ($\csc\theta$), and cotangent ($\cot\theta$) functions are defined as:

$$\sec\theta = \frac{1}{\cos\theta} \quad , \quad \csc\theta = \frac{1}{\sin\theta} \quad , \quad \cot\theta = \frac{1}{\tan\theta}$$

To evaluate the secant, cosecant, or cotangent function at a particular angle, first evaluate the corresponding trig function at the specified angle and then use the appropriate formula from the list above. For example, $\sec 60° = \frac{1}{\cos 60°} = \frac{1}{1/2} = 1 \div \frac{1}{2} = 1 \times \frac{2}{1} = 2$.

Note: The co's **don't** go together: Secant goes with cosine, whereas cosecant goes with sine.

Derivatives of Trig Functions

The first derivatives of the basic trig functions are:

$$\frac{d}{d\theta}\sin\theta = \cos\theta \quad , \quad \frac{d}{d\theta}\cos\theta = -\sin\theta \quad , \quad \frac{d}{d\theta}\tan\theta = \sec^2\theta$$

$$\frac{d}{d\theta}\sec\theta = \sec\theta\tan\theta \quad , \quad \frac{d}{d\theta}\csc\theta = -\csc\theta\cot\theta \quad , \quad \frac{d}{d\theta}\cot\theta = -\csc^2\theta$$

Integrals of Trig Functions

The integrals of the basic trig functions are anti-derivatives. If the integral is indefinite (meaning that it specifies no limits), the integral includes a constant of integration:

$$\int \sin\theta \, d\theta = -\cos\theta + c \quad , \quad \int \cos\theta \, d\theta = \sin\theta + c$$

$$\int \tan\theta \, d\theta = \ln|\sec\theta| + c \quad , \quad \int \sec\theta \, d\theta = \ln|\sec\theta + \tan\theta| + c$$

$$\int \csc\theta \, d\theta = -\ln|\csc\theta + \cot\theta| + c \quad , \quad \int \cot\theta \, d\theta = \ln|\sin\theta| + c$$

To perform a definite integral, first find the anti-derivative of the function in the integrand, then evaluate the anti-derivative at the upper and lower limits of integration, and subtract these two results. This method is illustrated by the following example.

Strategy for Performing a Definite Integral of a Basic Trig Function

To find the definite integral of a single, basic trig function, follow these steps:

1. First find the anti-derivative from the above list of indefinite integrals (but ignore the constant c which will cancel out in the subtraction).
2. Plug the upper limit of integration into the anti-derivative function, and do the same with the lower limit.
3. Subtract the two results from Step 2 as in the example that follows.

Example: Perform the following definite integral: $\int_{\theta=90°}^{180°} \cos\theta \, d\theta$.

First find the anti-derivative of $\cos\theta$:

$$\int_{\theta=90°}^{180°} \cos\theta \, d\theta = [\sin\theta]_{\theta=90°}^{180°}$$

The notation $[\sin\theta]_{\theta=90°}^{180°}$ means to evaluate the function $\sin\theta$ at $\theta = 180°$, then evaluate the function $\sin\theta$ at $\theta = 90°$, and subtract these results:

$$\int_{\theta=90°}^{180°} \cos\theta \, d\theta = [\sin\theta]_{\theta=90°}^{180°} = \sin 180° - \sin 90° = 0 - 1 = -1$$

See page 88 for help evaluating trig functions outside of Quadrant I.

73. Perform the following definite integral.

$$\int_{\theta=180°}^{360°} \sin\theta \, d\theta =$$

74. Perform the following definite integral.

$$\int_{\theta=120°}^{240°} \cos\theta \, d\theta =$$

Want help? Check the hints section at the back of the book.

Answers: -2, $-\sqrt{3}$

10 VECTOR ADDITION

Relevant Terminology

Magnitude – a number with units that indicates how much of a quantity there is.
Scalar – a quantity that has a magnitude, but which does **not** have a direction.
Vector – a quantity that has both a magnitude and a direction.
Component – the projection of a vector onto an axis. The components of a vector are shown visually on the following page.
Resultant – the combination of two or more vectors added together.

Examples of Scalars and Vectors

Velocity is one example of a vector. Velocity is a combination of speed and direction. Speed is a scalar, since it has only a magnitude (how fast). Velocity is a vector because it includes both a magnitude (how fast) and a direction (which way). Another example of a vector is force. A force includes both a magnitude (how much) and direction (which way). Unlike force, mass is a scalar. Mass has only a magnitude (how much), but no direction.

Notation

An arrow over a quantity indicates that it is a vector quantity. (Many textbooks use boldface to indicate a vector instead of an arrow, but an arrow stands out better. In the classroom or on homework, an arrow is used since boldface isn't possible.) For example, $\vec{\mathbf{A}}$ and $\vec{\mathbf{B}}$ represent vectors, whereas C and D do not. When a vector quantity lacks the arrow, it refers only to its magnitude. For example, A represents the magnitude of $\vec{\mathbf{A}}$. A subscript x or y indicates the component of a vector. For example, A_x is the x-component of $\vec{\mathbf{A}}$.

Graphical Representation

A vector can be represented visually by drawing an arrow. The length of the arrow represents the magnitude of the vector. A variety of vectors are illustrated below.

$\vec{\mathbf{A}}$ $\vec{\mathbf{B}}$ $\vec{\mathbf{v}}$ $\vec{\mathbf{g}}$ $\vec{\mathbf{p}}$ $\vec{\mathbf{A}}$ $\vec{\mathbf{F}}$

A vector can be moved around so long as the length and direction of the vector remain unchanged. See the same vector $\vec{\mathbf{A}}$ in two different places in the diagram above. It's sometimes helpful to move a vector when working with vectors visually. (An example of when this is useful appears on the next page under Graphical Vector Addition.)

Components

A vector $\vec{\mathbf{A}}$ can be resolved into Cartesian components A_x and A_y by projecting the vector onto each axis as illustrated below.

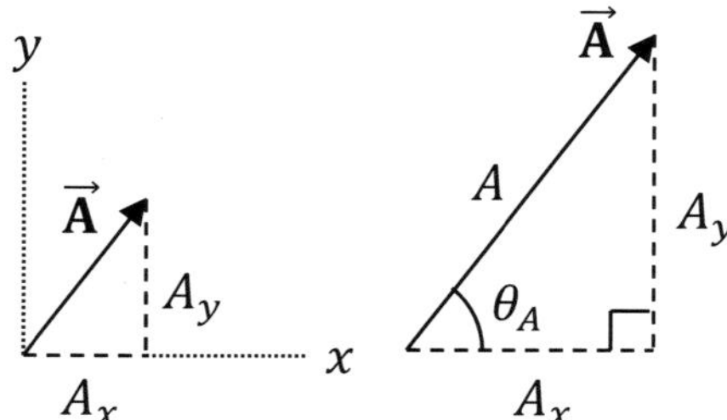

We can use trig to relate the components of a vector to its magnitude and direction. Use the following equations to find the **components** (A_x and A_y) of a vector given its magnitude (A) and direction (θ_A):

$$A_x = A\cos\theta_A \quad , \quad A_y = A\sin\theta_A$$

The x-component has a cosine because it is adjacent to θ_A, while the y-component has a sine because it is opposite to θ_A. Use the following equations to find the **magnitude** (A) and **direction** (θ_A) of a vector from its components (A_x and A_y):

$$A = \sqrt{A_x^2 + A_y^2} \quad , \quad \theta_A = \tan^{-1}\left(\frac{A_y}{A_x}\right)$$

The first equation follows from the Pythagorean theorem.

Graphical Vector Addition

Vector addition can be represented graphically: Join two vectors **tip-to-tail** in order to find the **resultant** vector graphically. In the illustration below, the vectors $\vec{\mathbf{A}}$ and $\vec{\mathbf{B}}$ are joined tip-to-tail to form the resultant vector, $\vec{\mathbf{R}}$.

It doesn't matter whether the tip of $\vec{\mathbf{A}}$ is joined to the tail of $\vec{\mathbf{B}}$ or if the tip of $\vec{\mathbf{B}}$ is joined to the tail of $\vec{\mathbf{A}}$, since either way $\vec{\mathbf{R}}$ will be the diagonal of the parallelogram, as shown below.

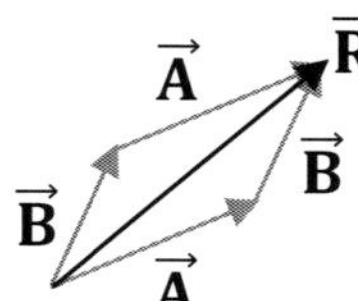

The resultant vector extends from the tail of the first vector to the tip of the second vector.

Vector Addition Strategy

Given the magnitudes (A, B, C,...) and directions (θ_A, θ_B, θ_C,...) of vectors, $\vec{\mathbf{A}}$, $\vec{\mathbf{B}}$, $\vec{\mathbf{C}}$,..., the goal is to find the magnitude (R) and direction (θ_R) of the resultant vector ($\vec{\mathbf{R}}$), where

$$\vec{\mathbf{R}} = \vec{\mathbf{A}} + \vec{\mathbf{B}} + \vec{\mathbf{C}} + \cdots$$

Follow these steps:

1. Make a sketch of the vector addition. Join the given vectors tip-to-tail. Draw the resultant vector, $\vec{\mathbf{R}}$, from the tail of the first vector to the tip of the last vector.
2. Find the components of the given vectors:
$$A_x = A\cos\theta_A$$
$$A_y = A\sin\theta_A$$
$$B_x = B\cos\theta_B$$
$$B_y = B\sin\theta_B$$
$$C_x = C\cos\theta_C$$
$$C_y = C\sin\theta_C$$
If there are more than three given vectors, do the same for the additional vectors. (If the problem only involves adding two vectors, just ignore C_x and C_y.)
3. Check the signs of your components:
   - In Quadrant I, the x- and y-components are both positive.
   - In Quadrant II, the x-component is negative and the y-component is positive.
   - In Quadrant III, the x- and y-components are both negative.
   - In Quadrant IV, the x-component is positive and the y-component is negative.
4. Add the respective components of the given vectors together to find the components of the resultant vector:
$$R_x = A_x + B_x + C_x + \cdots$$
$$R_y = A_y + B_y + C_y + \cdots$$
5. Use the following equations to determine the magnitude and direction of the resultant vector:
$$R = \sqrt{R_x^2 + R_y^2}$$
$$\theta_R = \tan^{-1}\left(\frac{R_y}{R_x}\right)$$
6. Be sure to put θ_R in the right Quadrant based on the signs of R_x and R_y:
   - If $R_x > 0$ and $R_y > 0$, then θ_R lies between 0° and 90°.
   - If $R_x < 0$ and $R_y > 0$, then θ_R lies between 90° and 180°.
   - If $R_x < 0$ and $R_y < 0$, then θ_R lies between 180° and 270°.
   - If $R_x > 0$ and $R_y < 0$, then θ_R lies between 270° and 360°.

Example: The simian vector, $\vec{\mathbf{S}}$, has a magnitude of 4.0 m and direction of 150°. The primate vector, $\vec{\mathbf{P}}$, has a magnitude of 4.0 m and direction of 90°. Find the magnitude and direction of the resultant of these two vectors.

Begin by sketching the vector addition. Draw the given vectors tip-to-tail.

$\vec{\mathbf{S}}$ + $\vec{\mathbf{P}}$ = $\vec{\mathbf{P}}$ $\vec{\mathbf{R}}$ $\vec{\mathbf{S}}$

Use trig to find the components of the given vectors:

$$S_x = S\cos\theta_S = 4\cos 150° = -4\cos 30° = -4\left(\frac{\sqrt{3}}{2}\right) = -2\sqrt{3}\text{ m}$$

$$S_y = S\sin\theta_S = 4\sin 150° = +4\sin 30° = 4\left(\frac{1}{2}\right) = 2\text{ m}$$

$$P_x = P\cos\theta_P = 4\cos 90° = 4(0) = 0$$

$$P_y = P\sin\theta_B = 4\sin 90° = 4(1) = 4\text{ m}$$

Add the respective components together in order to find the components of the resultant vector:

$$R_x = S_x + P_x = -2\sqrt{3} + 0 = -2\sqrt{3}\text{ m}$$

$$R_y = S_y + P_y = 2 + 4 = 6\text{ m}$$

Use the Pythagorean theorem to determine the magnitude of the resultant vector:

$$R = \sqrt{R_x^2 + R_y^2} = \sqrt{\left(-2\sqrt{3}\right)^2 + (6)^2} = \sqrt{12 + 36} = \sqrt{48} = \sqrt{(16)(3)} = 4\sqrt{3}\text{ m}$$

Use trig to determine the direction of the resultant vector:

$$\theta_R = \tan^{-1}\left(\frac{R_y}{R_x}\right) = \tan^{-1}\left(\frac{6}{-2\sqrt{3}}\right) = \tan^{-1}\left(-\frac{3}{\sqrt{3}}\right) = \tan^{-1}(-\sqrt{3})$$

First, ignore the sign to determine the reference angle. The reference angle is 60° since $\tan 60° = \sqrt{3}$. Tangent is negative in Quadrants II and IV. Therefore, the answer lies in Quadrant II or Quadrant IV. Which is it? We can deduce that the resultant vector lies in Quadrant II because $R_x < 0$ and $R_y > 0$. Use the equation (from Chapter 9) to determine θ_R (in Quadrant II) from the reference angle:

$$\theta_R = 180° - \theta_{ref} = 180° - 60° = 120°$$

Therefore, the direction of the resultant vector is 120°.

75. The monkey vector, $\vec{\mathbf{M}}$, has a magnitude of 36 ☺ and direction of 240°. The banana vector, $\vec{\mathbf{B}}$, has a magnitude of 18 ☺ and direction of 120°.

(A) Find the x- and y-components of the given vectors, $\vec{\mathbf{M}}$ and $\vec{\mathbf{B}}$:

______________ ______________ ______________ ______________

(B) Find the x- and y-components of the resultant vector, $\vec{\mathbf{R}}$:

______________ ______________

(C) Find the magnitude and direction of the resultant vector, $\vec{\mathbf{R}}$:

the magnitude of $\vec{\mathbf{R}}$ = ______________

the direction of $\vec{\mathbf{R}}$ = ______________ in Quad ____

Want help? Check the hints section at the back of the book.

Answers: $18\sqrt{3}$ ☺, 210°, III

76. The gorilla vector, $\vec{\boldsymbol{\Psi}}$, has a magnitude of $3\sqrt{2}$ N and direction of 135°. The chimpanzee vector, $\vec{\boldsymbol{\Phi}}$, has a magnitude of $6\sqrt{2}$ N and direction of 225°. The orangutan vector, $\vec{\boldsymbol{\Omega}}$, has a magnitude of 12 N and direction of 0°.

(A) Find the x- and y-components of the given vectors, $\vec{\boldsymbol{\Psi}}$, $\vec{\boldsymbol{\Phi}}$, and $\vec{\boldsymbol{\Omega}}$:

________ ________ ________ ________ ________ ________

(B) Find the x- and y-components of the resultant vector, $\vec{\mathbf{R}}$:

________ ________

(C) Find the magnitude and direction of the resultant vector, $\vec{\mathbf{R}}$:

the magnitude of $\vec{\mathbf{R}}$ = ________

the direction of $\vec{\mathbf{R}}$ = ________ in Quad ____

Want help? Check the hints section at the back of the book.

Answers: $3\sqrt{2}$ N, 315°, IV

Vector Subtraction Strategy

Given the magnitude (A and B) and direction (θ_A and θ_B) of two vectors, $\vec{\mathbf{A}}$ and $\vec{\mathbf{B}}$, the goal is to find the magnitude (C) and direction (θ_C) of a third vector ($\vec{\mathbf{C}}$), where

$$\vec{\mathbf{C}} = \vec{\mathbf{A}} - \vec{\mathbf{B}}$$

Follow these steps:

1. Find the components of the given vectors:

$$A_x = A\cos\theta_A$$
$$A_y = A\sin\theta_A$$
$$B_x = B\cos\theta_B$$
$$B_y = B\sin\theta_B$$

2. Check the signs of your components:
   - In Quadrant I, the x- and y-components are both positive.
   - In Quadrant II, the x-component is negative and the y-component is positive.
   - In Quadrant III, the x- and y-components are both negative.
   - In Quadrant IV, the x-component is positive and the y-component is negative.
3. Subtract the respective components of the given vectors to find the components of the resultant vector:

$$C_x = A_x - B_x$$
$$C_y = A_y - B_y$$

   Notes:
   - If you're trying to find $\vec{\mathbf{C}} = \vec{\mathbf{B}} - \vec{\mathbf{A}}$ instead of $\vec{\mathbf{C}} = \vec{\mathbf{A}} - \vec{\mathbf{B}}$, instead use the equations $C_x = B_x - A_x$ and $C_y = B_y - A_y$.
   - If you're instead solving a problem like $\vec{\mathbf{C}} = 2\vec{\mathbf{A}} + 3\vec{\mathbf{B}}$, change the component equations to match. In this example, you would use $C_x = 2A_x + 3B_x$ and $C_y = 2A_y + 3B_y$.
4. Use the following equations to determine the magnitude and direction of $\vec{\mathbf{C}}$:

$$C = \sqrt{C_x^2 + C_y^2}$$

$$\theta_C = \tan^{-1}\left(\frac{C_y}{C_x}\right)$$

5. Be sure to put θ_C in the right Quadrant based on the signs of C_x and C_y:
   - If $C_x > 0$ and $C_y > 0$, then θ_C lies between 0° and 90°.
   - If $C_x < 0$ and $C_y > 0$, then θ_C lies between 90° and 180°.
   - If $C_x < 0$ and $C_y < 0$, then θ_C lies between 180° and 270°.
   - If $C_x > 0$ and $C_y < 0$, then θ_C lies between 270° and 360°.

Example: The banana vector, $\vec{\mathbf{B}}$, has a magnitude of 20 m and direction of 150°. The coconut vector, $\vec{\mathbf{C}}$, has a magnitude of 10 m and direction of 210°. The watermelon vector, $\vec{\mathbf{W}}$, is defined according to $\vec{\mathbf{W}} = \vec{\mathbf{B}} - \vec{\mathbf{C}}$. Find the magnitude and direction of the watermelon vector, $\vec{\mathbf{W}}$.

Use trig to find the components of the given vectors:

$$B_x = B \cos \theta_B = 20 \cos 150° = -20 \cos 30° = -20\left(\frac{\sqrt{3}}{2}\right) = -10\sqrt{3} \text{ m}$$

$$B_y = B \sin \theta_B = 20 \sin 150° = +20 \sin 30° = 20\left(\frac{1}{2}\right) = 10 \text{ m}$$

$$C_x = C \cos \theta_C = 10 \cos 210° = -10 \cos 30° = -10\left(\frac{\sqrt{3}}{2}\right) = -5\sqrt{3} \text{ m}$$

$$C_y = C \sin \theta_C = 10 \sin 210° = -10 \sin 30° = -10\left(\frac{1}{2}\right) = -5 \text{ m}$$

Subtract the respective components in order to find the components of $\vec{\mathbf{W}}$:

$$W_x = B_x - C_x = -10\sqrt{3} - (-5\sqrt{3}) = -10\sqrt{3} + 5\sqrt{3} = -5\sqrt{3} \text{ m}$$

$$W_y = B_y - C_y = 10 - (-5) = 15 \text{ m}$$

Use the Pythagorean theorem to determine the magnitude of $\vec{\mathbf{W}}$:

$$W = \sqrt{W_x^2 + W_y^2} = \sqrt{\left(-5\sqrt{3}\right)^2 + (15)^2} = \sqrt{75 + 225} = \sqrt{300} = \sqrt{(100)(3)} = 10\sqrt{3} \text{ m}$$

Use trig to determine the direction of $\vec{\mathbf{W}}$:

$$\theta_W = \tan^{-1}\left(\frac{W_y}{W_x}\right) = \tan^{-1}\left(\frac{15}{-5\sqrt{3}}\right) = \tan^{-1}\left(-\frac{3}{\sqrt{3}}\right) = \tan^{-1}\left(-\sqrt{3}\right)$$

First, ignore the sign to determine the reference angle. The reference angle is 60° since $\tan 60° = \sqrt{3}$. Tangent is negative in Quadrants II and IV. Therefore, the answer lies in Quadrant II or Quadrant IV. Which is it? We can deduce that the vector $\vec{\mathbf{W}}$ lies in Quadrant II because W_x is negative while and W_y is positive. Use the equation (from Chapter 9) to determine θ_W (in Quadrant II) from the reference angle:

$$\theta_W = 180° - \theta_{ref} = 180° - 60° = 120°$$

Therefore, the direction of $\vec{\mathbf{W}}$ is 120°.

77. The monkey fur vector, $\vec{\mathbf{F}}$, has a magnitude of 16 m and direction of 270°. The monkey tail vector, $\vec{\mathbf{T}}$, has a magnitude of $8\sqrt{2}$ m and direction of 225°. The monkey belly vector, $\vec{\mathbf{B}}$, is defined according to $\vec{\mathbf{B}} = \vec{\mathbf{F}} - \vec{\mathbf{T}}$.

(A) Find the x- and y-components of the given vectors, $\vec{\mathbf{F}}$ and $\vec{\mathbf{T}}$:

____________ ____________ ____________ ____________

(B) Find the x- and y-components of $\vec{\mathbf{B}}$:

____________ ____________

(C) Find the magnitude and direction of $\vec{\mathbf{B}}$:

the magnitude of $\vec{\mathbf{B}}$ = ____________

the direction of $\vec{\mathbf{B}}$ = ____________ in Quad ____

Want help? Check the hints section at the back of the book.

Answers: $8\sqrt{2}$ m, 315°, IV

78. The math vector, $\vec{\mathbf{M}}$, has a magnitude of 4.0 N and direction of 210°. The science vector, $\vec{\mathbf{S}}$, has a magnitude of 3.0 N and direction of 150°. The physics vector, $\vec{\mathbf{P}}$, is defined according to $\vec{\mathbf{P}} = 3\vec{\mathbf{M}} - 2\vec{\mathbf{S}}$.

(A) Find the x- and y-components of the given vectors, $\vec{\mathbf{M}}$ and $\vec{\mathbf{S}}$:

_______________ _______________ _______________ _______________

(B) Find the x- and y-components of $\vec{\mathbf{P}}$:

_______________ _______________

(C) Find the magnitude and direction of $\vec{\mathbf{P}}$:

the magnitude of $\vec{\mathbf{P}}$ = _______________

the direction of $\vec{\mathbf{P}}$ = _______________ in Quad ____

Want help? Check the hints section at the back of the book.

Answers: $6\sqrt{3}$ N, 240°, III

Unit Vectors

Unit vectors are vectors which have a magnitude of exactly one unit. The Cartesian unit vectors $\hat{\mathbf{x}}$ and $\hat{\mathbf{y}}$ point one unit along the x- and y-axes, respectively, as illustrated below.

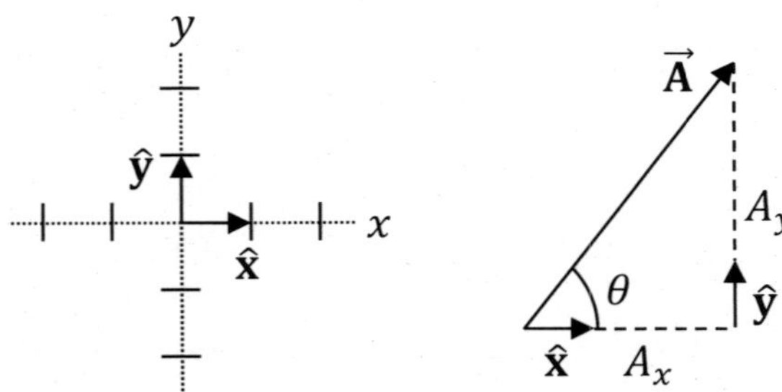

Any vector ($\vec{\mathbf{A}}$) can be expressed in terms of the Cartesian unit vectors ($\hat{\mathbf{x}}$ and $\hat{\mathbf{y}}$) via its components (A_x and A_y) in the following form:

$$\vec{\mathbf{A}} = A_x\hat{\mathbf{x}} + A_y\hat{\mathbf{y}}$$

Strategy for Working with Unit Vectors

If a problem involves the Cartesian unit vectors ($\hat{\mathbf{x}}$ and $\hat{\mathbf{y}}$), what to do depends on what you're given and what you're looking for, as described below:

- If you're given a vector in the form $\vec{\mathbf{A}} = A_x\hat{\mathbf{x}} + A_y\hat{\mathbf{y}}$ and want to find the magnitude and direction of that vector, first identify its components (A_x and A_y) as the coefficients of the unit vectors and then use the following equations.
$$A = \sqrt{A_x^2 + A_y^2} \quad , \quad \theta_A = \tan^{-1}\left(\frac{A_y}{A_x}\right)$$
- If you know the magnitude (A) and direction (θ_A) of a vector and want to express the vector in terms of Cartesian unit vectors ($\hat{\mathbf{x}}$ and $\hat{\mathbf{y}}$), first find the components (A_x and A_y) with trig.
$$A_x = A\cos\theta_A \quad , \quad A_y = A\sin\theta_A$$
Next substitute the components into the following equation:
$$\vec{\mathbf{A}} = A_x\hat{\mathbf{x}} + A_y\hat{\mathbf{y}}$$
- If you need to perform vector addition in the context of unit vectors, it's easy: You just group like terms together as follows to find the resultant vector ($\vec{\mathbf{R}}$):
$$\vec{\mathbf{R}} = \vec{\mathbf{A}} + \vec{\mathbf{B}} = (A_x + B_x)\hat{\mathbf{x}} + (A_y + B_y)\hat{\mathbf{y}}$$
Note that $\vec{\mathbf{R}}$ can be expressed in the form $\vec{\mathbf{R}} = R_x\hat{\mathbf{x}} + R_y\hat{\mathbf{y}}$ where
$$R_x = A_x + B_x \quad , \quad R_y = A_y + B_y$$
The magnitude and direction of the resultant can then be found like usual:
$$R = \sqrt{R_x^2 + R_y^2} \quad , \quad \theta_R = \tan^{-1}\left(\frac{R_y}{R_x}\right)$$

Example: Find the magnitude and direction of the following vector: $\vec{\mathbf{F}} = 3\,\hat{\mathbf{x}} - 3\,\hat{\mathbf{y}}$.
Compare $\vec{\mathbf{F}} = 3\,\hat{\mathbf{x}} - 3\,\hat{\mathbf{y}}$ to the general form $\vec{\mathbf{F}} = F_x\hat{\mathbf{x}} + F_y\hat{\mathbf{y}}$ to identify the components of $\vec{\mathbf{F}}$:

$$F_x = 3 \quad , \quad F_y = -3$$

Use the Pythagorean theorem to find the magnitude of $\vec{\mathbf{F}}$.

$$F = \sqrt{F_x^2 + F_y^2} = \sqrt{(3)^2 + (-3)^2} = \sqrt{9+9} = \sqrt{18} = \sqrt{(9)(2)} = 3\sqrt{2}$$

Use trig to find the direction of $\vec{\mathbf{F}}$.

$$\theta_F = \tan^{-1}\left(\frac{F_y}{F_x}\right) = \tan^{-1}\left(\frac{-3}{3}\right) = \tan^{-1}(-1)$$

Ignore the sign to find the reference angle. The reference angle is 45° since $\tan 45° = 1$. Tangent is negative in Quadrants II and IV. The vector $\vec{\mathbf{F}}$ lies in Quadrant IV since $F_x > 0$ and $F_y < 0$. Use the equation from Chapter 9 to determine the direction of $\vec{\mathbf{F}}$.

$$\theta_F = 360° - \theta_{ref} = 360° - 45° = 315°$$

Example: Vector $\vec{\mathbf{P}}$ has a magnitude of 8 and a direction of 180°. Express the vector $\vec{\mathbf{P}}$ in terms of Cartesian unit vectors.
First find the x- and y-components of $\vec{\mathbf{P}}$ using trig.

$$P_x = P\cos\theta_P = 8\cos 180° = 8(-1) = -8$$
$$P_y = P\sin\theta_P = 8\sin 180° = 8(0) = 0$$

Substitute these components into the equation for $\vec{\mathbf{P}}$.

$$\vec{\mathbf{P}} = P_x\hat{\mathbf{x}} + P_y\hat{\mathbf{y}} = -8\,\hat{\mathbf{x}} + 0\,\hat{\mathbf{y}} = -8\,\hat{\mathbf{x}}$$

Example: Find the magnitude and direction of the resultant of the following two vectors.

$$\vec{\mathbf{A}} = 2\,\hat{\mathbf{x}} - 4\,\hat{\mathbf{y}}$$
$$\vec{\mathbf{B}} = -6\,\hat{\mathbf{x}} + 8\,\hat{\mathbf{y}}$$

Add the respective components together to find the resultant vector.

$$\vec{\mathbf{R}} = \vec{\mathbf{A}} + \vec{\mathbf{B}} = 2\,\hat{\mathbf{x}} - 4\,\hat{\mathbf{y}} - 6\,\hat{\mathbf{x}} + 8\,\hat{\mathbf{y}} = 2\,\hat{\mathbf{x}} - 6\,\hat{\mathbf{x}} - 4\,\hat{\mathbf{y}} + 8\,\hat{\mathbf{y}} = (2-6)\hat{\mathbf{x}} + (-4+8)\,\hat{\mathbf{y}}$$
$$\vec{\mathbf{R}} = -4\,\hat{\mathbf{x}} + 4\,\hat{\mathbf{y}}$$

Use the Pythagorean theorem to find the magnitude of $\vec{\mathbf{R}}$.

$$R = \sqrt{R_x^2 + R_y^2} = \sqrt{(-4)^2 + (4)^2} = \sqrt{16+16} = \sqrt{32} = \sqrt{(16)(2)} = 4\sqrt{2}$$

Use trig to find the direction of $\vec{\mathbf{R}}$.

$$\theta_R = \tan^{-1}\left(\frac{R_y}{R_x}\right) = \tan^{-1}\left(\frac{4}{-4}\right) = \tan^{-1}(-1)$$

Ignore the sign to find the reference angle. The reference angle is 45° since $\tan 45° = 1$. Tangent is negative in Quadrants II and IV. The vector $\vec{\mathbf{R}}$ lies in Quadrant II since $R_x < 0$ and $R_y > 0$ (as $R_x = -4$ and $R_y = 4$). Therefore, $\theta_R = 180° - \theta_{ref} = 180° - 45° = 135°$.

79. Find the magnitude and direction of the following vector:

$$\vec{\mathbf{J}} = \sqrt{3}\,\hat{\mathbf{x}} - \hat{\mathbf{y}}$$

Answers: $2, 330°$

80. Vector $\vec{\mathbf{K}}$ has a magnitude of 4 and a direction of 120°. Express the vector $\vec{\mathbf{K}}$ in terms of Cartesian unit vectors.

Want help? Check the hints section at the back of the book.

Answers: $\vec{\mathbf{K}} = -2\,\hat{\mathbf{x}} + 2\sqrt{3}\,\hat{\mathbf{y}}$

81. Find the magnitude and direction of the resultant of the following two vectors.

$$\vec{\mathbf{V}} = 5\sqrt{3}\,\hat{\mathbf{x}} - 32\,\hat{\mathbf{y}}$$
$$\vec{\mathbf{W}} = 3\sqrt{3}\,\hat{\mathbf{x}} + 8\,\hat{\mathbf{y}}$$

Answers: $16\sqrt{3}, 300°$

82. Two vectors are defined as follows.

$$\vec{\mathbf{L}} = 7\sqrt{2}\,\hat{\mathbf{x}} + 3\sqrt{2}\,\hat{\mathbf{y}}$$
$$\vec{\mathbf{R}} = 2\sqrt{2}\,\hat{\mathbf{x}} - \sqrt{2}\,\hat{\mathbf{y}}$$

Determine the magnitude and direction of a third vector $\vec{\mathbf{Q}}$ where $\vec{\mathbf{Q}} = 3\vec{\mathbf{L}} - 4\vec{\mathbf{R}}$.

Want help? Check the hints section at the back of the book.

Answers: $26, 45°$

11 PROJECTILE MOTION

Essential Concepts

The motion of a projectile is easier to understand when you break the motion down into components:

1. The vertical component of the motion is like one-dimensional uniform acceleration. There is **constant acceleration vertically**. We will apply the equations of one-dimensional uniform acceleration to the y-component of the motion.
2. The horizontal component of velocity remains constant. There is **no horizontal acceleration**. That's because no forces are pushing or pulling horizontally on the projectile. The idea behind this is called inertia, which we'll explore in more detail in Chapter 13.

We use x to represent the **horizontal** component and y for the **vertical** component:

- $a_x = 0$ such that v_x remains constant: $v_x = v_{x0}$.
- $a_y = -9.81$ m/s^2 near the surface of the earth (this rounds to ≈ -10 m/s^2).

Symbols and Units

Symbol	Name	Units
Δx	net horizontal displacement	m
Δy	net vertical displacement	m
v_0	initial speed	m/s
θ_0	direction of initial velocity	°
v_{x0}	initial horizontal component of velocity	m/s
v_{y0}	initial vertical component of velocity	m/s
v_y	final vertical component of velocity	m/s
v	final speed	m/s
θ	direction of final velocity	°
a_y	acceleration (which is vertical)	m/s^2
t	time	s

Projectile Motion Strategy

To solve a projectile motion problem, follow these steps:

1. Draw a diagram of the path. Label the initial position (i), final position (f), the horizontal coordinate (x), and the vertical coordinate (y). Choose $+x$ to be forward. Choose $+y$ to be upward regardless of the motion of the object.
2. Use the following equations to determine the x- and y-components of the **initial velocity**:
$$v_{x0} = v_0 \cos\theta_0$$
$$v_{y0} = v_0 \sin\theta_0$$
3. The **acceleration** will equal $a_y = -9.81 \text{ m/s}^2 \approx -10 \text{ m/s}^2$ unless the problem specifically states that it is not falling near earth's surface. Near the moon, use $a_y = -1.62 \text{ m/s}^2 \approx -\frac{8}{5} \text{ m/s}^2$ instead.
4. Identify the unknown symbol and four known symbols. You must know four of the following symbols: $\Delta x, \Delta y, v_{x0}, v_{y0}, v_y, a_y,$ and t. See the previous note regarding a_y. You should also know v_{x0} and v_{y0} from Step 2 unless you're solving for initial speed.
5. Use the following equations to solve for the unknown. Think about which symbol you're solving for and which symbols you know to help you choose the right equations. Usually, you need to use the x-equation and one y-equation together.
$$\Delta x = v_{x0}t$$
$$\Delta y = v_{y0}t + \frac{1}{2}a_y t^2$$
$$v_y = v_{y0} + a_y t$$
$$v_y^2 = v_{y0}^2 + 2a_y\Delta y$$
6. If you're solving for the **final speed** or the direction of the final velocity, first find v_y using the equations above and then use the following equations:
$$v = \sqrt{v_{x0}^2 + v_y^2}$$
$$\theta = \tan^{-1}\left(\frac{v_y}{v_{x0}}\right)$$
(It's really v_x, but in projectile motion, $v_x = v_{x0}$, so we can write v_{x0} instead.)

Getting the Signs Right

Use the following sign conventions:

- a_y is negative for all projectile motion problems. (Draw $+y$ upward.)
- Δy is negative if the final position (f) is below the initial position (i).
- v_{y0} is negative if the object is moving downward in the initial position.
- v_y is negative if the object is moving downward in the final position.

Important Distinctions

There are two different components of net displacement:

- Δx is the **horizontal** component of the net displacement.
- Δy is the **vertical** component of the net displacement.

There are several different v's:

- v_0 is the **initial speed**, but only appears in the two trig equations.
- v_{x0} and v_{y0} are **components** of the **initial** velocity. Neither is the initial speed.
- v_y is the **final** y-component of velocity.
- v is the **final** speed.

Important Notes

In projectile motion, y is **vertical** and x is **horizontal**.

If an object makes an impact in the final position, the final velocity means just before impact (not after it lands). In this case, the final velocity will **not** be zero. (Final velocity only equals zero if the final position is at the very top of the trajectory.)

Similarly, if the initial position is where an object is launched or thrown, the initial velocity means just after it is released (not before it is launched). In this case, the initial velocity will **not** be zero. (Initial velocity only equals zero if the object is dropped from rest.)

Avoid using the two equations with v_y **unless** one of the following applies:

- You're solving for the maximum height (that is, the height at the very top). In this case, set $v_y = 0$ and solve for Δy. However, if you're solving for height, but the problem doesn't tell you that it's the top of the trajectory (or the maximum height during the trajectory), then **don't** set v_y equal to zero.
- You're solving for the final speed or the direction of the final velocity. In this case, you need to first solve for v_y and then use the equations from Step 6 of the strategy.
- The problem gives you the final speed and the final angle. In this case, you must use trig to solve for the components of the final velocity, and then you may use v_y as one of your four knowns. Most problems **don't** give you the final speed as a known.

Make the following assumptions except when a problem explicitly states otherwise:

- Neglect any effects of air resistance. Assume that all objects fall in vacuum.
- Assume that all objects are near the surface of the earth.
- Assume that the change in altitude is small enough that gravitational acceleration is approximately uniform throughout the motion.

Example: A monkey standing at the edge of the roof of a building throws a banana with an initial speed of 20 m/s at an angle of 30° above the horizontal. The banana travels $30\sqrt{3}$ m horizontally before landing on horizontal ground below. Approximately, how tall is the building?

Begin with a labeled diagram.

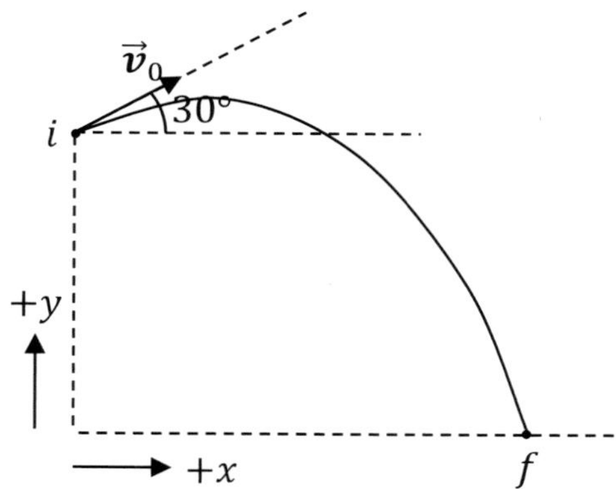

Use the two trig equations to find the components of the initial velocity:

$$v_{x0} = v_0 \cos\theta_0 = 20\cos 30° = 20\left(\frac{\sqrt{3}}{2}\right) = 10\sqrt{3}\text{ m/s}$$

$$v_{y0} = v_0 \sin\theta_0 = 20\sin 30° = 20\left(\frac{1}{2}\right) = 10\text{ m/s}$$

The unknown we are looking for is Δy. List the four knowns.

$$\Delta y = ?\ , v_{x0} = 10\sqrt{3}\text{ m/s}\, , v_{y0} = 10\text{ m/s}\, , \Delta x = 30\sqrt{3}\text{ m}\, , a_y = -9.81\text{ m/s}^2 \approx -10\text{ m/s}^2$$

We rounded 9.81 m/s² to 10 m/s² so that we can solve the problem without a calculator. Based on the list above and the unknown we are looking for, we should first use the Δx equation to solve for time:

$$\Delta x = v_{x0} t$$

$$30\sqrt{3} = (10\sqrt{3})t$$

$$t = 3.0\text{ s}$$

Now use the Δy equation to solve for Δy:

$$\Delta y = v_{y0} t + \frac{1}{2} a_y t^2$$

$$\Delta y = 10(3) + \frac{1}{2}(-9.81)(3)^2 \approx 10(3) + \frac{1}{2}(-10)(3)^2$$

$$\Delta y \approx 30 - 45$$

$$\Delta y \approx -15\text{ m}$$

The building is approximately 15 m tall, neglecting the height of the monkey. (The reason that Δy is negative is that the final position lies below the initial position.)

Example: A monkey throws a banana horizontally off the top of a 45-m tall building with an initial speed of 30 m/s. How far does the banana travel horizontally before it lands on horizontal ground below? Neglect the height of the monkey.

Begin with a labeled diagram.

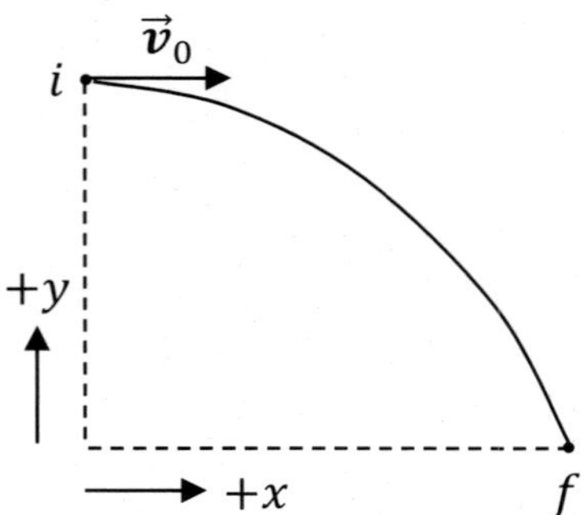

Use the two trig equations to find the components of the initial velocity. Since the banana is thrown horizontally, the launch angle is $\theta_0 = 0°$.

$$v_{x0} = v_0 \cos\theta_0 = 30\cos 0° = 30(1) = 30 \text{ m/s}$$
$$v_{y0} = v_0 \sin\theta_0 = 30\sin 0° = 30(0) = 0$$

The unknown we are looking for is Δx. List the four knowns.

$$\Delta x = ? \quad , \quad v_{x0} = 30 \text{ m/s} \quad , \quad v_{y0} = 0 \quad , \quad \Delta y = -45 \text{ m} \quad , \quad a_y = -9.81 \text{ m/s}^2 \approx -10 \text{ m/s}^2$$

The reason that Δy is negative is because the final position (f) lies below the initial position (i). As usual, we have rounded 9.81 m/s² to 10 m/s² so that we can solve the problem without a calculator. Based on the list above and the unknown we are looking for, we should first use the Δy equation to solve for time:

$$\Delta y = v_{y0}t + \frac{1}{2}a_y t^2$$
$$-45 = 0(t) + \frac{1}{2}(-9.81)t^2 \approx \frac{1}{2}(-10)t^2$$
$$-45 \approx -5t^2$$
$$9 \approx t^2$$
$$t \approx \sqrt{9} = 3.0 \text{ s}$$

Now use the Δx equation to solve for Δx:

$$\Delta x = v_{x0}t$$
$$\Delta x \approx (30)(3) = 90 \text{ m}$$

83. An ape standing atop a 60-m tall cliff launches a banana with an initial speed of 40 m/s at an angle of 30° above the horizontal. The banana lands on horizontal ground below.

Draw/label a diagram with the path, initial (i), final (f), and the $+x$- and $+y$-directions.

Carry out the trig step here.

Based on the question below, list the four symbols that you know (based on your labeled diagram) along with their values and SI units. **Hint**: Two of these numbers are **negative**.

_______________ _______________ _______________ _______________

How far does the banana travel horizontally? (Neglect the height of the ape.)

Want help? Check the hints section at the back of the book.

Answer: $120\sqrt{3}$ m

84. A monkey throws a banana horizontally off the top of an 80-m tall building with an initial speed of 40 m/s. Neglect the height of the monkey.

Draw/label a diagram with the path, initial (i), final (f), and the $+x$- and $+y$-directions.

Carry out the trig step here.

Based on the question below, list the four symbols that you know (based on your labeled diagram) along with their values and SI units. **Hint**: Two of these numbers are **negative**.

__________ __________ __________ __________

What is the final velocity of the banana, just before striking the (horizontal) ground below?

Want help? Check the hints section at the back of the book.

Answer: $40\sqrt{2}$ m/s, 315°

85. A monkey throws your physics textbook with an initial speed of 20 m/s at an angle of 30° below the horizontal from the roof of a 75-m tall building. Neglect the monkey's height.

Draw/label a diagram with the path, initial (i), final (f), and the $+x$- and $+y$-directions.

Carry out the trig step here.

Based on the question below, list the four symbols that you know (based on your labeled diagram) along with their values and SI units. **Hint**: Three of these numbers are **negative**.

______________ ______________ ______________ ______________

How far does the textbook travel horizontally before striking the (horizontal) ground below?

Want help? Check the hints section at the back of the book.

Answer: $30\sqrt{3}$ m

12 TWO-DIMENSIONAL MOTION WITH CALCULUS

The Position Vector

The position vector ($\vec{\mathbf{r}}$) extends from the origin to the location of the object.

$$\vec{\mathbf{r}} = x\,\hat{\mathbf{x}} + y\,\hat{\mathbf{y}}$$

The position vector is unique in that the x- and y-coordinates are also its x- and y-components. (Compare $\vec{\mathbf{r}} = x\,\hat{\mathbf{x}} + y\,\hat{\mathbf{y}}$ to the general form a vector $\vec{\mathbf{A}} = A_x\hat{\mathbf{x}} + A_y\hat{\mathbf{y}}$.) The initial position is given by $\vec{\mathbf{r}}_0 = x_0\,\hat{\mathbf{x}} + y_0\,\hat{\mathbf{y}}$. If you subtract $\vec{\mathbf{r}}_0$ from $\vec{\mathbf{r}}$, you get the net displacement ($\Delta\vec{\mathbf{r}}$), which is a straight line from the initial position to the final position.

$$\Delta\vec{\mathbf{r}} = \vec{\mathbf{r}} - \vec{\mathbf{r}}_0 = (x - x_0)\hat{\mathbf{x}} + (y - y_0)\hat{\mathbf{y}}$$

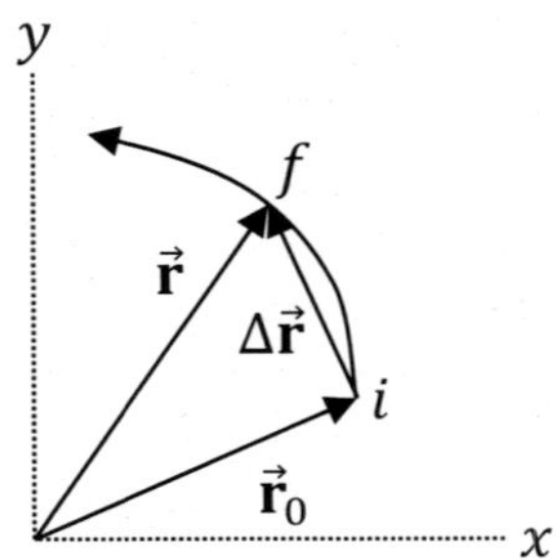

Calculus-based Motion Equations

Velocity ($\vec{v}$) is a first-derivative of position ($\vec{\mathbf{r}}$) with respect to time (t).

$$\vec{v} = \frac{d\vec{\mathbf{r}}}{dt}$$

Acceleration ($\vec{\mathbf{a}}$) is a first-derivative of velocity ($\vec{v}$) with respect to time (t). Acceleration ($\vec{\mathbf{a}}$) is also a second-derivative of position ($\vec{\mathbf{r}}$) with respect to time (t).

$$\vec{\mathbf{a}} = \frac{d\vec{v}}{dt} = \frac{d^2\vec{\mathbf{r}}}{dt^2}$$

Net displacement ($\Delta\vec{\mathbf{r}}$) is the definite integral of velocity ($\vec{v}$) over time (t).

$$\Delta\vec{\mathbf{r}} = \int_{t=t_0}^{t} \vec{v}\,dt$$

The change in velocity ($\vec{v} - \vec{v}_0$) is the definite integral of acceleration ($\vec{\mathbf{a}}$) over time (t). Therefore, final velocity ($\vec{v}$) equals initial velocity ($\vec{v}_0$) plus the definite integral of acceleration ($\vec{\mathbf{a}}$) over time (t).

$$\vec{v} = \vec{v}_0 + \int_{t=t_0}^{t} \vec{\mathbf{a}}\,dt$$

Note: When doing calculus with vectors, treat the unit vectors ($\hat{\mathbf{x}}$ and $\hat{\mathbf{y}}$) as constants.

Symbols and SI Units

Symbol	Name	SI Units
$\hat{\mathbf{x}}$	unit vector along the x-axis	unitless
$\hat{\mathbf{y}}$	unit vector along the y-axis	unitless
$\vec{\mathbf{r}}_0$	initial position vector	m
$\vec{\mathbf{r}}$	position vector	m
$\Delta\vec{\mathbf{r}}$	net displacement	m
x	horizontal position coordinate	m
Δx	horizontal displacement	m
y	vertical position coordinate	m
Δy	vertical displacement	m
$\vec{\boldsymbol{v}}_0$	initial velocity vector	m/s
$\vec{\boldsymbol{v}}$	velocity vector	m/s
v_{x0}	horizontal component of initial velocity	m/s
v_x	horizontal component of velocity	m/s
v_{y0}	vertical component of initial velocity	m/s
v_y	vertical component of velocity	m/s
$\vec{\mathbf{a}}$	acceleration vector	m/s^2
a_x	horizontal component of acceleration	m/s^2
a_y	vertical component of acceleration	m/s^2
t_0	initial time	s
t	time	s

Two-dimensional Calculus-based Motion Strategy

If a problem gives you a position vector ($\vec{\mathbf{r}}$), velocity vector ($\vec{\boldsymbol{v}}$), or acceleration vector ($\vec{\mathbf{a}}$) as a function of time in terms of unit vectors ($\hat{\mathbf{x}}$ and $\hat{\mathbf{y}}$), follow this strategy. How you solve the problem depends on what you're given and what you're looking for.

If you're given **position** ($\vec{\mathbf{r}}$) as a function of time (t):

1. To find net displacement, simply plug the final time into the position function to find $\vec{\mathbf{r}}(t)$, plug the initial time into the position function to find $\vec{\mathbf{r}}(t_0)$, and subtract the two results using vector subtraction (Chapter 10). Notation: $\vec{\mathbf{r}}(t)$ means "$\vec{\mathbf{r}}$ evaluated at t." It is **<u>not</u>** $\vec{\mathbf{r}}$ times t. Note that $\vec{\mathbf{r}}(t_0)$ is the same as the initial position vector $\vec{\mathbf{r}}_0$.
$$\Delta\vec{\mathbf{r}} = \vec{\mathbf{r}}(t) - \vec{\mathbf{r}}(t_0)$$
2. To find velocity, first take a derivative of the position function with respect to time and then plug the desired time into the velocity function.
$$\vec{\boldsymbol{v}} = \frac{d\vec{\mathbf{r}}}{dt}$$
3. To find acceleration, take two successive derivatives of the position function with respect to time and then plug the desired time into the acceleration function.
$$\vec{\mathbf{a}} = \frac{d^2\vec{\mathbf{r}}}{dt^2}$$

If you're given **velocity** ($\vec{\boldsymbol{v}}$) as a function of time (t):

4. To find net displacement, find the definite integral of the velocity function over the desired time interval.
$$\Delta\vec{\mathbf{r}} = \int_{t=t_0}^{t} \vec{\boldsymbol{v}}\,dt$$
5. To find velocity, simply plug the specified time into the velocity function.
6. To find acceleration, first take a derivative of the velocity function with respect to time and then plug the desired time into the acceleration function.
$$\vec{\mathbf{a}} = \frac{d\vec{\boldsymbol{v}}}{dt}$$

If you're given **acceleration** ($\vec{\mathbf{a}}$) as a function of time (t):

7. To find net displacement, first do Step 8 below and then do Step 4 above.
8. To find velocity, first find the definite integral of the acceleration function over the desired time interval and then add this result to the initial velocity.
$$\vec{\boldsymbol{v}} = \vec{\boldsymbol{v}}_0 + \int_{t=t_0}^{t} \vec{\mathbf{a}}\,dt$$
9. To find acceleration, simply plug the specified time into the acceleration function.

Note: You may also need to apply the unit vector strategy from page 105.

Tip: It may be helpful to review the examples from Chapter 4 before proceeding. Those examples are very similar to these. The difference is that these examples involve vectors and the motion is two-dimensional rather than one-dimensional.

Example: The position of a monkey is given according to the following equation, where SI units have been suppressed: $\vec{\mathbf{r}} = 3t^2\,\hat{\mathbf{x}} - 24t\,\hat{\mathbf{y}}$. Determine the monkey's speed at $t = 4.0$ s.

Find velocity in order to determine speed. Velocity is the first derivative of position with respect to time.

$$\vec{\boldsymbol{v}} = \frac{d\vec{\mathbf{r}}}{dt} = \frac{d}{dt}(3t^2\,\hat{\mathbf{x}} - 24t\,\hat{\mathbf{y}})$$

The unit vectors ($\hat{\mathbf{x}}$ and $\hat{\mathbf{y}}$) are constants. In the calculus, treat $\hat{\mathbf{x}}$ and $\hat{\mathbf{y}}$ the same way that you would treat any other constants: You can factor them out of the derivative.

$$\vec{\boldsymbol{v}} = \hat{\mathbf{x}}\frac{d}{dt}(3t^2) - \hat{\mathbf{y}}\frac{d}{dt}(24t) = 6t\,\hat{\mathbf{x}} - 24\,\hat{\mathbf{y}}$$

Evaluate the velocity function at $t = 4.0$ s:

$$\vec{\boldsymbol{v}}(\text{at } t = 4\text{ s}) = 6(4)\,\hat{\mathbf{x}} - 24\,\hat{\mathbf{y}} = 24\,\hat{\mathbf{x}} - 24\,\hat{\mathbf{y}}$$

The question asked us to find the speed at $t = 4.0$ s. We can find the speed from the components of velocity. As we learned in Chapter 10, we can express any vector in the form $\vec{\boldsymbol{v}} = v_x\hat{\mathbf{x}} + v_y\hat{\mathbf{y}}$. Compare $\vec{\boldsymbol{v}} = v_x\hat{\mathbf{x}} + v_y\hat{\mathbf{y}}$ to $\vec{\boldsymbol{v}} = 24\,\hat{\mathbf{x}} - 24\,\hat{\mathbf{y}}$ to realize that the x- and y-components of velocity are:

$$v_x = 24\text{ m/s} \quad , \quad v_y = -24\text{ m/s}$$

Speed (v) is the magnitude of velocity ($\vec{\boldsymbol{v}}$). Therefore, we can use the Pythagorean theorem to determine the speed from the components of velocity:

$$v = \sqrt{v_x^2 + v_y^2} = \sqrt{(24)^2 + (-24)^2} = \sqrt{(24)^2(1+1)} = 24\sqrt{2}\text{ m/s}$$

Example: The acceleration of an orangutan is given according to the following equation, where SI units have been suppressed: $\vec{\mathbf{a}} = 9t^2\,\hat{\mathbf{x}} - 18\,\hat{\mathbf{y}}$. The initial velocity of the orangutan is $\vec{\boldsymbol{v}}_0 = 36\,\hat{\mathbf{x}}$. Determine the velocity of the orangutan at $t = 2.0$ s.

Velocity equals initial velocity plus the definite integral of acceleration over time.

$$\vec{\boldsymbol{v}} = \vec{\boldsymbol{v}}_0 + \int_{t=t_0}^{t} \vec{\mathbf{a}}\,dt = 36\,\hat{\mathbf{x}} + \int_{t=0}^{2} (9t^2\,\hat{\mathbf{x}} - 18\,\hat{\mathbf{y}})\,dt$$

Find the anti-derivative of each term and then evaluate the definite integral over the limits following the strategy from Chapter 2. The unit vectors ($\hat{\mathbf{x}}$ and $\hat{\mathbf{y}}$) are constants.

$$\vec{\boldsymbol{v}} = 36\,\hat{\mathbf{x}} + [3t^3\,\hat{\mathbf{x}} - 18t\,\hat{\mathbf{y}}]_{t=0}^{2}$$

$$\vec{\boldsymbol{v}} = 36\,\hat{\mathbf{x}} + 3(2)^3\,\hat{\mathbf{x}} - (3)(0)^3\hat{\mathbf{x}} - 18(2)\,\hat{\mathbf{y}} - [-18(0)\hat{\mathbf{y}}] = 36\,\hat{\mathbf{x}} + 24\,\hat{\mathbf{x}} - 36\,\hat{\mathbf{y}}$$

$$\vec{\boldsymbol{v}} = 60\,\hat{\mathbf{x}} - 36\,\hat{\mathbf{y}}$$

86. The position of a bananamobile is given according to the following equation, where SI units have been suppressed.

$$\vec{\mathbf{r}} = 2t^4\,\hat{\mathbf{x}} - 4t^3\,\hat{\mathbf{y}}$$

Write an equation in symbols to determine the velocity of the bananamobile.

Determine the velocity of the bananamobile at $t = 3.0$ s.

Write an equation in symbols to determine the acceleration of the bananamobile.

Determine the acceleration of the bananamobile at $t = 5.0$ s.

Want help? Check the hints section at the back of the book.

Answers: $\vec{\boldsymbol{v}} = 216\,\hat{\mathbf{x}} - 108\,\hat{\mathbf{y}}$, $\vec{\mathbf{a}} = 600\,\hat{\mathbf{x}} - 120\,\hat{\mathbf{y}}$

87. The velocity of a bananaplane is given according to the following equation, where SI units have been suppressed.

$$\vec{v} = 8t^3\ \hat{\mathbf{x}} - 9t^2\ \hat{\mathbf{y}}$$

Write an equation in symbols to determine the net displacement of the bananaplane.

Determine the net displacement of the bananaplane from $t = 0$ to $t = 3.0$ s.

Write an equation in symbols to determine the acceleration of the bananaplane.

Determine the acceleration of the bananaplane at $t = 6.0$ s.

Want help? Check the hints section at the back of the book.

Answers: $\Delta\vec{\mathbf{r}} = 162\ \hat{\mathbf{x}} - 81\ \hat{\mathbf{y}}$, $\vec{\boldsymbol{a}} = 864\ \hat{\mathbf{x}} - 108\ \hat{\mathbf{y}}$

13 NEWTON'S LAWS OF MOTION

Relevant Terminology

Velocity – a combination of speed and direction.
Acceleration – the instantaneous rate at which velocity is changing.
Momentum – mass times velocity. All moving objects have momentum.
Inertia – the natural tendency of all objects to maintain constant momentum.
Mass – a measure of inertia. More massive objects are more difficult to accelerate as they have greater inertia to overcome.
Weight – the gravitational force that an object experiences.
Force – a push or a pull.
Vacuum – a region of space completely devoid of matter. There is not even air.

Important Distinction

The terms **mass** and **weight** are not interchangeable, although weight turns out to be equal to mass times gravitational acceleration ($W = mg$).

- Mass is a scalar quantity, whereas weight is a vector quantity: Weight has a direction, whereas mass does not. The direction of weight is toward the center of gravity. Mass doesn't have a direction because it's just as difficult to accelerate an object in any direction (the object resists acceleration equally in all directions).
- The SI unit of mass is the kilogram (kg), whereas the SI unit of weight is the Newton (N). A Newton equals a kilogram times a m/s^2 since $W = mg$.
- The mass of an object is the same regardless of its location in the universe, whereas weight varies with location. If you were on the moon, your mass would be the same, but you would weight about six times less.

Essential Concepts

An object must **change velocity** in order to **accelerate**. An object that has constant velocity has zero acceleration.

Any object that is moving has **momentum**, since momentum equals mass times velocity.

Every **force** is a kind of a push or a pull. Following are a few examples. Weight is a gravitational pull. Tension in a cord pulls on the objects attached to either end of the cord. Friction is a resistive force that attempts to decelerate two objects sliding against one another. When standing on the floor, a normal force supports you with an upward push.

Newton's Laws of Motion

Isaac Newton postulated several laws of motion. Three are very fundamental toward understanding the motion of an object:

1. According to Newton's first law of motion, every object has a natural tendency (called **inertia**) to maintain **constant momentum**.
2. According to Newton's second law of motion, the **net force** exerted on an object of constant mass equals the object's **mass times its acceleration**:
   $$\sum \vec{\mathbf{F}} = m\vec{\mathbf{a}}$$
   For an object with variable mass, like a rocket (Chapter 25), Newton's second law states that the net force equals a derivative of momentum ($\vec{\mathbf{p}}$) with respect to time.
   $$\sum \vec{\mathbf{F}} = \frac{d\vec{\mathbf{p}}}{dt}$$
   Notation: The uppercase Greek sigma (Σ) is a summation symbol. It states that the left-hand side is the sum of multiple forces (involving vector addition).
3. According to Newton's third law of motion, if one object (call it object A) exerts a force on a second object (call it object B), then object B exerts a force on object A that is **equal** in magnitude and **opposite** in direction to the force that object A exerts on object B. Newton's third law can be expressed concisely in an equation:
   $$\vec{\mathbf{F}}_{AB} = -\vec{\mathbf{F}}_{BA}$$

Essential Concepts

Newton's first law is sometimes called the law of **inertia**, since inertia is the natural tendency of an object to maintain constant momentum. If mass is constant, this implies a natural tendency to maintain **constant velocity**, which is equivalent to saying a natural tendency to maintain **zero acceleration**, or to saying a natural tendency to maintain both constant speed and travel in a straight line.

Newton's third law is sometimes stated as: For every **action**, there simultaneously occurs an equal and oppositely directed **reaction**. Most students are better able to reason out conceptual physics problems that relate to the third law if they think in terms of equal and opposite forces than if they think in terms of an action and reaction. Really, though, the action is the force that object A exerts on object B, and the equal and oppositely directed reaction is the force that object B exerts back on object A.

The main idea behind Newton's third law is that all **forces are mutual**, meaning that all **forces come in pairs**. If you can identify objects A and B, it's easy to apply Newton's third law, as we'll see in the examples and exercises.

Strategy for Conceptual Exercises Relating to Newton's Laws of Motion

To answer a conceptual exercise that applies Newton's laws of motion (or which involves related concepts like velocity, acceleration, momentum, mass, weight, force, or inertia), consider these tips:

1. Refrain from guessing. Most students find the concepts associated with Newton's laws of motion counterintuitive, meaning that their guesses are often incorrect. It usually takes practice applying Newton's laws to conceptual problems to refine one's intuition and to reconcile the laws with everyday experience.
2. **Let laws, concepts, and definitions guide your reasoning.**
3. If you feel stuck or unsure, remember that the correct answers can all be worked out through logic and reasoning. Make your best effort to reason out the answer, then check the Hints section at the back of the book for a full explanation of every conceptual exercise.

Example: A monkey is riding a skateboard. The skateboard crashes into a curb at the side of the road, which stops the skateboard. The monkey ____________________.

The answer is "keeps on moving." This exercise involves Newton's first law. The monkey has inertia such that the monkey continues moving forward with constant momentum even when the skateboard stops abruptly.

Example: If you double the mass of a box of bananas, it would take ____________________ force to achieve the same acceleration.

The answer is "twice as much." This exercise involves Newton's second law, which says that net force equals mass times acceleration: $\sum \vec{\mathbf{F}} = m\vec{\mathbf{a}}$. If you double mass ($m$) and want acceleration ($\vec{\mathbf{a}}$) to be unchanged, you must also double the net force ($\sum \vec{\mathbf{F}}$):

$$\vec{\mathbf{a}} = \frac{\sum \vec{\mathbf{F}}}{m} = \frac{2\sum \vec{\mathbf{F}}}{2m}$$

Example: A banana exerts a force of ____________________ on a monkey when the monkey stomps on the banana with a force of 200 N.

The answer is 200 N. This exercise involves Newton's third law. When applying the third law, identify the two objects, A and B. In this question, the two objects exerting forces on one another are the monkey (object A) and the banana (object B): $\vec{\mathbf{F}}_{AB} = -\vec{\mathbf{F}}_{BA}$. The force that the monkey exerts on the banana is equal in magnitude and opposite in direction to the force that the banana exerts on the monkey.

88. Fill in each blank with a word or phrase.

(A) An object is accelerating if it is ____________________.

(B) A banana will land ____________________ if a passenger riding inside of a train traveling horizontally to the east with constant velocity drops the banana directly above an X marked on the floor of the train.

(C) A banana will land ____________________ if a passenger riding inside of a train traveling horizontally to the east while decelerating drops the banana directly above an X marked on the floor of the train.

(D) A football player exerts ____________________ force on a cheerleader compared to the force that the cheerleader exerts on the football player when a 150-kg football player running 8 m/s collides with a 60-kg cheerleader who was at rest prior to the collision.

(E) A football player experiences ____________________ acceleration during the collision compared to a cheerleader when a 150-kg football player running 8 m/s collides with a 60-kg cheerleader who was at rest prior to the collision.

(F) All objects have a natural tendency to maintain ____________________ acceleration.

(G) A shooter experiences recoil due to Newton's ____________________ law of motion when firing a rifle. (Definition: A recoil is a push backwards.)

(H) A cannonball will land ____________________ when a monkey sailing with a constant velocity of 25 m/s to the south fires a cannonball straight upward relative to the ship.

(I) A 5-g mosquito exerts a force of ____________________ on a 500-g flyswatter when a monkey swats the mosquito with a force of 25 N.

For full explanations, consult the hints section at the back of the book.

Answers: (A) changing velocity (B) on the X (C) east of the X (D) the same (E) less (F) zero (G) third (H) in the cannon (I) 25 N

89. Fill in each blank with a word or phrase.

(A) A 24-kg monkey on earth has a ____________________ mass and ____________________ weight on the moon, where gravity is reduced by a factor of six compared to the earth.

(B) Objects tend to resist changing their ____________________.

(C) A 600-kg gorilla experiences ____________________ force and ____________________ acceleration as a 200-kg monkey when the gorilla collides head-on with the monkey.

(D) You get ____________________ when you multiply mass times velocity.

(E) You get ____________________ when you multiply mass times acceleration.

(F) The force that the earth exerts on the moon is ____________________ the force that the moon exerts on the earth.

(G) A monkey can ____________________ in order to have nonzero acceleration while running with constant speed.

(H) A monkey will have zero ____________________ and constant ____________________ if the net force acting on the monkey is zero.

(I) A banana reaches the ground ____________________ as a feather when the banana and feather are released from the same height above horizontal ground at the same time in a perfect vacuum.

(J) A monkey can get home by ____________________ when the monkey is stranded on horizontal frictionless ice with nothing but a banana.

For full explanations, consult the hints section at the back of the book.

Answers: (A) 24 kg, 40 N (B) momentum (C) the same, less (D) momentum (E) net force (F) equal to (G) change direction (H) acceleration, velocity (I) at the same time (J) throwing the banana directly away from his house

90. Select the best answer to each question.

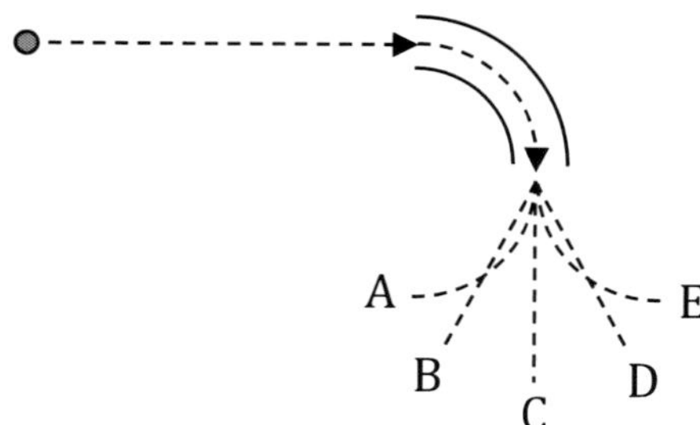

(A) The diagram above shows the top view of a miniature golf hole. A golf ball rolls along a circular metal arc. The dashed line represents the path of the golf ball. When the golf ball loses contact with the metal arc, which path will it follow?

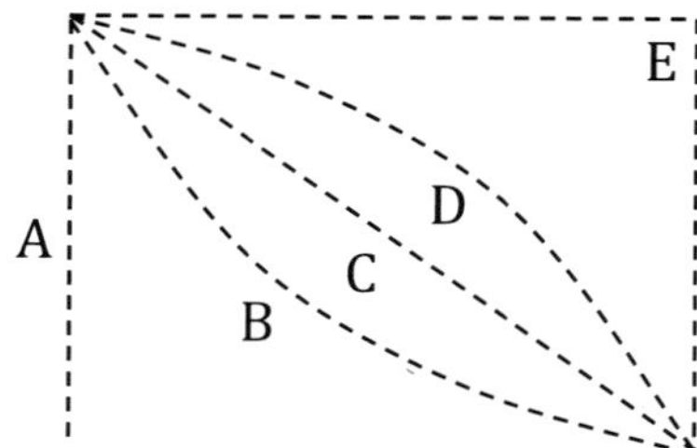

(B) The diagram above illustrates an airplane flying horizontally to the right. The airplane releases a box of bananas from rest (relative to the airplane) when it is at the top left position of the picture above. Which path will the box of bananas follow?

(C) A 0.2-kg banana is released from rest at the same time as a 20-kg box of bananas from the same height above horizontal ground. Which object strikes the ground first?

(D) One bullet is released from rest and falls straight down at the same time as a second bullet with identical mass is shot horizontally to the right from the same height above horizontal ground. Which bullet strikes the ground first?

(E) A necklace dangles from the rearview mirror of a car. Which way does the necklace lean when the car speeds up? slows down? travels with constant velocity along a level road? rounds a turn to the left with constant speed?

For full explanations, consult the hints section at the back of the book.

Answers: (A) C (B) D (C) same (D) same (E) backward, forward, straight down, right

14 APPLICATIONS OF NEWTON'S SECOND LAW

Essential Concepts

According to Newton's second law of motion, the net force ($\sum \vec{\mathbf{F}}$) acting on an object equals the mass (m) of the object times the object's acceleration ($\vec{\mathbf{a}}$):

$$\sum \vec{\mathbf{F}} = m\vec{\mathbf{a}}$$

The left-hand side of the above equation involves the vector addition (Chapter 10) of the forces. The main idea behind vector addition is to resolve the vectors into x- and y-components. Therefore, rather than work with the above equation, when we apply Newton's second law to solve a problem, we will work with the following equations in component form:

$$\sum F_x = ma_x \quad , \quad \sum F_y = ma_y$$

The first equation states that the sum of the x-components of the forces ($\sum F_x$) acting on an object equals the mass (m) of the object times the x-component of its acceleration (a_x). The second equation states the same thing in terms of the y-components.

Symbols and SI Units

Symbol	Name	SI Units
m	mass	kg
mg	weight	N
F	force	N
N	normal force	N
f	friction force	N
μ	coefficient of friction	unitless
T	tension	N
P, D, L	other kinds of forces	N
a_x	x-component of acceleration	m/s^2
a_y	y-component of acceleration	m/s^2

Notes Regarding Units

A Newton (N) is equivalent to a kilogram (kg) times a meter per second squared (m/s^2):

$$1 \text{ N} = 1 \text{ kg} \cdot \text{m/s}^2$$

This follows from Newton's second law ($\sum \vec{\mathbf{F}} = m\vec{\mathbf{a}}$), which states that the net force (in Newtons) must equal mass (in kilograms) times acceleration (in meters per second squared). When performing calculations, put the mass in kilograms and the acceleration in meters per second squared in order to get force in Newtons. It's a good habit to convert to SI units, which include kilograms (kg), meters (m), and seconds (s).

Types of Forces

Following is a list of forces commonly encountered in physics problems:

- Any object in the presence of a gravitational field experiences **weight**. Weight (W) equals mass (m) times gravitational acceleration (g).
  $$W = mg$$
- When an object is in contact with a surface, the surface exerts a **normal force** (N) on the object. The word "normal" means perpendicular: Normal force is perpendicular to the surface.
- When one object slides against another, the force of **kinetic friction** (f_k) equals the coefficient of kinetic friction (μ_k) times normal force (N):
  $$f_k = \mu_k N$$
- When a stationary object is in contact with a surface, the force of **static friction** (f_s) is less than or equal to the coefficient of static friction (μ_s) times normal force (N):
  $$f_s \leq \mu_s N$$
  The reason for the inequality is that friction can't cause an object to gain speed, but can only resist acceleration. Friction can cause deceleration, and it can limit the rate at which an object gains speed. Most physics problems involve using μN for the friction force, whether or not the problem is static or kinetic, so in practice, the inequality is usually not an issue.
- There is a **tension** force (T) whenever an object is connected to a cord. When two objects are connected to the same cord, the objects exert tension forces on one another that are equal in magnitude and opposite in direction according to Newton's third law.
- A **spring** exerts a **restoring force** ($k\Delta x$) on an object. We'll consider springs in detail in Chapter 15.
- Other common forces include a person's push or pull (P), a drive force (D) in a car, or a helicopter's lift force (L).

Strategy for Applying Newton's Second Law

To solve a problem involving Newton's second law (which relates forces to acceleration), follow these steps:

1. Draw a free-body diagram (FBD) for each object. Label the forces acting on each object. Consider each of the following forces:
   - Every object has **weight** (mg). Draw mg **straight down**. If there are multiple objects, distinguish their masses with subscripts: $m_1 g$, $m_2 g$, etc.
   - Is the object in contact with a surface? If it is, draw **normal** force (N) **perpendicular to the surface**. If there are two or more normal forces in the problem, use N_1, N_2, etc.
   - If the object is in contact with a surface, there will also be a **friction** force (f), unless the problem declares the surface to be frictionless. Draw the friction force **opposite to the velocity** (or to the potential velocity if the object is at rest). If there is more than one friction force, use f_1, f_2, etc.
   - Is the object connected to a cord, rope, thread, or string? (But not a spring: See the next note regarding springs.) If so, there will be a tension (T) **along the cord**. If two objects are connected to one cord, draw **equal and oppositely directed** forces acting on the two objects in accordance with Newton's third law. If there are two separate cords, then there will be two different pairs of tensions (T_1 and T_2), one pair for each cord.
   - Is the object connected to a **spring**? If so, there will be a restoring force ($k\Delta x$). We will discuss the restoring force in Chapter 12.
   - Does the problem describe or involve any other forces? If so, draw and label these forces. Examples include a monkey's pull (P), the drive force of a car (D), or the lift force of a helicopter (L).
2. Label the $+x$ and $+y$-axes in each FBD. Choose $+x$ for each object to be in the **direction of that object's acceleration**. This choice will let you set $a_y = 0$. (If the object is accelerating horizontally, for example, it won't have vertical acceleration.)
3. Write Newton's second law in component form for each object:

$$\sum F_{1x} = m_1 a_x \quad , \quad \sum F_{1y} = m_1 a_y \quad , \quad \sum F_{2x} = m_2 a_x \quad , \quad \sum F_{2y} = m_2 a_y \quad \cdots$$

4. Rewrite the left-hand side of each sum in terms of the x- and y-components of the forces acting on each object. Consider each force one at a time. Ask yourself if the force lies on an axis:
   - If a force lies on a positive or negative coordinate axis, the force only goes in that sum (x or y) with no trig.
   - If a force lies in the middle of a Quadrant, the force goes in both the x- and y-sums using trig. One component will involve cosine, while the other will involve sine. Whichever axis is adjacent to the angle gets the cosine.

5. Check the signs of each term in your sum. If the force has a component pointing along the $+x$-axis, it should be positive in the x-sum, but if it has a component pointing along the $-x$-axis, it should be negative in the x-sum. Apply similar reasoning for the y-sum.
6. Solve for the desired unknowns. Here are a few tips to help with the algebra:
   - Is there friction in the problem? If so, follow these steps: (A) Isolate normal force in the y-sum. (B) Plug this expression for normal force into the equation $f = \mu N$. (C) Substitute this expression for friction into the x-sum.
   - Does the problem involve a cord connecting two different objects? If so, you should see a $+T$ in one x-sum and a $-T$ in the other x-sum. Add the x-sums together so that the unknown tension will cancel out.
7. When plugging in numbers, note that gravitational acceleration (g) is always positive, whereas a component of acceleration (such as a_x) may be negative.

Important Notes

There are equations for weight ($W = mg$), friction ($f = \mu N$), and restoring force in a spring (see Chapter 15 for a spring).

However, there are **no** magic equations for normal force (N) or tension (T). The equations for normal force and tension will vary from one problem to another. For example, in one problem, normal force may turn out to equal mg, in another problem it may be $mg \cos\theta$, and in yet another problem it may be $mg - P\sin\theta$. The important thing is not to guess the equation for normal force: Solve for it instead. Don't use the equation for normal force from one problem in a different problem, as it likely won't be the same.

It's okay that there isn't a magic equation for normal force or tension. You will always be able to solve for these forces using algebra. For example, to find the expression for normal force, solve for N in the y-sum.

The way to become adept at applying Newton's second law is to study the main features. Some problems feature friction, some involve a cord, some have a pulley, some are on an incline, some have a spring, and some have two objects sharing a mutual surface. Every problem is a different combination of these features. Study the examples: Each example shows you how to deal with one of these features. Deal with each feature the same way in the problems as in the examples. The other way to become fluent in applying Newton's second law is through practice, which the exercises will help with.

Example: Atwood's machine, illustrated below, is constructed by suspending a 75-g banana (on the right) and a 25-g banana (on the left) from the two ends of a cord that passes over a pulley. The cord slides over the pulley without friction. Determine the acceleration of the system and the tension in the cord.

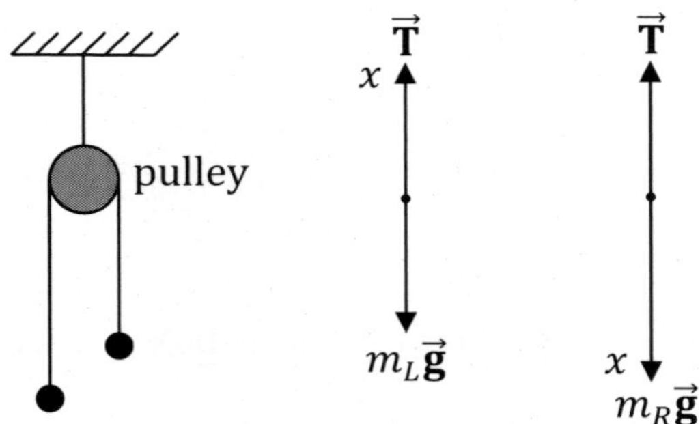

Draw and label a FBD (free-body diagram) for each banana. Each banana has weight pulling downward and tension pulling upward. The tensions are the same due to Newton's third law. Since the right banana falls down while the left banana rises up, we choose $+x$ to be down for the right banana and up for the left banana. Think of the pulley as bending the x-axis around it. The reason behind this is that this way the two accelerations will be the same (not only in magnitude, but also in sign, since positive will be down for the right banana and up for the left banana), since they are connected by a cord.

Apply Newton's second law to each banana:

$$\sum F_{Lx} = m_L a_x \quad , \quad \sum F_{Rx} = m_R a_x$$

$$T - m_L g = m_L a_x \quad , \quad m_R g - T = m_R a_x$$

The signs are different in the two equations because (as discussed above) $+x$ is down for the right banana and up for the left banana. As expected from Newton's third law, tension has opposite sign in the two equations. Add the two equations together to cancel tension:

$$T - m_L g + m_R g - T = m_R g - m_L g = m_L a_x + m_R a_x$$

$$(m_R - m_L)g = (m_L + m_R)a_x$$

$$a_x = \frac{m_R - m_L}{m_L + m_R} g = \frac{75 - 25}{75 + 25}(9.81) \approx \frac{75 - 25}{75 + 25}(10) = \frac{50}{100}(10) = 5.0 \text{ m/s}^2$$

Plug the acceleration into one of the original equations to solve for tension:

$$T - m_L g = m_L a_x$$

Convert the mass from grams to kilograms (since the SI unit of kg is needed to make a Newton for force):

$$25 \text{ g} = 25 \text{ g} \times \frac{1 \text{ kg}}{1000 \text{ g}} = 0.025 \text{ kg}$$

Plug the mass and acceleration into the tension equation:

$$T = m_L(a_x + g) = (0.025)(5 + 9.81) \approx (0.025)(5 + 10) = (0.025)(15) = 0.375 \text{ N} = \frac{3}{8} \text{ N}$$

Example: As illustrated below, a monkey pulls a 25-kg box of bananas with a force of 200 N at an angle of 30° above the horizontal. The coefficient of friction between the box of bananas and the ground is $\frac{1}{\sqrt{3}}$. Find the acceleration of the box, which begins from rest.

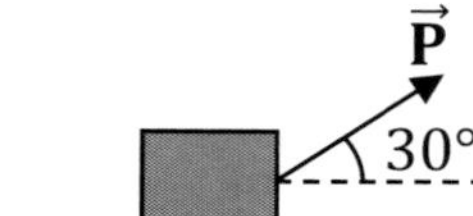

Draw and label a FBD for the box of bananas. Weight pulls downward. Normal force is upward since it is perpendicular to the surface. Friction acts to the left, opposite to the velocity. The monkey's pull acts 30° above the horizontal. Choose $+x$ to be to the right (along acceleration). Then y is upward and $a_y = 0$ (the box won't accelerate up or down).

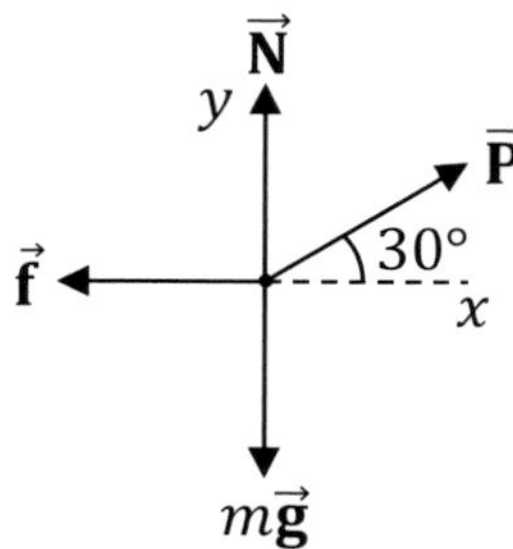

Apply Newton's second law to the box of bananas. Since $\vec{\mathbf{P}}$ doesn't lie on an axis, it appears in both the x- and y-sums with trig. The x-component of $\vec{\mathbf{P}}$ has a cosine since x is adjacent to 30°, while the y-component of $\vec{\mathbf{P}}$ has a sine since y is opposite to 30°. Each of the other forces only appears in one sum (x or y only) because the other forces all lie on an axis. Friction gets a minus sign because it is left (since $+x$ is right), and weight has a minus sign because it is down (since $+y$ is up).

$$\sum F_x = ma_x \quad , \quad \sum F_y = ma_y$$

$$P\cos 30° - f = ma_x \quad , \quad P\sin 30° + N - mg = 0$$

$$\frac{P\sqrt{3}}{2} - f = ma_x \quad , \quad \frac{P}{2} + N - mg = 0$$

When there is friction in a problem, first solve for normal force in the y-sum:

$$N = mg - \frac{P}{2} = (25)(9.81) - \frac{200}{2} \approx (25)(10) - \frac{200}{2} = 250 - 100 = 150 \text{ N}$$

Plug normal force into the friction equation:

$$f = \mu N \approx \frac{1}{\sqrt{3}}(150) = \frac{150}{\sqrt{3}}\frac{\sqrt{3}}{\sqrt{3}} = \frac{150\sqrt{3}}{3} = 50\sqrt{3} \text{ N}$$

Now substitute friction into the x-equation:

$$\frac{P\sqrt{3}}{2} - f = \frac{200\sqrt{3}}{2} - 50\sqrt{3} = 100\sqrt{3} - 50\sqrt{3} = 50\sqrt{3} = ma_x$$

$$50\sqrt{3} = 25a_x$$

$$a_x = 2\sqrt{3} \text{ m/s}^2$$

Example: As illustrated below, a box of bananas slides down a frictionless incline. If the incline makes a 30° angle with the horizontal, what is the acceleration of the box?

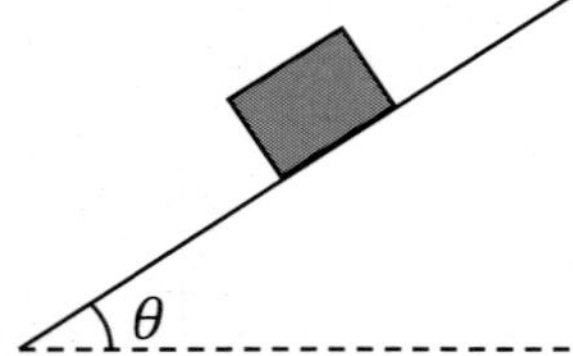

Draw and label a FBD for the box of bananas. Weight pulls downward, while normal force pushes perpendicular to the surface. According to the strategy for applying Newton's second law, we choose $+x$ to be along the direction of the acceleration, which in this case is down the incline (x is **not** horizontal in this problem). Since y must be perpendicular to x, we choose $+y$ to be along the normal force (y is **not** vertical in this problem). The reason for this is that this choice of coordinates makes $a_y = 0$, since the object won't accelerate perpendicular to the incline. The benefit is that it makes the math much simpler.

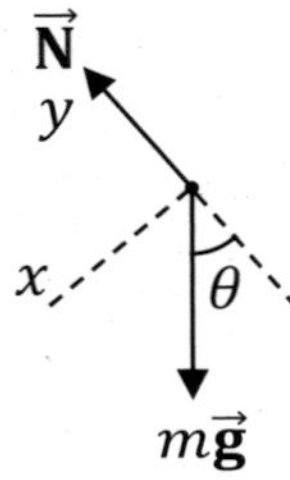

Apply Newton's second law to the box of bananas. Since weight doesn't lie on an axis, it appears in both the x- and y-sums with trig. Study the diagram below.

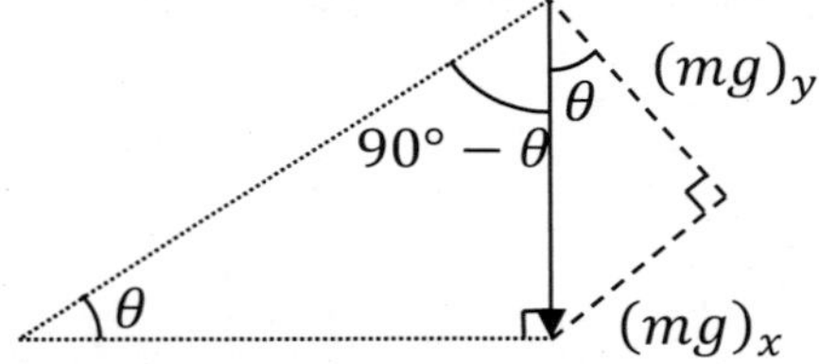

The x-component of weight has a sine since x is opposite to 30°, while the y-component of weight has a cosine since y is adjacent to 30° (see the figure above). Normal force only appears in the y-sum because it lies on the y-axis.

$$\sum F_x = ma_x \quad , \quad \sum F_y = ma_y$$

$$mg \sin\theta = ma_x \quad , \quad N - mg\cos\theta = 0$$

Divide both sides of the x-sum by mass to solve for acceleration:

$$a_x = g\sin\theta = (9.81)\sin 30° \approx 10 \sin 30° = 10\left(\frac{1}{2}\right) = 5.0 \text{ m/s}^2$$

Example: As illustrated below, a box of bananas rests in the back of a truck. The coefficient of friction between the box and the truck is $\frac{1}{5}$. What is the maximum acceleration that the truck can have without the box of bananas sliding backward relative to the truck?

Draw and label a FBD for each object. The box has weight pulling downward, normal force pushing upward (perpendicular to the surface), and friction pushing to the right (to oppose it from sliding backward).

The FBD for the truck is somewhat tricky. Weight pulls downward on the truck. The truck is propelled by a drive force (D). A resistive force (F_R) acting on the truck may include air resistance. By Newton's third law, the box exerts an equal and oppositely directed friction force on the truck (so it acts to the left on the truck). The truck experiences not one, but two, normal forces: an upward normal force (N_t) from the ground and a downward normal force (N_b) from the box. The second normal force is also associated with Newton's third law (equal and opposite N_b's).

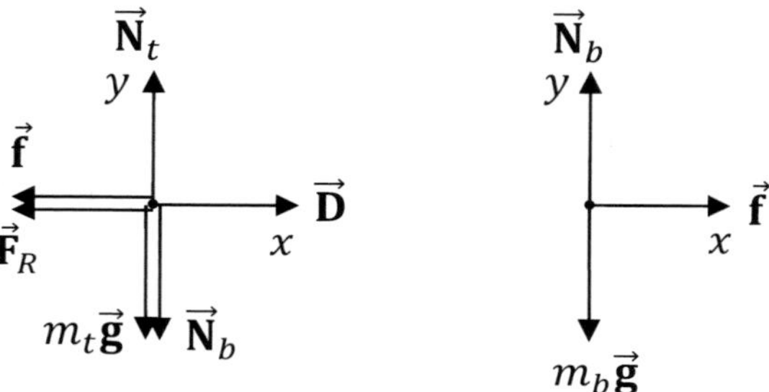

Apply Newton's second law to object:

$$\sum F_{tx} = m_t a_x \quad , \quad \sum F_{ty} = m_t a_x \quad , \quad \sum F_{bx} = m_b a_x \quad , \quad \sum F_{by} = m_b a_y$$

$$D - f - F_R = m_t a_x \quad , \quad N_t - N_b - m_t g = 0 \quad , \quad f = m_b a_x \quad , \quad N_b - m_b g = 0$$

Since there is friction in the problem, first isolate normal force in a y-sum:

$$N_b = m_b g$$

Plug this expression into the equation for friction force:

$$f = \mu N_b = \frac{1}{5} m_b g$$

Now substitute this expression for friction into an x-sum:

$$f = \frac{1}{5} m_b g = m_b a_x$$

Divide both sides by the mass of the box to isolate the unknown:

$$a_x = \frac{g}{5} = \frac{9.81}{5} \approx \frac{10}{5} = 2.0 \text{ m/s}^2$$

91. As shown below, a 600-kg helicopter flies straight upward with a lift force of 12,000-N. A 150-kg spy (00π) hangs onto a rope that is connected to the helicopter. A 50-kg physics student hangs onto a second rope, which is held by the spy.

(A) Draw a FBD for each object (three in all). Label each force and the x- and y-coordinates.

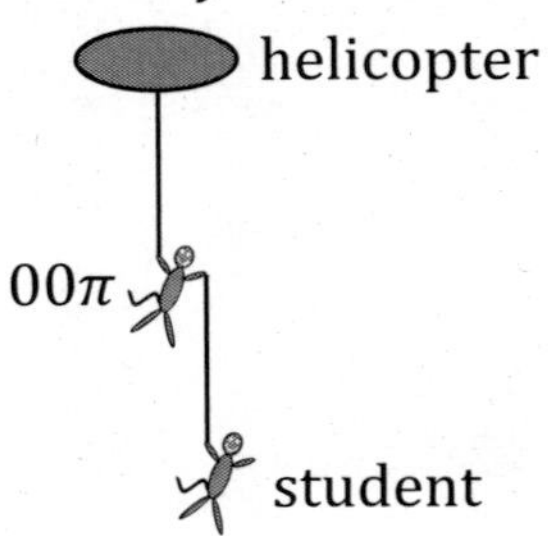

(B) Write the vertical sums for the forces acting on each object. There will be three sums. On the line immediately below each sum, rewrite the left-hand side in terms of the forces.

(C) Determine the acceleration of the system and the tension in each cord.

Want help? Check the hints section at the back of the book.

Answers: 5.0 m/s^2, 3000 N, 750 N

92. As illustrated below, a 30-kg monkey is connected to a 20-kg box of bananas by a cord that passes over a pulley. The coefficient of friction between the box and ground is 0.50. The system is initially at rest.

(A) Draw a FBD for each object (two in all). Label each force and the x- and y-coordinates.

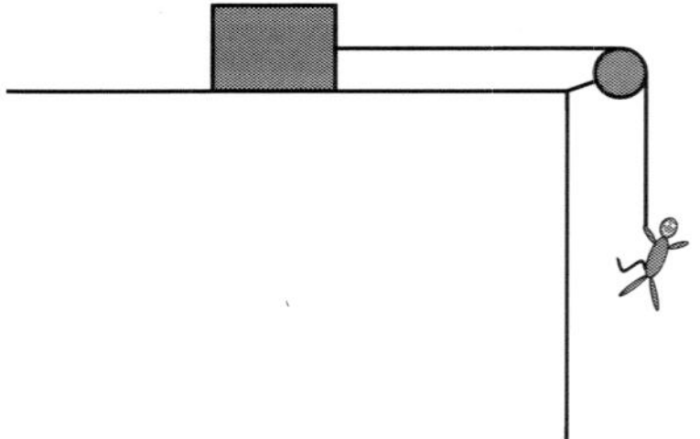

(B) Write the x- and y-sums for the forces acting on each object. There will be three sums. On the line immediately below each sum, rewrite the left-hand side in terms of the forces.

(C) Determine the acceleration of the system and the tension in the cord.

Want help? Check the hints section at the back of the book.

Answers: 4.0 m/s^2, 180 N

93. Two boxes of bananas connected by a cord are pulled by a monkey with a force of $300\sqrt{3}$ N as illustrated below. The left box of bananas has a mass of $10\sqrt{3}$ kg and the right box of bananas has a mass of $30\sqrt{3}$ kg. The coefficient of friction between the boxes of bananas and the horizontal is $\frac{\sqrt{3}}{5}$. The system is initially at rest.

(A) Draw a FBD for each box (two in all). Label each force and the x- and y-coordinates.

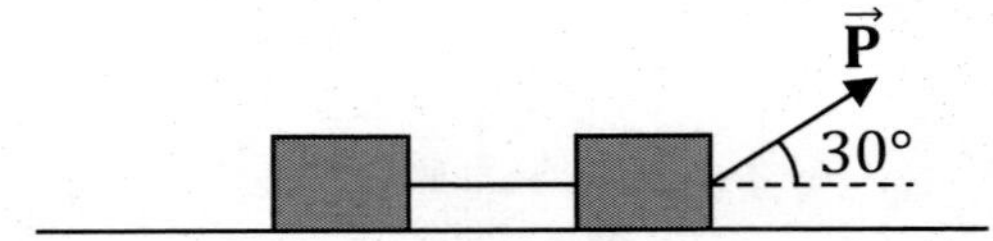

(B) Write the x- and y-sums for the forces acting on each box. There will be four sums. On the line immediately below each sum, rewrite the left-hand side in terms of the forces.

(C) Determine the acceleration of the system and the tension in the cord.

Want help? Check the hints section at the back of the book.

Answers: $\frac{5\sqrt{3}}{2}$ m/s^2, 135 N

94. A box of bananas slides down an inclined plane that makes an angle of 30° with the horizontal. The coefficient of friction between the box and the incline is $\frac{\sqrt{3}}{5}$.

(A) Draw a FBD for the box. Label each force and the x- and y-coordinates.

(B) Write the x- and y-sums for the forces acting on the box. There will be two sums. On the line immediately below each sum, rewrite the left-hand side in terms of the forces.

(C) Determine the acceleration of the box of bananas.

Want help? Check the hints section at the back of the book.

Answer: 2.0 m/s^2

95. As illustrated below, a 40-kg box of bananas on a frictionless incline is connected to a 60-kg monkey by a cord that passes over a pulley. The system is initially at rest.

(A) Draw a FBD for each object (two in all). Label each force and the x- and y-coordinates.

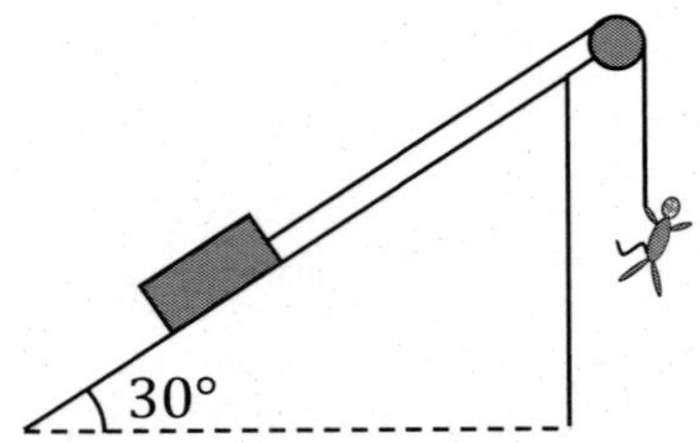

(B) Write the x- and y-sums for the forces acting on each object. There will be three sums. On the line immediately below each sum, rewrite the left-hand side in terms of the forces.

(C) Determine the acceleration of each object.

Want help? Check the hints section at the back of the book.

Answer: 4.0 m/s^2

96. As illustrated below, a 40-kg chest filled with bananas is on top of an 80-kg chest filled with coconuts. The coefficient of static friction between the chests is $\frac{1}{4}$, but the ground is frictionless. A monkey exerts a 200-N horizontal force on the top chest.

(A) Draw a FBD for each object (two in all). Label each force and the x- and y-coordinates.

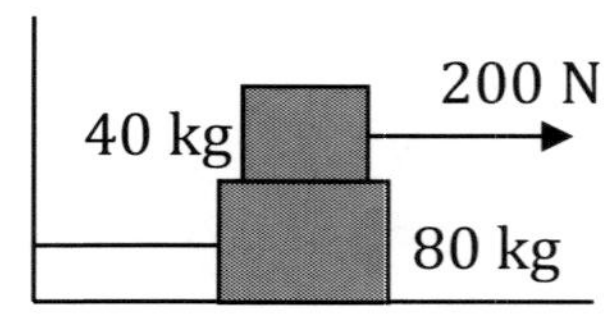

(B) Write the x- and y-sums for the forces acting on each object. There will be four sums. On the line immediately below each sum, rewrite the left-hand side in terms of the forces.

(C) Determine the acceleration of the top chest.

(D) What is the tension in the cord?

Want help? Check the hints section at the back of the book.

Answers: $\frac{5}{2}$ m/s^2, 100 N

15 HOOKE'S LAW

Relevant Terminology

Equilibrium – the natural position of a spring. A spring displaced from its equilibrium position tends to oscillate about its equilibrium position.
Displacement from equilibrium – the net displacement of a spring from its equilibrium position. It is a directed distance, telling how far from equilibrium a spring is.
Spring constant – a measure of the stiffness of a spring. A stiffer spring has a higher value for its spring constant.
Restoring force – a force that a spring exerts to return toward its equilibrium position.

Hooke's Law

If a **spring** is displaced from its equilibrium position, according to Hooke's law the spring exerts a **restoring** force (F_r) directed toward the equilibrium position, such that the restoring force is proportional to the displacement from equilibrium (Δx):

$$F_r = -k\Delta x$$

The proportionality constant in Hooke's law is the spring constant (k). The significance of the minus sign is that the restoring force is opposite to the displacement from equilibrium, which means that the restoring force always points toward the equilibrium position:

- If the spring is compressed from equilibrium, the restoring force pushes to stretch the spring back to its equilibrium position.
- If the spring is stretched from equilibrium, the restoring forces pulls to compress the spring back to its equilibrium position.

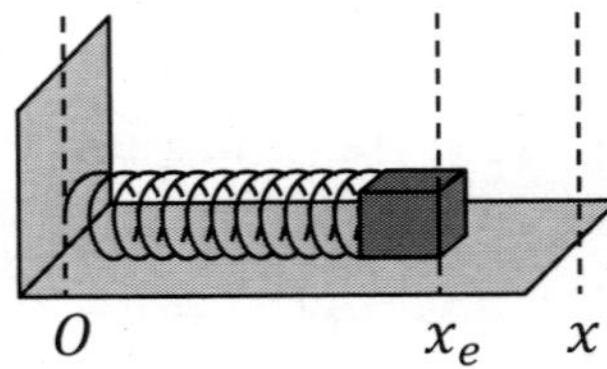

Symbols and SI Units

Symbol	Name	SI Units
k	spring constant	N/m (or kg/s^2)
Δx	displacement from equilibrium	m
F_r	restoring force	N

Notes Regarding Units

The SI units for the spring constant follow naturally from Hooke's law:

$$F_r = -k\Delta x$$

If you solve for the spring constant, you get:

$$k = -\frac{F_r}{\Delta x}$$

Thus, spring constant must have SI units of a Newton (the SI unit of force) per meter (the SI unit of displacement). Recall from Chapter 14 how a Newton relates to a kilogram, meter, and second:

$$1 \text{ N} = 1 \text{ kg} \cdot \text{m/s}^2$$

Using this relationship, we see that a Newton per meter equals:

$$1\frac{\text{N}}{\text{m}} = 1\frac{\text{kg}}{\text{s}^2}$$

Therefore, the SI units of a spring constant can alternatively be expressed as a Newton per meter (N/m) or as a kilogram per second squared (kg/s^2).

Strategy for Applying Newton's Second Law to Problems with Springs

To apply Newton's second law to a problem involving a spring, follow the Newton's second law strategy described in detail in Chapter 14, while following these steps to treat the spring:

1. In the FBD, draw a **restoring force** directed **towards equilibrium** and label this force as $k|\Delta x|$ (spring constant times the absolute value of the displacement of the spring from its equilibrium position). The reason for the absolute values, along with the reason that we're not including the minus sign from Hooke's law, is that the sign is already factored into the direction. (The sign is directional. Once you draw the arrow, the sign has already been used. The sign is why you draw the arrow towards the equilibrium position.)
2. Choose the $+x$-axis to be along the length of the spring.
3. When you sum the components of the forces, include $k|\Delta x|$ for the spring. It will appear only in the x-sum (if you followed the advice in Step 2 above), with no trig (though other forces in the problem may involve trig). In the sum, write $k|\Delta x|$ if the restoring force is along $+x$ in your FBD, and write $-k|\Delta x|$ if your restoring force is instead along $-x$ in your FBD.
4. Follow the strategy from Chapter 14.

Example: As illustrated below, a spring with a spring constant of 80 N/m on a frictionless horizontal surface has its left end attached to a wall while its right end is attached to a 12-kg box of bananas. What is the acceleration of the spring when it is stretched 6.0 cm from its equilibrium position?

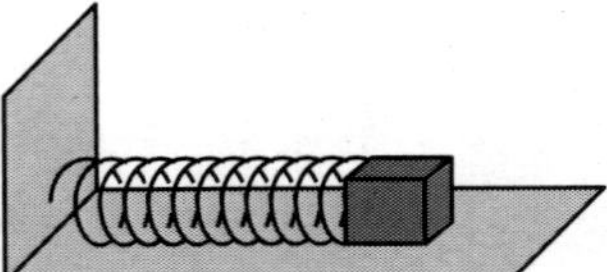

Draw and label a FBD for the box. Weight pulls downward, normal force pushes upward (perpendicular to the surface), and the spring exerts a restoring force ($k\Delta x$) to the left (towards equilibrium).

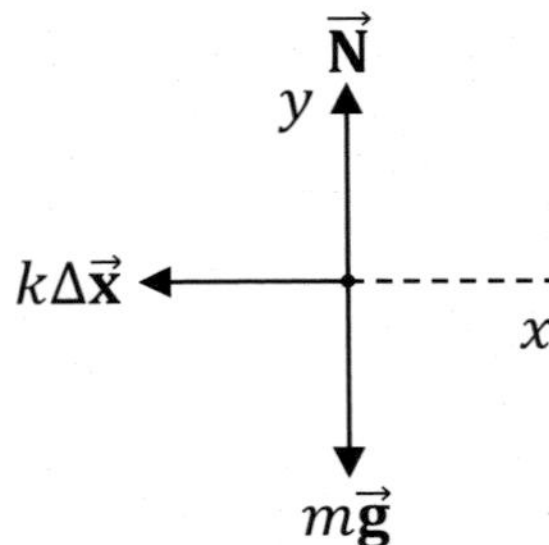

Apply Newton's second law to the box of bananas. The restoring force will be negative since it is directed opposite to the $+x$-axis in the diagram above. As usual, $a_y = 0$ because the box has only horizontal acceleration (it doesn't accelerate up or down).

$$\sum F_x = ma_x \quad , \quad \sum F_y = ma_y$$

$$-k|\Delta x| = ma_x \quad , \quad N - mg = 0$$

Solve for acceleration in the x-sum:

$$-\frac{k|\Delta x|}{m} = a_x$$

Convert Δx from centimeters (cm) to meters (m) since cm isn't an SI unit:

$$6.0 \text{ cm} = 6.0 \text{ cm} \times \frac{1 \text{ m}}{100 \text{ cm}} = \frac{6}{100} \text{ m}$$

Plug the values of the spring constant ($k = 80$ N/m), displacement from equilibrium ($|\Delta x| = 6.0$ cm), and mass ($m = 12$ kg) into the above equation for acceleration:

$$a_x = -\frac{k|\Delta x|}{m} = -\frac{80}{12} \cdot \frac{6}{100} = -\frac{(80)(6)}{(12)(100)} = -\frac{(80)(6)}{(100)(12)} = -\left(\frac{4}{5}\right)\left(\frac{1}{2}\right) = -\frac{2}{5} \text{ m/s}^2$$

The minus sign here represents that the box is accelerating to the left (since we chose the $+x$-axis to point to the right).

Example: As illustrated below, a spring is stretched by fixing its upper end and connecting a 5.0-kg box of bananas to its lower end. When the spring is in its new vertical equilibrium position, it is 25 cm longer than when lying in its natural horizontal equilibrium position. Determine the spring constant.

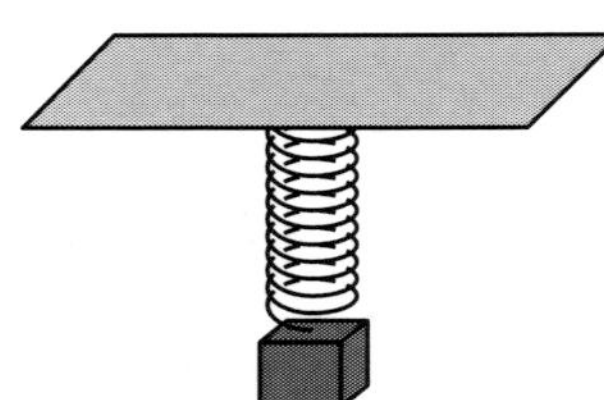

Draw and label a FBD for the box. Weight pulls downward, while the spring exerts a restoring force ($k\Delta x$) upward. The restoring force is upward as the stretched spring tries (unsuccessfully) to return to its original length.

Apply Newton's second law to the box of bananas.

$$\sum F_x = ma_x$$

$$k|\Delta x| - mg = ma_x$$

Since the problem states that the spring is in equilibrium, the acceleration is zero:

$$k|\Delta x| - mg = 0$$

Convert Δx from centimeters (cm) to meters (m) since cm isn't an SI unit:

$$25 \text{ cm} = 25 \text{ cm} \times \frac{1 \text{ m}}{100 \ cm} = \frac{25}{100} \text{ m} = \frac{1}{4} \text{ m}$$

Solve for the spring constant:

$$k = \frac{mg}{|\Delta x|} = \frac{5(9.81)}{1/4} \approx \frac{5(10)}{1/4} = \frac{50}{1/4}$$

Recall from math that the rule for dividing by a fraction is to multiply by its reciprocal:

$$k \approx 50 \div \frac{1}{4} = 50 \times \frac{4}{1} = 200 \text{ N/m}$$

The spring constant is approximately equal to 200 N/m.

97. As illustrated below, a spring with a spring constant of 45 N/m on a horizontal surface has its left end attached to a wall while its right end is attached to a 3.0-kg box of bananas. The coefficient of friction between the box and the horizontal surface is $\frac{1}{4}$. What is the acceleration of the spring when it is stretched 50 cm to the right of its equilibrium position while moving to the left (heading back to equilibrium)?

(A) Draw a FBD for the box. Label each force and the x- and y-coordinates.

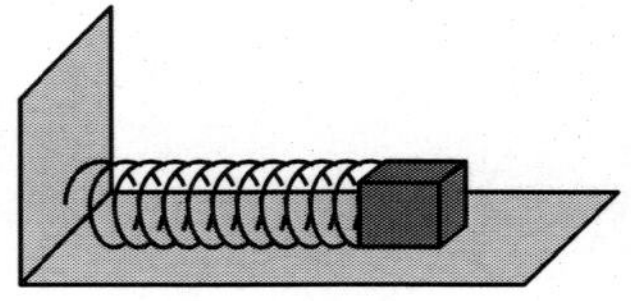

(B) Write the x- and y-sums for the forces acting on the box. There will be two sums. On the line immediately below each sum, rewrite the left-hand side in terms of the forces.

(C) Determine the acceleration of the box of bananas.

Want help? Check the hints section at the back of the book.

Answers: -5.0 m/s^2 (where the $-$ sign means to the left)

98. As illustrated below, a spring is stretched on a frictionless incline by fixing its upper end and connecting a 36-kg box of bananas to its lower end. When the spring is in its new inclined equilibrium position, it is 4.0 m longer than when lying in its natural horizontal equilibrium position.

(A) Draw a FBD for the box. Label each force and the x- and y-coordinates.

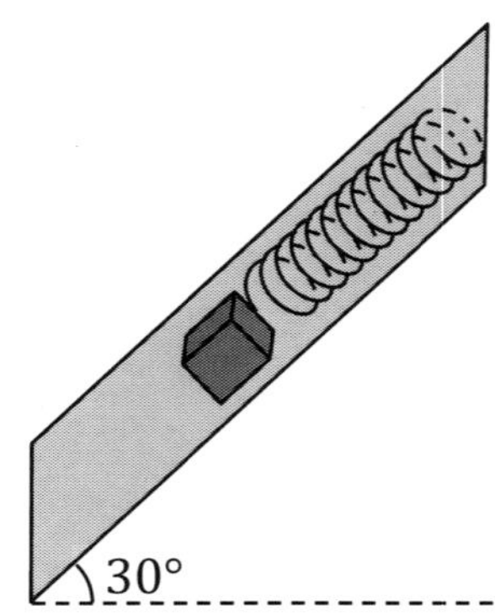

(B) Write the x- and y-sums for the forces acting on the box. There will be two sums. On the line immediately below each sum, rewrite the left-hand side in terms of the forces.

(C) Determine the spring constant.

Want help? Check the hints section at the back of the book.

Answers: 45 N/m

16 UNIFORM CIRCULAR MOTION

Relevant Terminology

Uniform circular motion – the motion of an object that travels in a circle with constant speed. The word 'uniform' specifically refers to the speed.
Centripetal – towards the center of the circle (inward). In uniform circular motion, the acceleration is centripetal and at least one force is centripetal or has an inward component.
Centrifugal – away from the center of the circle (outward). A person traveling in a circle has a perception of being pushed outward. This perception is called centrifugal force, but what the person perceives is really just a consequence of inertia.
Angular speed – the instantaneous rate at which the angle changes (as measured from the center of the circle, as shown in the diagram on page 152).
Period – the time it takes to complete exactly one revolution.
Frequency – the number of oscillations completed per second.

Equations of Uniform Circular Motion

Centripetal acceleration (a_c) depends on the speed (v) and the radius (R):

$$a_c = \frac{v^2}{R}$$

The **speed** (v) is proportional to the **angular speed** (ω):

$$v = R\omega$$

The **speed** (v) equals the total distance traveled (s) over the total time (t) or the circumference ($C = 2\pi R$) divided by the period (T):

$$v = \frac{s}{t} = \frac{C}{T} = \frac{2\pi R}{T}$$

The **angular speed** (ω) equals the angle (θ) over the total time (t) or the angle for one revolution (2π) divided by the period (T) or 2π times the frequency (f).

$$\omega = \frac{\theta}{t} = \frac{2\pi}{T} = 2\pi f$$

Frequency (f) and **period** (T) share a reciprocal relationship:

$$f = \frac{1}{T}$$

The **total distance traveled** (s) is proportional to the angle (θ) in radians:

$$s = R\theta$$

The centripetal force (the net inward force) can be found by multiplying mass by centripetal acceleration according to Newton's second law:

$$\sum F_{in} = ma_c$$

Symbols and SI Units

Symbol	Name	SI Units
a_c	centripetal acceleration	m/s^2
v	speed	m/s
ω	angular speed	rad/s
s	total distance traveled (arc length)	m
C	circumference	m
R	radius	m
D	diameter	m
θ	angle	rad
t	time	s
T	period	s
f	frequency	Hz
m	mass	kg
F	force	N

Note: The symbol for angular speed (ω) is the lowercase Greek letter omega.

Essential Concepts

Although an object traveling in uniform circular motion has constant speed, it does have acceleration because the direction of its velocity is changing. This type of acceleration is called **centripetal** acceleration. Centripetal acceleration is toward the center of the circle.

From Newton's second law, centripetal acceleration must be caused by centripetal force.

$$\sum F_{in} = ma_c$$

There is at least one real force either pointing inward or which has an inward component.

If you travel in a circle, you feel like you're being pushed outward. We call this centrifugal force, but it is really just your **inertia**. Your body has a natural tendency to go in a straight line, so you feel like you're trying to go off on a tangent.

Important Distinctions

There are several pairs of similar symbols used to describe uniform circular motion. It's important not to get these mixed up:

- The symbol t represents the **total time** that the object has been moving, whereas the symbol T (called the **period**) is the time it takes to go around exactly once.
- The symbol s represents the **total distance** that the object has traveled, whereas the symbol C (the **circumference**) is the distance that the object would travel if it went around the circle exactly once.
- The symbol v is the **speed** of the object in m/s, whereas the symbol ω is the **angular speed** of the object in rad/s.
- The symbol f can either mean **frequency** or friction force. You must pay attention to the context to distinguish between them. The units can also help: The SI unit of frequency is the Hertz (Hz) and the SI unit of friction force is the Newton (N).
- The symbol T can either mean **period** or tension. You must pay attention to the context to distinguish between them. The units can also help: The SI unit of period is the second (s) and the SI unit of tension is the Newton (N).

Notes Regarding Units

Angle (θ) can be expressed in radians, revolutions, or degrees, but only the radian is the SI unit for angle. The following equation only works if the angle (θ) is expressed in radians. If you have degrees or revolutions, you must convert to radians first.

$$s = R\theta$$

Similarly, the following equation only works if the angular speed (ω) is expressed in rad/s. If you have rev/s, you must convert to rad/s first.

$$v = R\omega$$

The conversion between **revolutions**, degrees, and **radians** is:

$$1 \text{ rev} = 360° = 2\pi \text{ rad}$$

To convert between degrees and radians, you can also use:

$$180° = \pi \text{ rad}$$

The units can help you determine which information is given in a problem. For example, the number 0.25 rad/s must be the angular speed (ω) because it's the only symbol that can be expressed in rad/s, whereas the number 6.0 m/s must be the speed (v). However, a number like 30 s could be either t or T, and a number like 4.0 m could be s, C, R, or D, so in these cases you must read the problem carefully and choose wisely. The SI unit of frequency, the Hertz (Hz), is equal to an inverse second:

$$\text{Hz} = \frac{1}{\text{s}}$$

This follow from the equation $f = 1/T$.

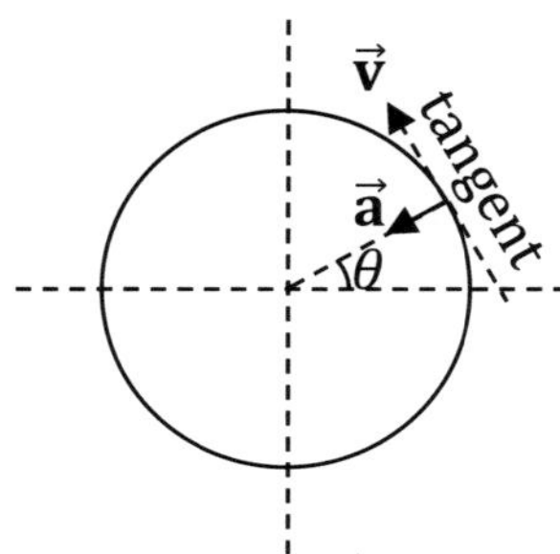

Uniform Circular Motion Strategy

To solve a problem involving uniform circular motion (where an object travels in a circle with constant speed), follow these steps:

1. Make a list of the symbols that you know along with their numerical values and SI units. Look at the units and wording to determine which symbols you know. Also identify the symbol that you're solving for:
   - A value in m/s is the speed, v.
   - A value in rad/s is the angular speed, ω.
   - A value in rev/s is the angular speed, ω, but must be converted to rad/s.
   - A value in m/s^2 is the centripetal acceleration, a_c.
   - A value in Hz is the frequency, f.
   - A value in revolutions is the angle, θ, but must be converted to radians.
   - A value in seconds could be the total time, t, or the period, T.
   - A value in meters could be the total distance traveled, s, the circumference, C, the diameter, D, or the radius, R.
   - A value in kg is the mass, m.
   - A value in N is a force. It could be centripetal force or a specific force.
2. Look at the list of equations for uniform circular motion. Specifically, look at the equations which involve the known symbols and the symbol you're looking for.
3. Choose the relevant equation(s) and solve for the desired unknown.
4. If the problem asks you to find the **number of revolutions**, solve for θ and then convert the answer from radians to revolutions: 2π rad $= 1$ rev. Similarly, if the problem tells you how many revolutions are completed, then θ is one of your knowns (but in this case you must first convert it from revolutions to radians).
5. If the problem asks for centripetal force, simply multiply mass times centripetal acceleration: $\sum F_{in} = ma_c$. Similarly, if the problem gives you centripetal force, you can set that equal to mass times centripetal acceleration.
6. If the problem involves applying Newton's second law further than described in Step 5, see Chapter 17, which discusses how to apply Newton's second law to uniform circular motion problems.

Example: A monkey drives a bananamobile in a 40-m diameter circle with a constant speed of 30 m/s. What is the bananamobile's acceleration?

Make a list of the knowns and the desired unknown:

$$D = 40 \text{ m} \quad , \quad v = 30 \text{ m/s} \quad , \quad a_c = ?$$

First find the radius from the diameter:

$$R = \frac{D}{2} = \frac{40}{2} = 20 \text{ m}$$

Use the equation for centripetal acceleration:

$$a_c = \frac{v^2}{R} = \frac{(30)^2}{20} = \frac{900}{20} = 45 \text{ m/s}^2$$

Example: A monkey drives a bananamobile in a 30-m diameter circle with a constant speed of 60 m/s. What is the bananamobile's angular speed?

Make a list of the knowns and the desired unknown:

$$D = 30 \text{ m} \quad , \quad v = 60 \text{ m/s} \quad , \quad \omega = ?$$

First find the radius from the diameter:

$$R = \frac{D}{2} = \frac{30}{2} = 15 \text{ m}$$

Use the equation that relates speed (v) to angular speed (ω):

$$v = R\omega$$

$$\omega = \frac{v}{R} = \frac{60}{15} = 4.0 \text{ rad/s}$$

Example: A monkey skates in a 32-m diameter circle with a period of 4.0 s. Determine the monkey's acceleration.

Make a list of the knowns and the desired unknown:

$$D = 32 \text{ m} \quad , \quad T = 4.0 \text{ s} \quad , \quad a_c = ?$$

First find the radius from the diameter:

$$R = \frac{D}{2} = \frac{32}{2} = 16 \text{ m}$$

We need to determine the monkey's speed because the equation for acceleration involves the speed:

$$v = \frac{2\pi R}{T} = \frac{2\pi(16)}{4} = 8\pi \text{ m/s}$$

Use the equation for centripetal acceleration:

$$a_c = \frac{v^2}{R} = \frac{(8\pi)^2}{16} = \frac{64\pi^2}{16} = 4\pi^2 \text{ m/s}^2$$

Example: A monkey runs in a circle with a period of 0.50 s. Determine the frequency.

Make a list of the knowns and the desired unknown:

$$T = 0.50 \text{ s} \quad , \quad f = ?$$

The frequency is the reciprocal of the period:

$$f = \frac{1}{T} = \frac{1}{0.5} = 2.0 \text{ Hz}$$

Example: A monkey runs in a circle with a constant speed of 8.0 m/s for a total time of 2.0 minutes. How far does the monkey travel?

Make a list of the knowns and the desired unknown:

$$v = 8.0 \text{ m/s} \quad , \quad t = 2.0 \text{ min.} \quad , \quad s = ?$$

First convert the time from minutes to seconds:

$$t = 2.0 \text{ min.} \times \frac{60 \text{ s}}{1 \text{ min.}} = 120 \text{ s}$$

Use the equation that involves the two knowns and the desired unknown:

$$v = \frac{s}{t}$$

$$s = vt = (8)(120) = 960 \text{ m}$$

Example: A monkey runs in a circle, completing 12 revolutions in one minute. Determine the period and the frequency.

Make a list of the knowns and the desired unknowns. Note that the number of revolutions corresponds to θ, provided that we convert the angle to radians before using it.

$$\theta = 12 \text{ rev} \quad , \quad t = 1.0 \text{ min.} \quad , \quad T = ? \quad , \quad f = ?$$

We must convert the angle from revolutions to radians:

$$\theta = 12 \text{ rev} \times \frac{2\pi \text{ rad}}{1 \text{ rev}} = 24\pi \text{ rad}$$

We also need to convert the time from minutes to seconds:

$$t = 1.0 \text{ min.} \times \frac{60 \text{ s}}{1 \text{ min.}} = 60 \text{ s}$$

Use the two equations for angular speed:

$$\omega = \frac{\theta}{t} = \frac{2\pi}{T}$$

Now we can find the period with a little algebra (cross-multiply to solve for T):

$$T = \frac{2\pi t}{\theta} = \frac{2\pi(60)}{24\pi} = \frac{120}{24} = 5.0 \text{ s}$$

The frequency is the reciprocal of the period:

$$f = \frac{1}{T} = \frac{1}{5} \text{ Hz}$$

Example: A monkey drives 120π m in a circle with a radius of 15 m. How many revolutions does the monkey complete?

Make a list of the knowns and the desired unknown:

$$s = 120\pi \text{ m} \quad , \quad R = 15 \text{ m} \quad , \quad \theta = ?$$

Use the arc length equation to find the angle:

$$s = R\theta$$

$$\theta = \frac{s}{R} = \frac{120\pi}{15} = 8\pi \text{ rad}$$

Convert the angle from radians to revolutions:

$$\theta = 8\pi \text{ rad} \times \frac{1 \text{ rev}}{2\pi \text{ rad}} = 4.0 \text{ rev}$$

99. A monkey skates in a 12-m diameter circle with a constant angular speed of $\frac{1}{2}$ rad/s for a total time of 4.0 minutes.

List the symbols that you know along with their values and SI units.

________________ ________________ ________________

(A) What is the monkey's speed?

(B) What is the monkey's acceleration?

(C) How far does the monkey travel?

(D) What is the period?

(E) How many revolutions does the monkey complete?

Want help? Check the hints section at the back of the book.

Answers: 3.0 m/s, $\frac{3}{2}$ m/s^2, 720 m, 4π s, $\frac{60}{\pi}$ rev

100. A monkey runs in circle with a constant speed of 4.0 m/s and a period of 8π s, completing $\frac{20}{\pi}$ revolutions.

List the symbols that you know along with their values and SI units.

____________________ ____________________ ____________________

(A) What is the monkey's angular speed?

(B) What is the radius of the circle?

(C) What is the monkey's acceleration?

(D) How far does the monkey travel?

(E) What is the frequency?

Want help? Check the hints section at the back of the book.

Answers: $\frac{1}{4}$ rad/s, 16 m, 1.0 m/s^2, 640 m, $\frac{1}{8\pi}$ Hz

17 UNIFORM CIRCULAR MOTION WITH NEWTON'S SECOND LAW

Essential Concepts

When applying Newton's second law to uniform circular motion, rather than working with x- and y-axes, we work with three directions relating to the circle:

- The inward (in) direction points toward the center of the circle.
- The tangential (tan) direction points along a line that is tangent to the circle.
- The z-direction is perpendicular to the plane of the circle.

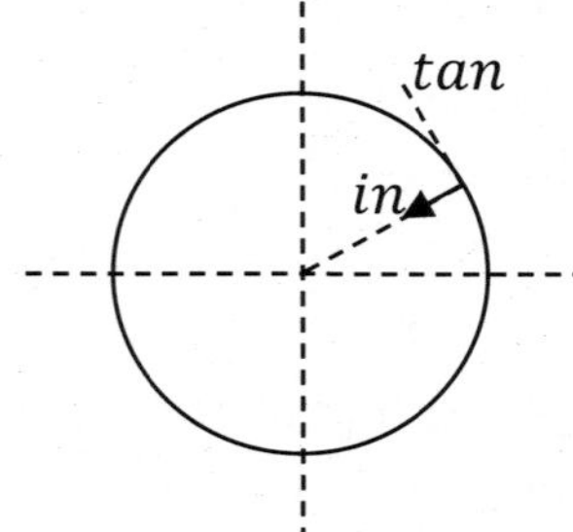

In uniform circular motion, the direction of the acceleration is centripetal, meaning that it is directed towards the center of the circle. Therefore, the sum of the inward components of the forces equals mass times centripetal acceleration, while the other sums equal zero:

$$\sum F_{in} = ma_c \quad , \quad \sum F_{tan} = 0 \quad , \quad \sum F_z = 0$$

Note: The label tan in the diagram above, in the FBD's that follow, and in the sum above stands for "tangential." This refers to the tangential direction, not to be confused with the tangent function. It's easy to tell the difference because the tangent function has an argument, like $\tan\theta$. So if you don't see an angle in an argument after the letters tan, it's just referring to the tangential direction.

Important Distinction

Depending on the context, the symbol θ may mean one of two different things in uniform circular motion:

- In Chapter 16, θ was an angle measured from the center of the circle corresponding to how far the object had traveled around the circle.
- In Chapter 17, θ will be a different angle appearing in the FBD, and appearing in the sums via trig. One such θ will appear in the first example, where θ is the angle between the cord and the vertical.

These two angles are not interchangeable.

Strategy for Applying Newton's Second Law to Uniform Circular Motion

To apply Newton's second law to a problem involving uniform circular motion, follow the Newton's second law strategy described in detail in Chapter 14, while following these steps to treat the uniform circular motion:

1. Draw the FBD according to the instructions in Chapter 14 (and in Chapter 15 if the problem involves a spring). There are no new forces to draw relating to the circular motion. For example, don't draw and label a "centripetal force." There will be at least one force with an inward component, but it will be one of the usual forces, like tension or friction. (What we call the "centripetal force" is really the sum of the inward components of the forces, $\sum F_{in}$.)
2. Label your coordinates as follows:
   - The **inward** (in) direction points toward the center of the circle. Find the center of the circle and draw the in-axis towards that point.
   - The **tangential** (tan) direction points along a line that is tangent to the circle. Draw a tangent line to label the tan-axis.
   - The z-direction is **perpendicular to the plane of the circle**. Draw a line perpendicular to the plane of the circle for the z-axis.
3. Write Newton's second law as follows for an object in uniform circular motion:
$$\sum F_{in} = ma_c \quad , \quad \sum F_{tan} = 0 \quad , \quad \sum F_z = 0$$
4. Follow the strategy from Chapter 14.
5. You can write centripetal acceleration as:
$$a_c = \frac{v^2}{R}$$
6. You may need other equations of uniform circular motion from Chapter 16:
$$v = R\omega$$
$$v = \frac{s}{t} = \frac{C}{T} = \frac{2\pi R}{T}$$
$$\omega = \frac{\theta}{t} = \frac{2\pi}{T} = 2\pi f$$
$$f = \frac{1}{T}$$
$$s = R\theta$$

Example: As illustrated below, a monkey ties a cord around a banana and whirls the cord in a horizontal circle with a constant speed of $\sqrt{10}$ m/s. The radius of the circle is 1.0 m. What angle does the cord make with the vertical?

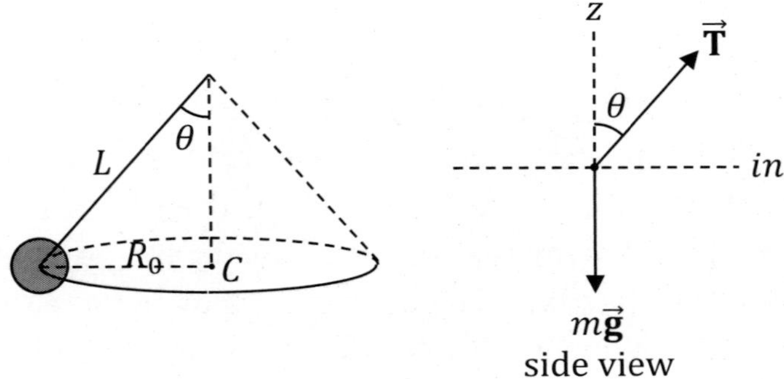

Note that while the banana travels in a horizontal circle, the cord sweeps out the shape of a cone. Draw and label a FBD for the banana. Tension pulls along the cord, while weight pulls down. The inward (in) direction is to the right for the position shown above (toward the center of the circle). The tangential direction is perpendicular to the page (coming out of the page), but isn't needed for this problem since none of the forces have a tangential component. The z-direction is upward (perpendicular to the plane of the circle).

Apply Newton's second law to the banana. For tension, the inward component receives a sine because it is opposite to θ, while the z-component receives a cosine because z is adjacent to θ. Weight appears only in the z-sum since it lies on the negative z-axis. The acceleration is centripetal (a_c) for a problem involving uniform circular motion.

$$\sum F_{in} = ma_c \quad , \quad \sum F_z = 0$$

$$T \sin\theta = ma_c \quad , \quad T\cos\theta - mg = 0$$

Bring weight to the right-hand side:

$$T \sin\theta = ma_c \quad , \quad T\cos\theta = mg$$

Now that we have moved weight over, we can divide the two equations to cancel tension:

$$\frac{T\sin\theta}{T\cos\theta} = \frac{ma_c}{mg}$$

$$\frac{\sin\theta}{\cos\theta} = \frac{a_c}{g}$$

Recall from trig that sine divided by cosines makes tangent:

$$\tan\theta = \frac{a_c}{g}$$

Use the equation for centripetal acceleration from Chapter 16 $\left(a_c = \frac{v^2}{R}\right)$:

$$\tan\theta = a_c\frac{1}{g} = \frac{v^2}{R}\frac{1}{g} = \frac{\left(\sqrt{10}\right)^2}{(1)(9.81)} \approx \frac{\left(\sqrt{10}\right)^2}{10} = 1$$

$$\theta = \tan^{-1}(1) = 45°$$

Example: A car rounds a turn with constant speed on horizontal ground. The coefficient of friction between the tires and road is $\frac{1}{2}$. The radius of the circle is 80 m. What maximum speed can the car have and still round the turn safely?

Draw and label a FBD for the car. Weight (mg) pulls down, normal force (N) pushes up, drive force (D) is forward, resistive forces (F_R) including air resistance push backward, and the force of static friction (f_s) pushes the car inward. Friction between the road and tires supplies the needed centripetal acceleration, allowing the car to make the turn: If you remove friction (like an icy road in the winter), you won't be able to make the turn. Newton's third law is at work: Just like the equal and opposite force that the ground exerts on your feet when you walk, when you turn the steering wheel, the force that the tires exert on the ground results in an equal and opposite force exerted on the car. The reason it's static friction is that the car isn't moving inward or outward (it's moving tangentially).

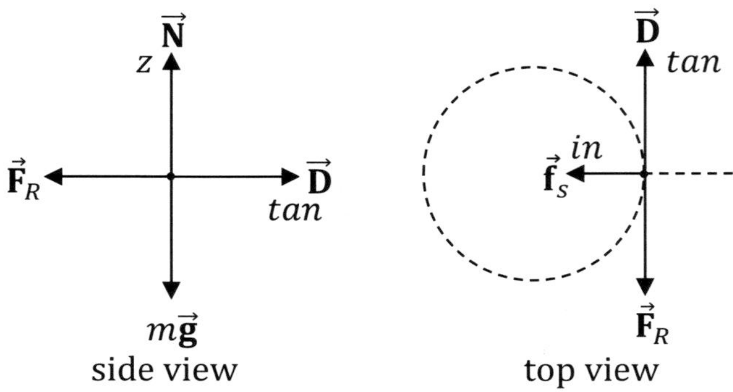

Apply Newton's second law to the car:

$$\sum F_{in} = ma_c \quad , \quad \sum F_{tan} = 0 \quad , \quad \sum F_z = 0$$

$$f_s = ma_c \quad , \quad D - F_R = 0 \quad , \quad N - mg = 0$$

Solve for normal force in the z-sum:

$$N = mg$$

Substitute the expression for normal force into the equation for static friction (Chapter 14):

$$f_s \leq \mu_s mg$$

Plug this into the equation from the inward sum:

$$f_s = ma_c \leq \mu_s mg$$

$$a_c \leq \mu_s g$$

Use the equation for centripetal acceleration (Chapter 16):

$$a_c = \frac{v^2}{R} \leq \mu_s g$$

$$v \leq \sqrt{\mu_s g R} = \sqrt{\frac{1}{2}(9.81)(80)} \approx \sqrt{\frac{1}{2}(10)(80)} = \sqrt{400} = 20 \text{ m/s}$$

The car must not travel faster than 20 m/s in order to round this turn safely.

101. An amusement park ride has a $48\sqrt{3}$-m diameter horizontal disc high above the ground. Several swings are suspended from the disc near its edge using $32\sqrt{3}$-m long chains. As the disc spins, the swings extend outward at an angle of 30° from the vertical, as illustrated below. A 25-kg monkey sitting in one of the swings travels in a horizontal circle.

(A) Draw a FBD for the monkey. Label each force and the *in* and *z*-coordinates.

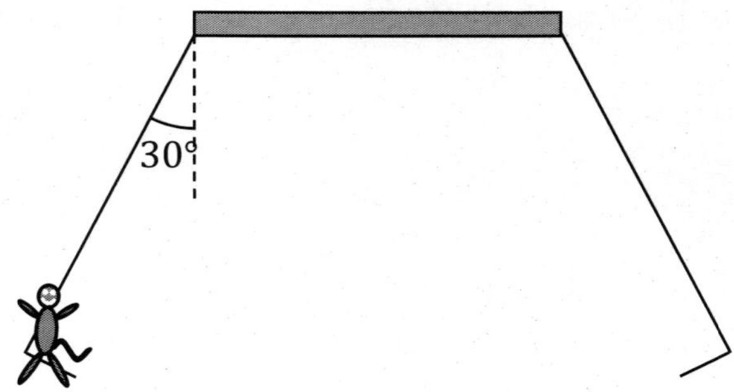

(B) Write the *in*- and *z*-sums for the forces acting on the monkey. On the line immediately below each sum, rewrite the left-hand side in terms of the forces.

(C) What is the tension in the chain of the monkey's swing, which has a 5.0-kg seat?

(D) What is the monkey's speed?

Want help? Check the hints section at the back of the book.

Answers: $200\sqrt{3}$ N, 20 m/s

102. A monkey grabs a 500-g mouse by the tail and whirls the mouse in a vertical circle with a constant speed of 4.0 m/s. The radius of the circle is 200 cm.

Note: Unlike the previous problem, which involved a horizontal circle, this problem involves a vertical circle. The two solutions appear much different.

(A) Draw a FBD for the mouse at the bottom of its circular arc. Label each force and the *in* and *tan*-coordinates.

(B) Write the *in*-sum for the forces acting on the mouse. (Just **one** sum is relevant here.) On the line immediately below the sum, rewrite the left-hand side in terms of the forces.

(C) What is the acceleration of the mouse?

(D) Determine the tension in the tail when the mouse is at the bottom of its arc.

Want help? Check the hints section at the back of the book.

Answers: 8.0 m/s^2, 9.0 N

103. A 50-kg monkey defies gravity on an amusement park centrifuge, which consists of a large cylindrical room, as illustrated below. The monkey stands against the wall. First, the centrifuge rotates with constant angular speed, then the floor disappears! Yet, the monkey does not slide down the wall. The coefficient of friction between the monkey and the wall is $\frac{1}{4}$. The radius of the centrifuge is 10 m.

(A) Draw a FBD for the monkey. Label each force and the *in* and *z*-coordinates.

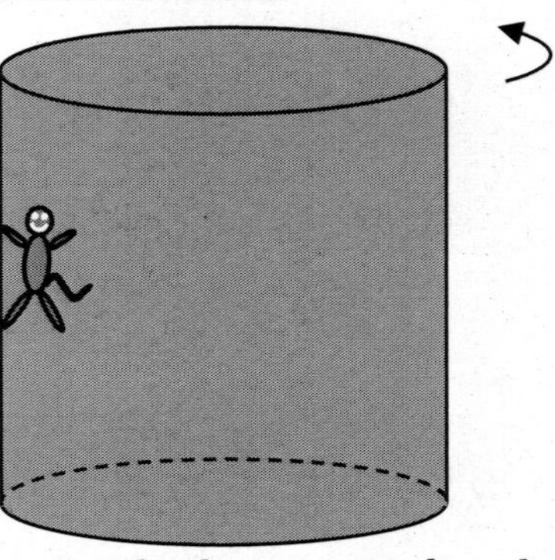

(B) Write the *in*- and *z*-sums for the forces acting on the monkey. On the line immediately below each sum, rewrite the left-hand side in terms of the forces.

(C) What minimum acceleration does the monkey need in order to not to slide downward?

(D) What minimum speed does the monkey need in order to not to slide downward?

Want help? Check the hints section at the back of the book.

Answers: 40 m/s^2, 20 m/s

104. A circular racetrack has a diameter of $180\sqrt{3}$ m (the radius R is labeled below) and constant banking angle of 30° (as shown below, with the racecar headed out of the page). The racetrack is frictionless.

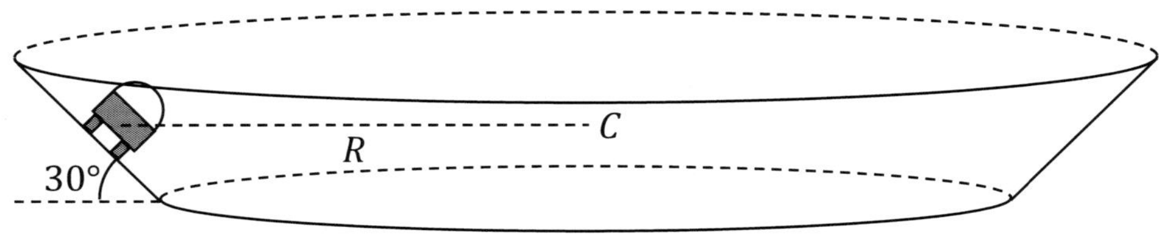

side view

(A) Draw a FBD for the racecar. Label each force and the *in* and z-coordinates.

(B) Write the *in*- and z-sums for the forces acting on the racecar. On the line immediately below each sum, rewrite the left-hand side in terms of the forces.

(C) What speed does the racecar need to have in order not to slide up or down the bank?

Want help? Check the hints section at the back of the book.

Answers: 30 m/s

18 NEWTON'S LAW OF GRAVITY

Relevant Terminology

Mass – a measure of inertia. More massive objects are more difficult to accelerate as they have greater inertia to overcome.
Weight – the gravitational force that an object experiences.
Gravitational acceleration – the acceleration of an object in free fall.

Newton's Law of Gravity

According to Newton's law of gravity, any two objects with mass attract one another with a **gravitational force** that is directly proportional to each mass and inversely proportional to the square of the separation between the two masses:

$$F_g = G\frac{m_1 m_2}{R^2}$$

The proportionality constant in Newton's law of gravity is called the gravitational constant:

$$G = 6.67 \times 10^{-11} \frac{\text{N}\cdot\text{m}^2}{\text{kg}^2} \approx \frac{20}{3} \times 10^{-11} \frac{\text{N}\cdot\text{m}^2}{\text{kg}^2} = \frac{2}{3} \times 10^{-10} \frac{\text{N}\cdot\text{m}^2}{\text{kg}^2}$$

Gravitational force is also called **weight**, and the formula for weight is mass times gravity:

$$F_g = mg$$

Therefore, if you divide both sides of Newton's law of gravity by one of the masses, you obtain an equation for **gravitational acceleration** created by a large astronomical body such as a planet or moon:

$$g = G\frac{m_p}{R^2}$$

Symbols and SI Units

Symbol	Name	SI Units
F_g	gravitational force	N
m	mass	kg
R	separation	m
G	gravitational constant	$\frac{\text{N}\cdot\text{m}^2}{\text{kg}^2}$ or $\frac{\text{m}^3}{\text{kg}\cdot\text{s}^2}$
g	gravitational acceleration	m/s^2

Important Distinctions

The symbol F_g represents gravitational **force** in N, whereas the symbol g represents gravitational **acceleration** in m/s^2.

The symbol g represents gravitational acceleration in m/s^2, whereas the symbol G represents the gravitational **constant** in $\frac{\text{N·m}^2}{\text{kg}^2}$. The value of g varies with location: For example, g equals 9.81 m/s^2 near earth's surface and 1.62 m/s^2 near the moon's surface. The value of G is universal: It equals 6.67×10^{-11} $\frac{\text{N·m}^2}{\text{kg}^2}$ everywhere.

Notes Regarding Units

The SI units of the gravitational constant (G) follow by solving for G in Newton's law of gravity:

$$G = \frac{F_g R^2}{m_1 m_2}$$

The SI units of G equal $\frac{\text{N·m}^2}{\text{kg}^2}$ since these are the SI units of $\frac{F_g R^2}{m_1 m_2}$. Recall from Chapter 14 that a Newton is equivalent to:

$$1\text{ N} = \frac{\text{kg·m}}{\text{s}^2}$$

Plugging this into $\frac{\text{N·m}^2}{\text{kg}^2}$, the SI units of G can alternatively be expressed as $\frac{\text{m}^3}{\text{kg·s}^2}$.

Algebra with Powers

It may be helpful to recall the following rules of algebra relating to powers (Chapter 1):

$$x^a x^b = x^{a+b}$$

$$\frac{x^a}{x^b} = x^{a-b}$$

$$x^{-a} = \frac{1}{x^a}$$

$$\frac{1}{x^{-a}} = x^a$$

$$x^0 = 1$$

$$(x^a)^b = x^{ab}$$

$$(ax)^b = a^b x^b$$

Gravitational Problems Strategy

How you solve a problem involving gravity depends on which kind of problem it is:

1. The simpler gravitational problems can be solved by making a list of the symbols that you know, choosing the appropriate equation, and solving for the desired unknown. These three equations apply to gravity problems:
   - The equation for gravitational force:
   $$F_g = G\frac{m_1 m_2}{R^2}$$
   - The equation for gravitational acceleration:
   $$g = G\frac{m_p}{R^2}$$
   - The equation for weight:
   $$F_g = mg$$

   Look at the units and wording to determine which symbols you know. Also identify the symbol that you're solving for:
   - A value in m/s^2 is an acceleration, such as gravitational acceleration, g.
   - A value in kg is the mass, m.
   - A value in N is a force, such as gravitational force (also called weight), F_g.

   You should know the following numerical values:
   - $G = 6.67 \times 10^{-11} \frac{\text{N}\cdot\text{m}^2}{\text{kg}^2} \approx \frac{20}{3} \times 10^{-11} \frac{\text{N}\cdot\text{m}^2}{\text{kg}^2} \approx \frac{2}{3} \times 10^{-10} \frac{\text{N}\cdot\text{m}^2}{\text{kg}^2}$ everywhere.
   - $g = 9.81 \text{ m/s}^2 \approx 10 \text{ m/s}^2$ near earth's surface.
   - $g = 1.62 \text{ m/s}^2 \approx \frac{8}{5} \text{ m/s}^2$ near the moon's surface.
2. If a problem gives you ratios of masses, radii, or gravitational accelerations, such as "with a mass five times greater than earth's," take a **ratio**, as shown in one of the examples that follow.
3. To find the point where the net gravitational field from two separate masses equals zero, set the expression for the gravitational field created by each mass equal to one another. Also write down an equation for R_1 and R_2 based on a picture. Then solve the system using algebra. This strategy is shown in one of the examples.
4. To find the **net gravitational field** created by two different masses (like the earth and the moon), use the formula for gravitational acceleration for each mass and then do vector addition (Chapter 10). This strategy is shown in one of the examples.
5. For a **satellite** problem, see Chapter 19.

Rounding Note

When we round G from $6.67 \times 10^{-11} \frac{\text{N}\cdot\text{m}^2}{\text{kg}^2}$ to $\frac{2}{3} \times 10^{-10} \frac{\text{N}\cdot\text{m}^2}{\text{kg}^2}$, it makes just a slight difference.

Example: Star Phy has a mass of 4.0×10^{30} kg. Star Phy has a planet named Six with a mass of 6.0×10^{24} kg. The distance between the center of Phy and Six is 2.0×10^{11} m. Determine the gravitational force of attraction between Phy and Six.

Make a list of the knowns and the desired unknown:

$$G \approx \frac{2}{3} \times 10^{-10} \frac{\text{N·m}^2}{\text{kg}^2} \quad , \quad m_1 = 4.0 \times 10^{30} \text{ kg} \quad , \quad m_2 = 6.0 \times 10^{24} \text{ kg}$$

$$R = 2.0 \times 10^{11} \text{ m} \quad , \quad F_g = ?$$

Plug these values into Newton's law of gravity:

$$F_g = G\frac{m_1 m_2}{R^2} \approx \left(\frac{2}{3} \times 10^{-10}\right)\frac{(4 \times 10^{30})(6 \times 10^{24})}{(2 \times 10^{11})^2}$$

It's convenient to separate the powers:

$$F_g \approx \frac{(2)(4)(6)}{(3)(2)^2} \times \frac{10^{-10}10^{30}10^{24}}{(10^{11})^2} = 4 \times \frac{10^{-10+30+24}}{10^{22}} = 4 \times \frac{10^{44}}{10^{22}} = 4.0 \times 10^{22} \text{ N}$$

Example: Planet Monk has a mass of 8.1×10^{24} kg and a radius of 3.0×10^6 m. What is gravitational acceleration near the surface of planet Monk?

Make a list of the knowns and the desired unknown:

$$G \approx \frac{2}{3} \times 10^{-10} \frac{\text{N·m}^2}{\text{kg}^2} \quad , \quad m_p = 8.1 \times 10^{24} \text{ kg} \quad , \quad R = 3.0 \times 10^6 \text{ m} \quad , \quad g = ?$$

Plug these values into the equation for gravitational acceleration:

$$g = G\frac{m_p}{R^2} \approx \left(\frac{2}{3} \times 10^{-10}\right)\frac{(8.1 \times 10^{24})}{(3 \times 10^6)^2}$$

It's convenient to separate the powers:

$$g \approx \frac{(2)(8.1)}{(3)(3)^2} \times \frac{10^{-10}10^{24}}{(10^6)^2} = 0.6 \times \frac{10^{-10+24}}{10^{12}} = 0.6 \times \frac{10^{14}}{10^{12}} = 0.6 \times 10^2 = 60 \text{ m/s}^2$$

Example: A planet has 15 times earth's mass and 5 times earth's radius. What is the value of gravitational acceleration near the surface of that planet?

Express the given ratios with symbols:

$$\frac{m_p}{m_e} = 15 \quad , \quad \frac{R_p}{R_e} = 5$$

Take a ratio of gravitational acceleration for each planet:

$$\frac{g_p}{g_e} = \frac{\frac{Gm_p}{R_p^2}}{\frac{Gm_e}{R_e^2}} = \frac{Gm_p}{R_p^2} \div \frac{Gm_e}{R_e^2} = \frac{Gm_p}{R_p^2} \times \frac{R_e^2}{Gm_e} = \frac{m_p}{m_e}\frac{R_e^2}{R_p^2} = \left(\frac{m_p}{m_e}\right)\left(\frac{R_e}{R_p}\right)^2 = (15)\left(\frac{1}{5}\right)^2 = \frac{3}{5}$$

Multiply both sides by earth's gravity:

$$g_p = \frac{3}{5}g_e = \frac{3}{5}(9.81) \approx \frac{3}{5}(10) = 6.0 \text{ m/s}^2$$

Example: Planet Ein has a mass of 9.0×10^{24} kg. Planet Ein has a moon named Stein with a mass of 4.0×10^{22} kg. The distance between the center of Ein and Stein is 8.0×10^{8} m. Find the point where the net gravitational field equals zero.

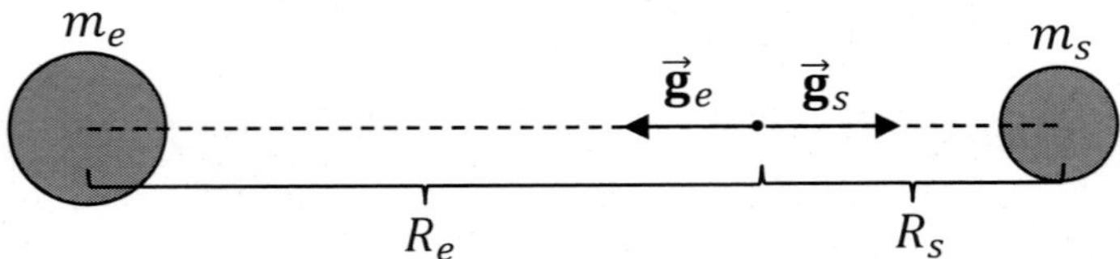

In between the planet and moon, the two gravitational fields will be in opposite directions. At some point in between, the two gravitational fields will also have equal magnitude and therefore cancel out. To figure out where this special point is, set the magnitudes of the two gravitational fields equal to one another ($g_e = g_s$):

$$G\frac{m_e}{R_e^2} = G\frac{m_s}{R_s^2}$$

$$\frac{m_e}{R_e^2} = \frac{m_s}{R_s^2}$$

Cross-multiply in order to remove the unknowns from the denominator:

$$R_s^2 m_e = R_e^2 m_s$$

We have one equation, but two unknowns: The unknowns are R_e and R_s. This means that we need a second equation. From the picture above, we can see that R_e and R_s add up to the distance between the centers of Ein and Stein, d, where $d = 8.0 \times 10^8$ m.

$$R_e + R_s = d$$

Isolate one of the unknowns in this equation:

$$R_s = d - R_e$$

Substitute this expression in place of R_s in the previous equation:

$$(d - R_e)^2 m_e = R_e^2 m_s$$

If you foil the left-hand side and distribute, you get a quadratic equation for R_e. You can't always avoid a quadratic, but in this case you can avoid the quadratic by taking the square-root of both sides:

$$(d - R_e)\sqrt{m_e} = \pm R_e\sqrt{m_s}$$

The $\pm$ is there because $\sqrt{x^2} = \pm x$. For example, $\sqrt{4}$ has two solutions, -2 and $+2$, since $(-2)^2 = 4$ and $(+2)^2 = 4$. We must consider both signs when solving for R_e.

$$d\sqrt{m_e} - R_e\sqrt{m_e} = \pm R_e\sqrt{m_s}$$

$$d\sqrt{m_e} = R_e\sqrt{m_e} \pm R_e\sqrt{m_s} = R_e\left(\sqrt{m_e} \pm \sqrt{m_s}\right)$$

$$R_e = \frac{d\sqrt{m_e}}{\sqrt{m_e} \pm \sqrt{m_s}}$$

Only the positive root yields an answer which lies in between the two masses:

$$R_e = \frac{(8 \times 10^8)\sqrt{9 \times 10^{24}}}{\sqrt{9 \times 10^{24}} + \sqrt{4 \times 10^{22}}} = \frac{(8 \times 10^8)(3 \times 10^{12})}{3 \times 10^{12} + 2.0 \times 10^{11}} = \frac{24 \times 10^{20}}{3.2 \times 10^{12}} = 7.5 \times 10^8 \text{ m}$$

Note: In the previous example and next example, R is **not** the radius of a planet or moon. Instead, R is the distance from the center of the body. R is only the radius of a planet or moon when you're trying to find g at its surface (which is **not** the case in the last example or the next example). In contrast, R was the radius in the second and third examples.

Example: Planet New has a mass of 5.4×10^{25} kg. Planet Ton has a mass of 1.8×10^{25} kg. In the diagram below, $R_n = 6.0 \times 10^8$ m and $R_t = 4.0 \times 10^8$ m. What is the magnitude of the net gravitational field at the point marked X in the diagram below?

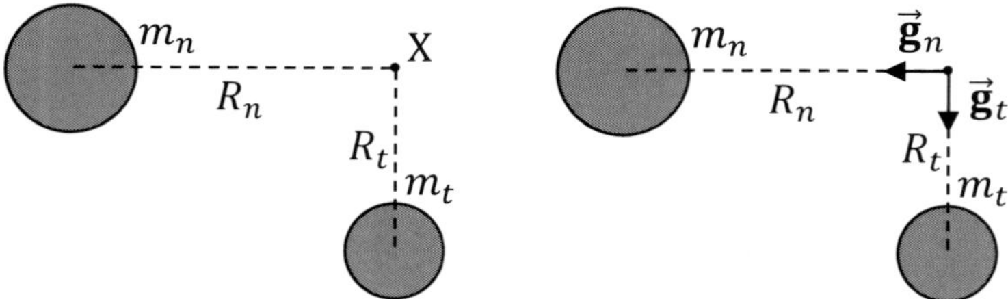

The first step is to determine the magnitude of each individual gravitational field at the specified point. Note that R_n and R_t are **not** the radii of the two planets because we're not finding gravitational acceleration on their surfaces, but at the point X instead.

$$g_n = G\frac{m_n}{R_n^2} \approx \left(\frac{2}{3} \times 10^{-10}\right)\frac{(5.4 \times 10^{25})}{(6 \times 10^8)^2} = \frac{(2)(5.4)}{(3)(6)^2} \times \frac{10^{-10}10^{25}}{(10^8)^2} = \frac{10.8}{108} \times 10^{-1} = \frac{1}{100} \text{ m/s}^2$$

$$g_t = G\frac{m_t}{R_t^2} \approx \left(\frac{2}{3} \times 10^{-10}\right)\frac{(1.8 \times 10^{25})}{(4 \times 10^8)^2} = \frac{(2)(1.8)}{(3)(4)^2} \times \frac{10^{-10}10^{25}}{(10^8)^2} = \frac{3.6}{48} \times 10^{-1} = \frac{3}{400} \text{ m/s}^2$$

We must now add the two gravitational fields considering them as vectors (Chapter 10), where $\vec{\mathbf{g}}_n$ has a magnitude of $\frac{1}{100}$ m/s^2 and points to the left in the picture above, while $\vec{\mathbf{g}}_t$ has a magnitude of $\frac{3}{400}$ m/s^2 and points down in the picture above. We could go through all of the steps of vector addition, but this problem is simpler if you realize that $\vec{\mathbf{g}}_n$ and $\vec{\mathbf{g}}_t$ are perpendicular. Because they form a right angle, we can simply use the Pythagorean theorem to determine the magnitude of the net gravitational field:

$$g_{net} = \sqrt{g_n^2 + g_t^2} \approx \sqrt{\left(\frac{1}{100}\right)^2 + \left(\frac{3}{400}\right)^2}$$

If not using a calculator, it's convenient to factor out $\left(\frac{1}{100}\right)^2$. Of course, you can reach the same final answer if you choose not to factor.

$$g_{net} = \sqrt{\left(\frac{1}{100}\right)^2}\sqrt{1^2 + \left(\frac{3}{4}\right)^2} = \frac{1}{100}\sqrt{1 + \frac{9}{16}} = \frac{1}{100}\sqrt{\frac{16}{16} + \frac{9}{16}}$$

$$g_{net} = \frac{1}{100}\sqrt{\frac{25}{16}} = \frac{1}{100}\frac{5}{4} = \frac{1}{80} \text{ m/s}^2$$

105. Planet Ban has a radius of 6.0×10^7 m. Gravitational acceleration near the surface of Ban is 20 m/s^2.

(A) What is the mass of Ban?

(B) Ban has a moon named Ana that has an orbital radius of 4.0×10^9 m and a mass of 2.0×10^{22} kg. What is the force of attraction between Ban and Ana?

Want help? Check the hints section at the back of the book.

Answers: 1.08×10^{27} kg, 9.0×10^{19} N

106. (A) A planet has 5 times earth's radius and 3 times the surface gravity of earth. How does that planet's mass compare to earth's?

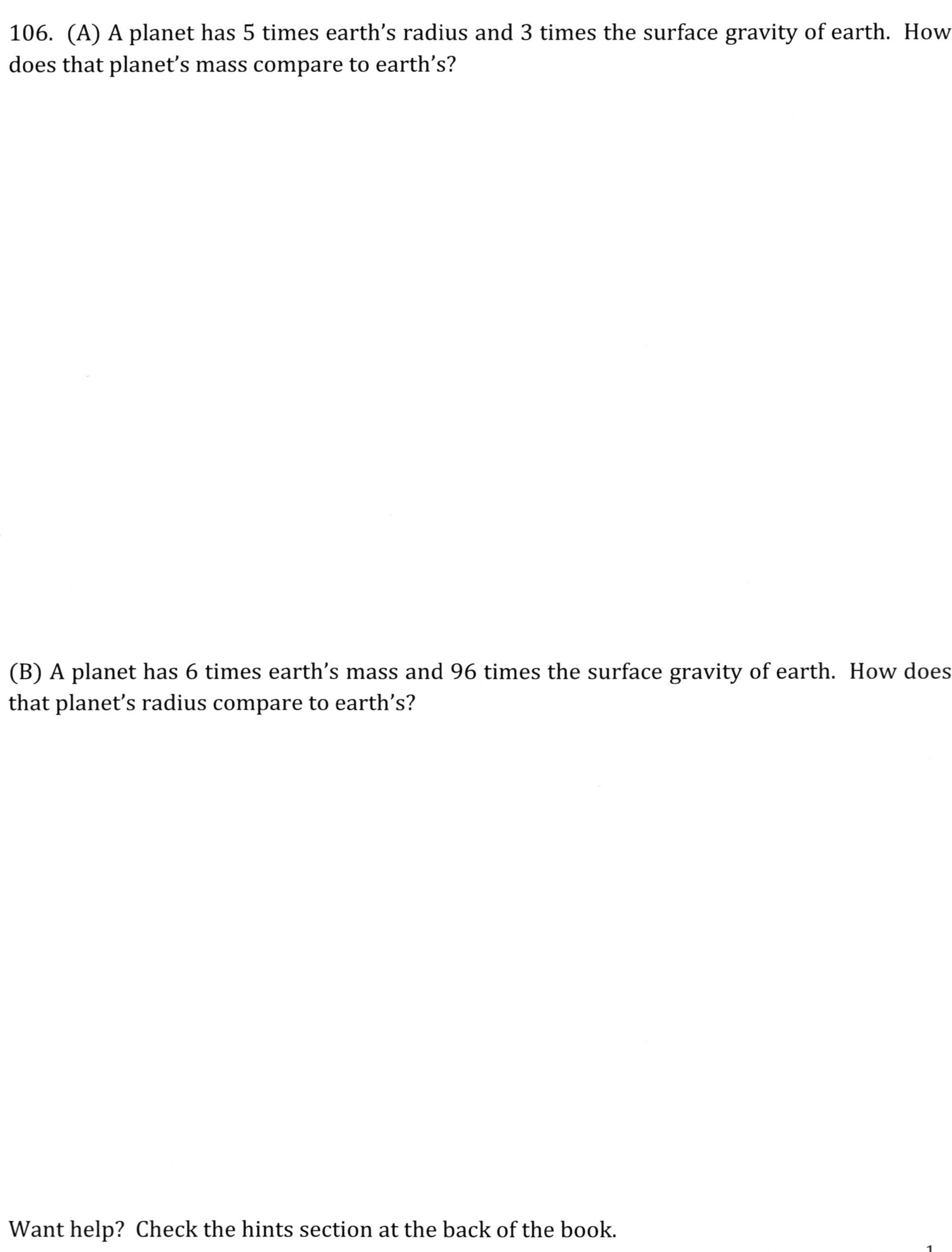

(B) A planet has 6 times earth's mass and 96 times the surface gravity of earth. How does that planet's radius compare to earth's?

Want help? Check the hints section at the back of the book.

Answers: $75\times, \frac{1}{4}\times$

107. Planet Mon has a mass of 1.6×10^{25} kg. Planet Mon has a moon named Key with a mass of 2.5×10^{23} kg. The distance between the center of Mon and Key is 9.0×10^{8} m. Find the point where the net gravitational field equals zero.

Want help? Check the hints section at the back of the book.

Answer: 8.0×10^{8} m from Mon

108. Planet Coco has a mass of 3.0×10^{24} kg. Planet Nut has a mass of 6.0×10^{24} kg. The distance between the center of Coco and Nut is 4.0×10^{8} m. What are the magnitude and direction of the net gravitational field at the point exactly halfway between the centers of these two planets?

Want help? Check the hints section at the back of the book.

Answer: $\frac{1}{200}$ m/s^2 towards Nut

19 SATELLITE MOTION

Relevant Terminology

Satellite – an object that orbits another body.
Geosynchronous – in an orbit synchronized with the earth's rotation. A geosynchronous satellite has a period of 24 hours (convert this to seconds before using it).

Satellite Equations

Consider a satellite traveling with constant speed in a circle. The net force is a gravitational force directed inward, which causes centripetal acceleration. Apply Newton's second law to uniform circular motion as described in Chapter 17.

$$\sum F_{in} = m_s a_c$$

Newton's law of gravity provides the equation $\left(F_g = G\frac{m_p m_s}{R^2}\right)$ for the force:

$$G\frac{m_p m_s}{R^2} = m_s a_c$$

The satellite's mass appears on the right-hand side because it is the satellite which is traveling in a large circle. Divide both sides by the mass of the satellite to solve for the **acceleration** (a_c) of the satellite:

$$a_c = G\frac{m_p}{R^2}$$

Use the equation for centripetal acceleration $\left(a_c = \frac{v^2}{R}\right)$ from Chapter 16:

$$\frac{v^2}{R} = G\frac{m_p}{R^2}$$

Multiply both sides by R to solve for the **speed** (v) of the satellite:

$$v^2 = G\frac{m_p}{R} \quad \text{or} \quad v = \sqrt{G\frac{m_p}{R}}$$

To find the period of the satellite, use an equation from Chapter 16 $\left(v = \frac{2\pi R}{T}\right)$:

$$\left(\frac{2\pi R}{T}\right)^2 = G\frac{m_p}{R} \quad \text{or} \quad \frac{4\pi^2 R^2}{T^2} = G\frac{m_p}{R}$$

Cross-multiply to solve for the **period** (T):

$$4\pi^2 R^3 = G m_p T^2$$

$$T^2 = \frac{4\pi^2 R^3}{G m_p} \quad \text{or} \quad T = 2\pi\sqrt{\frac{R^3}{G m_p}}$$

Note that R is the radius of the satellite's orbit. It is **<u>not</u>** the radius of the planet.

Satellite Strategy

To solve a problem with a satellite traveling in a circular orbit, follow these steps:

1. Make a list of the symbols that you know along with their numerical values and SI units. Also identify the symbol that you're solving for. If the term geosynchronous appears in the problem, one of the knowns will be $T = 24$ hrs (which you would need to convert to seconds). However, if the problem doesn't tell you that the satellite is geosynchronous, **<u>don't</u>** use 24 hrs for the period. You should also know that $G = 6.67 \times 10^{-11} \frac{\text{N}\cdot\text{m}^2}{\text{kg}^2} \approx \frac{20}{3} \times 10^{-11} \frac{\text{N}\cdot\text{m}^2}{\text{kg}^2} \approx \frac{2}{3} \times 10^{-10} \frac{\text{N}\cdot\text{m}^2}{\text{kg}^2}$.
2. Use the appropriate equation based on the values that you know and the unknown you are looking for. **Note**: Some physics instructors expect their students to derive the following equations before using them to solve a problem. If that's the case with you, first copy the steps from the previous page and then proceed to use the desired equation.
   - To find the **speed** of the satellite from m_p and R:
   $$v = \sqrt{G\frac{m_p}{R}}$$
   - To find the **speed** of the satellite from R and T:
   $$v = \frac{2\pi R}{T}$$
   - To find the **acceleration** of the satellite from m_p and R:
   $$a_c = G\frac{m_p}{R^2}$$
   - To find the **acceleration** of the satellite from v and R:
   $$a_c = \frac{v^2}{R}$$
   - To find the **period** of the satellite from m_p and R:
   $$T = 2\pi\sqrt{\frac{R^3}{Gm_p}}$$
   - To find the **period** of the satellite from v and R:
   $$T = \frac{2\pi R}{v}$$
3. You may need to solve for other quantities, like **angular speed** (ω), using the equations of uniform circular motion (Chapter 16). For example, either of these equation can help you find angular speed:
$$\omega = \frac{v}{R} \quad \text{or} \quad \omega = \frac{2\pi}{T}$$
4. To find **altitude** (h), subtract the planet's radius (R_p) from the orbital radius (R):
$$h = R - R_p$$

Symbols and SI Units

Symbol	Name	SI Units
m_p	mass of the large astronomical body near the center of the orbit	kg
m_s	mass of the satellite	kg
R	radius of the satellite's orbit	m
R_p	radius of the planet	m
h	altitude	m
G	gravitational constant	$\frac{\text{N}\cdot\text{m}^2}{\text{kg}^2}$ or $\frac{\text{m}^3}{\text{kg}\cdot\text{s}^2}$
a_c	centripetal acceleration	m/s^2
v	speed of the satellite	m/s
T	period of the satellite's revolution	s
ω	angular speed	rad/s

Example: Planet Ape has a mass of 3.6×10^{25} kg. A satellite orbits Ape in a circular orbit with a radius of 6.0×10^8 m. What is the satellite's orbital speed?

Make a list of the knowns and the desired unknown:

$$G \approx \frac{2}{3} \times 10^{-10} \frac{\text{N}\cdot\text{m}^2}{\text{kg}^2} \quad , \quad m_p = 3.6 \times 10^{25} \text{ kg} \quad , \quad R = 6.0 \times 10^8 \text{ m} \quad , \quad v = ?$$

Plug these values into the appropriate equation for speed:

$$v = \sqrt{G\frac{m_p}{R}} \approx \sqrt{\left(\frac{2}{3} \times 10^{-10}\right)\frac{(3.6 \times 10^{25})}{(6 \times 10^8)}} = \sqrt{\frac{(2)(3.6)}{(3)(6)}\frac{10^{-10}10^{25}}{10^8}} = \sqrt{\frac{7.2}{18}\frac{10^{15}}{10^8}} = \sqrt{\frac{2}{5}10^7}$$

We used the fact that $\frac{7.2}{18}$ reduces to $\frac{2}{5}$ (both equal 0.4 in decimal form). Now we'll rewrite 10^7 as $10^6 \times 10$ since 10^6 is a perfect square ($10^6 = 10^3 10^3$ such that $\sqrt{10^6} = 10^3$):

$$v \approx \sqrt{\frac{2}{5}10^7} = \sqrt{\frac{2}{5}10^6 10} = \sqrt{\frac{20}{5}10^6} = \sqrt{4 \times 10^6} = 2.0 \times 10^3 \text{ m/s} = 2.0 \text{ km/s}$$

(If you're using a calculator, this arithmetic is much simpler, of course. The math tricks applied above help if you're not using a calculator, which is sometimes a handy skill.)

Note: We will round G just slightly from $6.67 \times 10^{-11}\ \frac{\text{N}\cdot\text{m}^2}{\text{kg}^2}$ to $\frac{2}{3} \times 10^{-10}\ \frac{\text{N}\cdot\text{m}^2}{\text{kg}^2}$ in order to solve the problems without a calculator. This will cause just slight approximations in the examples that follow. (Note that $\frac{2}{3} \times 10^{-10} = \frac{20}{3} \times 10^{-11} \approx 6.67 \times 10^{-11}$.)

Example: Planet Chimp has a mass of 4.5×10^{25} kg. A satellite orbits Chimp in a circular orbit with a radius of 3.0×10^7 m. What is the satellite's orbital period?
Make a list of the knowns and the desired unknown:

$$G \approx \frac{2}{3} \times 10^{-10}\ \frac{\text{N}\cdot\text{m}^2}{\text{kg}^2} \quad , \quad m_p = 4.5 \times 10^{25}\text{ kg} \quad , \quad R = 3.0 \times 10^7\text{ m} \quad , \quad T = ?$$

Plug these values into the appropriate equation for period:

$$T = 2\pi\sqrt{\frac{R^3}{Gm_p}} \approx 2\pi\sqrt{\frac{(3 \times 10^7)^3}{\left(\frac{2}{3} \times 10^{-10}\right)(4.5 \times 10^{25})}} = 2\pi\sqrt{\frac{(3)^3}{\left(\frac{2}{3}\right)(4.5)} \times \frac{(10^7)^3}{10^{-10}10^{25}}} = 2\pi\sqrt{\frac{27}{3}\frac{10^{21}}{10^{15}}}$$

$$T \approx 2\pi\sqrt{9 \times 10^6} = 2\pi(3 \times 10^3) = 6\pi \times 10^3\text{ s} = 6000\pi\text{ s}$$

Example: Planet Lemur has a mass of 1.35×10^{26} kg and a radius of 3.0×10^6 m. A satellite orbits Lemur in a circular orbit with a speed of 30 km/s. What is the satellite's altitude?
Make a list of the knowns and the desired unknown:

$$G \approx \frac{2}{3} \times 10^{-10}\ \frac{\text{N}\cdot\text{m}^2}{\text{kg}^2} \quad , \quad m_p = 1.35 \times 10^{26}\text{ kg} \quad , \quad R_p = 3.0 \times 10^6\text{ m}$$

$$v = 30\text{ km/s} \quad , \quad h = ?$$

Unlike the other examples, this problem gave us the radius of the planet, R_p, and not the radius of the satellite's orbit, R. Also, since the given speed is in km/s, we need to convert it to SI units (m/s) before using it:

$$v = 30\text{ km/s} = 30 \times 10^3\text{ m/s} = 3.0 \times 10^4\text{ m/s}$$

We need to find the satellite's orbital radius, R, before we can find the altitude, h. Given what we know, we can find R from one of the speed equations:

$$v = \sqrt{G\frac{m_p}{R}}$$

Square both sides to eliminate the squareroot:

$$v^2 = G\frac{m_p}{R}$$

Multiply both sides by R and divide both sides by v^2:

$$R = G\frac{m_p}{v^2} \approx \left(\frac{2}{3} \times 10^{-10}\right)\frac{(1.35 \times 10^{26})}{(3 \times 10^4)^2} = \frac{(2)(1.35)}{(3)(3)^2}\frac{10^{-10}10^{26}}{(10^4)^2} = \frac{2.7}{27}\frac{10^{16}}{10^8} = \frac{1}{10} \times 10^8$$

$$R \approx 1.0 \times 10^7\text{ m}$$

Subtract the planet's radius, R_p, from the satellite's orbital radius, R, to find the altitude, h:

$$h = R - R_p \approx 1.0 \times 10^7 - 3.0 \times 10^6 = 10.0 \times 10^6 - 3.0 \times 10^6 = 7.0 \times 10^6\text{ m}$$

109. Planet Φ_6 has a mass of 1.2×10^{24} kg. A satellite orbits Φ_6 in a circular orbit with a radius of 2.0×10^7 m.

(A) What is the satellite's orbital speed?

(B) What is the satellite's orbital period?

Want help? Check the hints section at the back of the book.

Answers: 2.0 km/s, $2\pi \times 10^4$ s

110. A satellite orbits the earth in a geosynchronous circular orbit above the equator. Approximate earth's mass as $\approx 6.0 \times 10^{24}$ kg. What is the radius of the satellite's orbit?

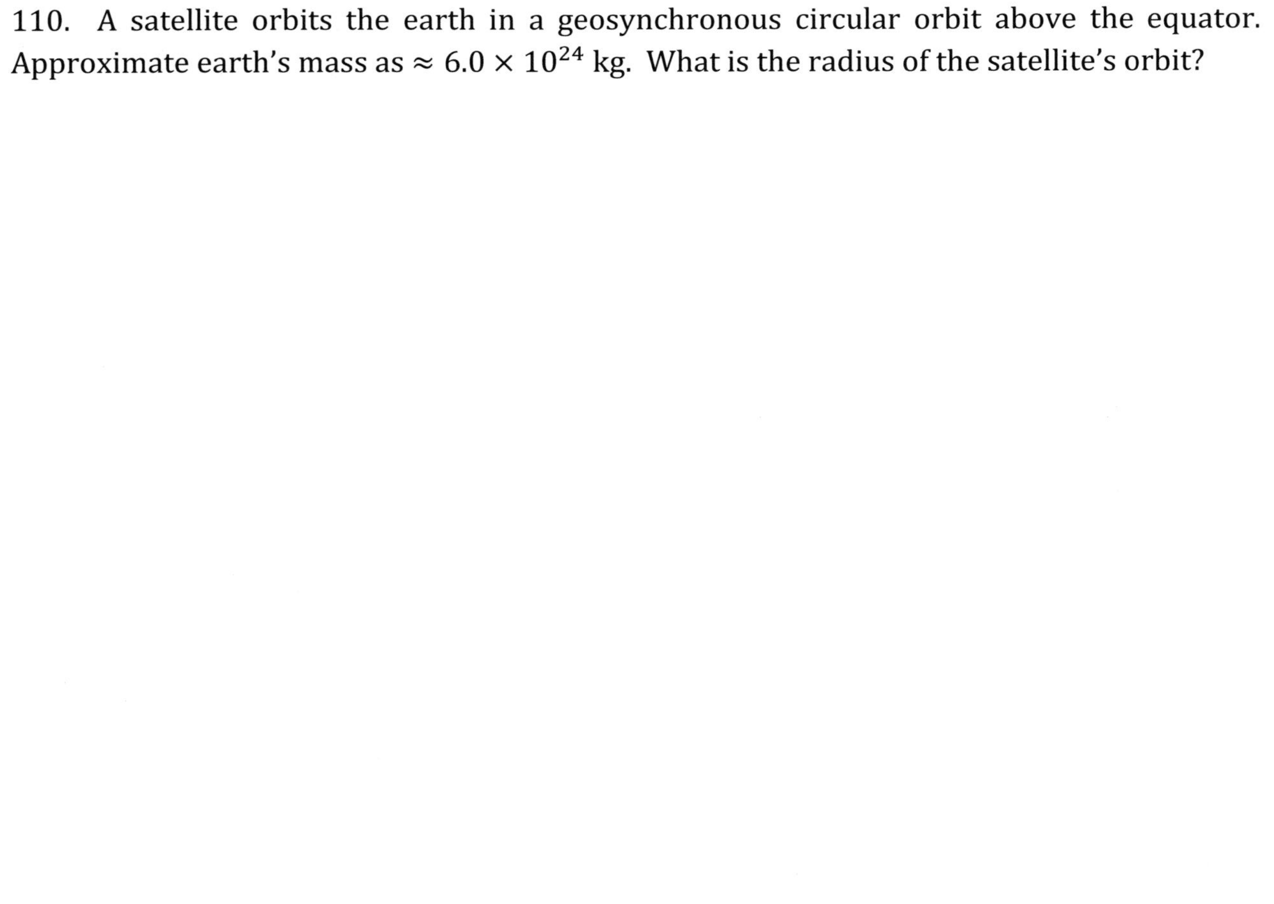

Want help? Check the hints section at the back of the book.

Answer: $72 \times \left(\frac{2}{\pi^2}\right)^{\frac{1}{3}} \times 10^6$ m (if you prefer to work with decimals, it is 4.23×10^7 m)

20 THE SCALAR PRODUCT

Essential Concepts

The **scalar** product (or **dot** product) $\vec{\mathbf{A}} \cdot \vec{\mathbf{B}}$ between two vectors $\vec{\mathbf{A}}$ and $\vec{\mathbf{B}}$ represents B times the projection of $\vec{\mathbf{A}}$ onto $\vec{\mathbf{B}}$ (or, equivalently, A times the projection of $\vec{\mathbf{B}}$ onto $\vec{\mathbf{A}}$). Whereas $\vec{\mathbf{A}}$ and $\vec{\mathbf{B}}$ are **vectors** (they each have both a magnitude and direction), the scalar product $\vec{\mathbf{A}} \cdot \vec{\mathbf{B}}$ results in a **scalar** (meaning that $\vec{\mathbf{A}} \cdot \vec{\mathbf{B}}$ has <u>no</u> direction, only a magnitude).

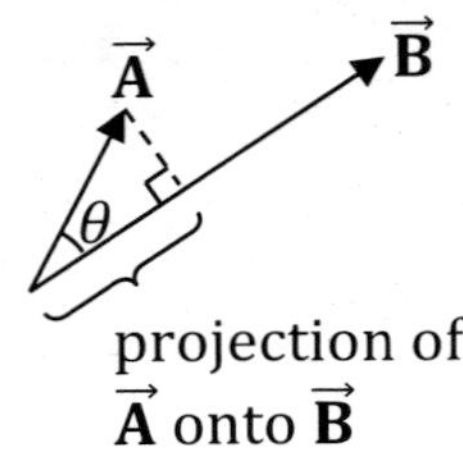

Scalar Product Equations

There are two equivalent ways of expressing the scalar product:

- Use the following equation to compute the scalar product when you know the **components** of the given vectors.
$$\vec{\mathbf{A}} \cdot \vec{\mathbf{B}} = A_x B_x + A_y B_y + A_z B_z$$
- Use the following equation to compute the scalar product when you know the magnitudes of $\vec{\mathbf{A}}$ and $\vec{\mathbf{B}}$ along with the angle between the two vectors.
$$\vec{\mathbf{A}} \cdot \vec{\mathbf{B}} = AB \cos\theta$$
- You can use both equations together to **find the angle** between two vectors, as shown in one of the examples that follow.

Recall from Chapter 10 that any vector can be expressed in terms of its components (A_x, A_y, and A_z) and Cartesian unit vectors ($\hat{\mathbf{x}}$, $\hat{\mathbf{y}}$, and $\hat{\mathbf{z}}$) in the following form:

$$\vec{\mathbf{A}} = A_x \hat{\mathbf{x}} + A_y \hat{\mathbf{y}} + A_z \hat{\mathbf{z}}$$

Strategy for Finding or Applying the Scalar Product

To compute the scalar product between two vectors, or to determine the angle between two vectors, follow these steps:

- If you know the components of the given vectors, use $\vec{\mathbf{A}} \cdot \vec{\mathbf{B}} = A_x B_x + A_y B_y + A_z B_z$.
- If you know the magnitudes of the given vectors along with the angle between the two vectors, use $\vec{\mathbf{A}} \cdot \vec{\mathbf{B}} = AB \cos\theta$.
- To find the angle between two vectors, first compute the scalar product as in Step 1 and then solve for θ using the equation from Step 2. Find the magnitudes of the given vectors using the Pythagorean theorem. For example, $A = \sqrt{A_x^2 + A_y^2 + A_z^2}$.

Commutativity of the Scalar Product

The scalar product is commutative because $\vec{\mathbf{B}} \cdot \vec{\mathbf{A}} = \vec{\mathbf{A}} \cdot \vec{\mathbf{B}}$.

Scalar Product between Unit Vectors

Recall from Chapter 10 that $\hat{\mathbf{x}}$ is a unit vector pointing along the $+x$-axis and $\hat{\mathbf{y}}$ is a unit vector pointing along the $+y$-axis. Similarly, $\hat{\mathbf{z}}$ is a unit vector pointing along the $+z$-axis. Since the angle between $\hat{\mathbf{x}}$ and itself is 0°, the scalar product $\hat{\mathbf{x}} \cdot \hat{\mathbf{x}}$ must equal 1 according to the formula $\vec{\mathbf{A}} \cdot \vec{\mathbf{B}} = AB\cos\theta$: Here is the math: $\hat{\mathbf{x}} \cdot \hat{\mathbf{x}} = (1)(1)\cos 0°$. In contrast, since the angle between $\hat{\mathbf{x}}$ and $\hat{\mathbf{y}}$ is 90°, the scalar product $\hat{\mathbf{x}} \cdot \hat{\mathbf{y}}$ equals 0: $\hat{\mathbf{x}} \cdot \hat{\mathbf{y}} = (1)(1)\cos 90° = 0$. Similar relations are tabulated below.

$$\hat{\mathbf{x}} \cdot \hat{\mathbf{x}} = 1 \quad , \quad \hat{\mathbf{y}} \cdot \hat{\mathbf{y}} = 1 \quad , \quad \hat{\mathbf{z}} \cdot \hat{\mathbf{z}} = 1$$
$$\hat{\mathbf{x}} \cdot \hat{\mathbf{y}} = 0 \quad , \quad \hat{\mathbf{y}} \cdot \hat{\mathbf{z}} = 0 \quad , \quad \hat{\mathbf{z}} \cdot \hat{\mathbf{x}} = 0$$
$$\hat{\mathbf{y}} \cdot \hat{\mathbf{x}} = 0 \quad , \quad \hat{\mathbf{z}} \cdot \hat{\mathbf{y}} = 0 \quad , \quad \hat{\mathbf{x}} \cdot \hat{\mathbf{z}} = 0$$

Example: (A) Find the scalar product between the vectors $\vec{\mathbf{A}} = 2\,\hat{\mathbf{x}} - 3\,\hat{\mathbf{y}} + 4\,\hat{\mathbf{z}}$ and $\vec{\mathbf{B}} = 5\,\hat{\mathbf{x}} - 6\,\hat{\mathbf{y}} - 7\,\hat{\mathbf{z}}$. (B) What is the angle between these two vectors?

(A) Compare $\vec{\mathbf{A}} = 2\,\hat{\mathbf{x}} - 3\,\hat{\mathbf{y}} + 4\,\hat{\mathbf{z}}$ to $\vec{\mathbf{A}} = A_x\hat{\mathbf{x}} + A_y\hat{\mathbf{y}} + A_z\hat{\mathbf{z}}$ to see that $A_x = 2$, $A_y = -3$, and $A_z = 4$. Similarly, $B_x = 5$, $B_y = -6$, and $B_z = -7$. Use the appropriate equation:

$$\vec{\mathbf{A}} \cdot \vec{\mathbf{B}} = A_xB_x + A_yB_y + A_zB_z = (2)(5) + (-3)(-6) + (4)(-7) = 10 + 18 - 28 = 0$$

(B) Now apply the other equation for the scalar product:

$$\vec{\mathbf{A}} \cdot \vec{\mathbf{B}} = AB\cos\theta$$
$$0 = AB\cos\theta$$

Since A and B are clearly nonzero, it follows that:

$$\cos\theta = 0$$
$$\theta = \cos^{-1}(0) = 90°$$

Example: Determine the angle between $\vec{\mathbf{A}} = 2\,\hat{\mathbf{x}} - \hat{\mathbf{z}}$ and $\vec{\mathbf{B}} = 5\,\hat{\mathbf{x}} - \sqrt{55}\,\hat{\mathbf{y}}$.

First work with the components of the given vectors in order to find the scalar product. Compare $\vec{\mathbf{A}} = 2\hat{\mathbf{x}} - \hat{\mathbf{z}}$ to $\vec{\mathbf{A}} = A_x\hat{\mathbf{x}} + A_y\hat{\mathbf{y}} + A_z\hat{\mathbf{z}}$ to see that $A_x = 2$, $A_y = 0$, and $A_z = -1$. Similarly, $B_x = 5$, $B_y = -\sqrt{55}$, and $B_z = 0$.

$$\vec{\mathbf{A}} \cdot \vec{\mathbf{B}} = A_xB_x + A_yB_y + A_zB_z = (2)(5) + (0)\left(-\sqrt{55}\right) + (-1)(0) = 10 + 0 + 0 = 10$$

Next we need to determine the magnitudes of the given vectors:

$$A = \sqrt{A_x^2 + A_y^2 + A_z^2} = \sqrt{2^2 + 0^2 + (-1)^2} = \sqrt{5}$$

$$B = \sqrt{B_x^2 + B_y^2 + B_z^2} = \sqrt{5^2 + (-\sqrt{55})^2 + 0^2} = \sqrt{80} = \sqrt{(16)(5)} = 4\sqrt{5}$$

Now use the second form of the scalar product to solve for the angle.

$$\vec{\mathbf{A}} \cdot \vec{\mathbf{B}} = AB\cos\theta$$

$$10 = \left(\sqrt{5}\right)\left(4\sqrt{5}\right)\cos\theta = (4)(5)\cos\theta = 20\cos\theta$$

$$\cos\theta = \frac{10}{20} = \frac{1}{2}$$

$$\theta = \cos^{-1}\left(\frac{1}{2}\right) = 60°$$

111. Find the scalar product between $\vec{\mathbf{A}} = 6\,\hat{\mathbf{x}} + \hat{\mathbf{y}} + 3\,\hat{\mathbf{z}}$ and $\vec{\mathbf{B}} = 2\,\hat{\mathbf{x}} + 8\,\hat{\mathbf{y}} + 4\,\hat{\mathbf{z}}$.

Answer: 32

112. Find the scalar product between $\vec{\mathbf{A}} = 2\,\hat{\mathbf{x}} - \hat{\mathbf{y}} + 6\,\hat{\mathbf{z}}$ and $\vec{\mathbf{B}} = -3\,\hat{\mathbf{x}} + 2\,\hat{\mathbf{y}} + 5\,\hat{\mathbf{z}}$.

Answer: 22

113. Find the scalar product between $\vec{\mathbf{A}} = 3\,\hat{\mathbf{y}} + 4\,\hat{\mathbf{z}}$ and $\vec{\mathbf{B}} = 9\,\hat{\mathbf{x}} - 3\,\hat{\mathbf{y}} - \hat{\mathbf{z}}$.

Want help? Check the hints section at the back of the book.

Answer: −13

114. Find the scalar product between $\vec{\mathbf{A}} = 12\,\hat{\mathbf{x}} + 4\,\hat{\mathbf{y}} - 16\,\hat{\mathbf{z}}$ and $\vec{\mathbf{B}} = 2\,\hat{\mathbf{x}} - 24\,\hat{\mathbf{y}} - 6\,\hat{\mathbf{z}}$.

Answer: 24

115. Determine the angle between $\vec{\mathbf{A}} = 3\,\hat{\mathbf{y}} - 3\,\hat{\mathbf{z}}$ and $\vec{\mathbf{B}} = -6\,\hat{\mathbf{y}} + 6\,\hat{\mathbf{z}}$.

Answer: 180°

116. Determine the angle between $\vec{\mathbf{A}} = 3\,\hat{\mathbf{x}} - \sqrt{3}\,\hat{\mathbf{y}} + 2\,\hat{\mathbf{z}}$ and $\vec{\mathbf{B}} = \sqrt{3}\,\hat{\mathbf{x}} - \hat{\mathbf{y}} + 2\sqrt{3}\,\hat{\mathbf{z}}$.

Want help? Check the hints section at the back of the book.

Answer: 30°

21 WORK AND POWER

Relevant Terminology

Displacement – a straight line from the initial position to the final position.
Work – work is done when there is not only a force acting on an object, but when the force also contributes toward the displacement of the object.
Power – the instantaneous rate at which work is done.

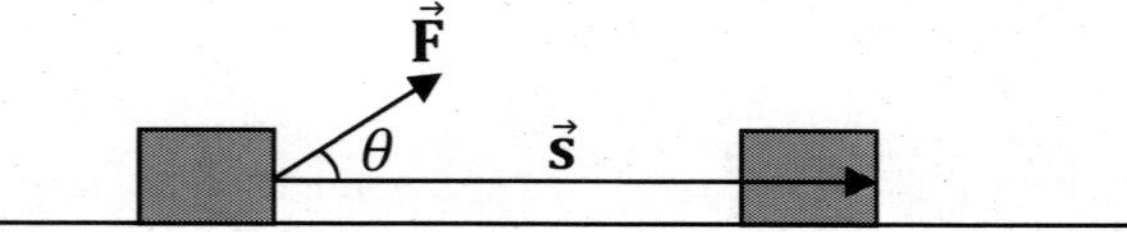

Work Equations

The work (W) done by a specified force equals the integral of the scalar product of the force ($\vec{\mathbf{F}}$) with the displacement ($d\vec{\mathbf{s}}$) from the initial position (i) to the final position (f):

$$W = \int_i^f \vec{\mathbf{F}} \cdot d\vec{\mathbf{s}} = \int_{x=x_0}^{x} F_x\, dx + \int_{y=y_0}^{y} F_y\, dy + \int_{z=z_0}^{z} F_z\, dz$$

For a **constant** force, this simplifies to the scalar product (Chapter 20) between the force ($\vec{\mathbf{F}}$) and the displacement ($\vec{\mathbf{s}}$). The displacement vector is a straight line from the initial position (i) to the final position (f) and can be expressed as $\vec{\mathbf{s}} = \Delta x\, \hat{\mathbf{x}} + \Delta y\, \hat{\mathbf{y}} + \Delta z\, \hat{\mathbf{z}}$ (it's no different than what we called $\Delta\vec{\mathbf{r}}$ in Chapter 12). The angle θ is the angle between $\vec{\mathbf{F}}$ and $\vec{\mathbf{s}}$.

$$W = \vec{\mathbf{F}} \cdot \vec{\mathbf{s}} = F_x \Delta x + F_y \Delta y + F_z \Delta z = F\, s \cos\theta$$

For the work done by **gravity** (over a non-astronomical change in altitude), the force is mg. The work done by gravity is proportional to the change in height (Δh).

$$W_g = -mg\Delta h$$

For an **astronomical** change in altitude, W_g involves the gravitational constant (G), masses of the planet and object (m_p and m), and distance from the center of the planet (R). The equation below comes from integrating Newton's law of gravity (Chapter 18) from R_0 to R.

$$W_g = Gm_p m \left(\frac{1}{R} - \frac{1}{R_0}\right)$$

For the work done by **friction** (which is **nonconservative**, so we call it W_{nc}), the force is μN (the coefficient of friction times normal force) and the angle is 180° (since friction is opposite to the displacement). The work done by friction is negative since $\cos 180° = -1$. For friction, s is the total distance traveled, not the net displacement.

$$W_{nc} = -\mu\, N\, s$$

Normal force does no work because it is perpendicular to the displacement ($\cos 90° = 0$):

$$W_N = 0$$

The work done by a **spring** involves the spring constant (k) and the displacement from equilibrium (Δx, which is simply x if we put the origin at equilibrium). It is positive if the spring displaced towards equilibrium and negative if displaced away from equilibrium. The equation below comes from integrating Hooke's law (Chapter 15) over x.

$$W_s = \pm\frac{1}{2}kx^2$$

The **net** work uses the net force, $\sum F_x$, which equals ma_x according to Newton's second law:

$$W_{net} = (\textstyle\sum F_x)s \quad \text{or} \quad W_{net} = ma_x s$$

Symbols and Units

Symbol	Name	Units
$\vec{\mathbf{F}}$	force	N
$\vec{\mathbf{s}}$	displacement	m
$d\vec{\mathbf{s}}$	differential displacement	m
θ	angle between $\vec{\mathbf{F}}$ and $\vec{\mathbf{s}}$	° or rad
W	work	J
m	mass	kg
g	gravitational acceleration	m/s^2
Δh	change in height	m
G	gravitational constant	$\frac{\text{N}\cdot\text{m}^2}{\text{kg}^2}$ or $\frac{\text{m}^3}{\text{kg}\cdot\text{s}^2}$
m_p	mass of large astronomical body	kg
R	distance from the center of the planet	m
μ	coefficient of friction	unitless
N	normal force	N
k	spring constant	N/m (or kg/s^2)
x	displacement of a spring from equilibrium	m
a_x	x-component of acceleration	m/s^2

Strategy for Finding Work Done

To find the work done, it depends on which force is specified in the problem. Choose the correct equation based on which force or which type of work is specified:

1. The work done by **gravity** for a small change in altitude, where Δh equals the change in altitude:
$$W_g = -mg\Delta h$$
This will turn out negative $(-)$ if the final position (f) is above the initial position (i), and positive $(+)$ if the final position (f) is below the initial position (i).
2. The work done by gravity for a **significant change in altitude** (like climbing a tall mountain or launching a rocket into space):
$$W_g = Gm_p m\left(\frac{1}{R} - \frac{1}{R_0}\right)$$
3. The work done by **friction** (nonconservative work), where s is the distance traveled:
$$W_{nc} = -\mu\, N\, s$$
To find normal force (N), draw a FBD and set $\sum F_y = 0$ (Chapter 14).
4. The work done by **normal** force is zero because $\theta = 90°$ and $\cos 90° = 0$:
$$W_N = 0$$
5. The work done by a **spring** is:
$$W_s = \pm\frac{1}{2}kx^2$$
Choose the positive $(+)$ sign if the spring is displaced towards equilibrium and the negative $(-)$ sign if it is displaced away from equilibrium.
6. The **net** work can be found two ways, depending upon whether you know the net force $(\sum F_x)$ or the acceleration (a_x):
$$W_{net} = \left(\sum F_x\right)s \quad \text{or} \quad W_{net} = ma_x s$$
See Chapter 14 regarding how to find $\sum F_x$ and Chapters 3-4 regarding a_x.
7. Given a constant force in the form $\vec{\mathbf{F}} = F_x\hat{\mathbf{x}} + F_y\hat{\mathbf{y}} + F_z\hat{\mathbf{z}}$ along with displacement in the form $\vec{\mathbf{s}} = \Delta x\,\hat{\mathbf{x}} + \Delta y\,\hat{\mathbf{y}} + \Delta z\,\hat{\mathbf{z}}$, find the scalar product (Chapter 20):
$$W = \vec{\mathbf{F}} \cdot \vec{\mathbf{s}} = F_x\Delta x + F_y\Delta y + F_z\Delta z$$
8. Given the magnitude of the force (F), the magnitude of the displacement (s), and the θ between $\vec{\mathbf{F}}$ and $\vec{\mathbf{s}}$, use the trig form of the scalar product (Chapter 20):
$$W = \vec{\mathbf{F}} \cdot \vec{\mathbf{s}} = F\, s\cos\theta$$
9. Given force as a **function** of the Cartesian coordinates x, y, and z, perform the following integral as described on page 196.
$$W = \int_{x=x_0}^{x} F_x\, dx + \int_{y=y_0}^{y} F_y\, dy + \int_{z=z_0}^{z} F_z\, dz$$

Notes Regarding Units

The SI unit of work is the Joule (J). It follows from the work equation,

$$W = F\, s \cos\theta$$

that a Joule is equivalent to a Newton times a meter:

$$1\ \text{J} = 1\ \text{N} \cdot \text{m}$$

Recall from Chapter 14 how a Newton relates to a kilogram, meter, and second:

$$1\ \text{N} = 1\ \text{kg} \cdot \text{m/s}^2$$

Using this relationship, we see that a Joule equals:

$$1\ \text{J} = 1\frac{\text{kg} \cdot \text{m}^2}{\text{s}^2}$$

Example: How much work must a monkey do against gravity in order to lift a 30-kg box of bananas 2.0 m off the ground?

Find the work done by gravity:

$$W_g = -mg\Delta h = -(30)(9.81)(2 - 0) \approx -(30)(10)(2) = -600\ \text{J}$$

The work done by gravity is negative ($-$) because the final position (f) is above the initial position (i). However, this is a tricky question because it asked for work done "against gravity" instead of the work done "by gravity." The word "against" makes it the opposite, so the correct answer to this question is $+600$ J. (The distinction is that -600 J is done "by" gravity, whereas $+600$ J is done "against" gravity. In this example, the work done "by" gravity is negative because gravity wants the object to fall, and this box instead went up, while the work done "against" gravity is positive because it was displaced in a direction opposite to, meaning against, the direction of gravity.)

Example: A 30-kg box of bananas uniformly accelerates for a distance of 4.0 m along an incline. The acceleration of the box is $6.0\ \text{m/s}^2$. Find the net work done by the system.

Use the equation for net work involving acceleration.

$$W_{net} = ma_x s = (30)(6)(4) = 720\ \text{J}$$

(The work done "by" the system is positive since the acceleration is parallel to the net displacement. If the question had asked for the work done "on" the system, instead of the work done "by" the system, then the answer would have been -720 J.)

Example: A monkey pushes a 15-kg box of bananas against a spring, compressing the spring 2.0 m from equilibrium. The spring constant is 30 N/m. How much work is done by the spring?

The work is negative because the spring went away from equilibrium:

$$W_s = -\frac{1}{2}kx^2$$

$$W_s = -\frac{1}{2}(30)(2)^2 = -60 \text{ J}$$

(If the question had asked for the work done "against" the spring, or "by" the "monkey," instead of the actual question which states "by" the spring, then the sign of the answer would have been opposite. That is, the work done against the spring or by the monkey would be +60 J, whereas the work done by the spring is −60 J.)

Example: A chimpanzee carries a 2.0-kg bunch of bananas 8.0 m horizontally. How much work is done by gravity?

Gravity doesn't do any work because the height doesn't change: $W_g = 0$ since $\Delta h = 0$.

Example: Planet ν_{10} has a mass of 9.0×10^{24} kg. An 8000-kg satellite orbits ν_{10} in an elliptical orbit. How much work is done by gravity when the satellite's orbital distance is reduced from 4.0×10^7 m to 3.0×10^7 m?

Since this change in altitude is astronomical (unlike the previous example), we use the other equation for the work done by gravity, where $R_0 = 4.0 \times 10^7$ m is the initial position and $R = 3.0 \times 10^7$ m is the final position (it went *from* 4.0×10^7 m *to* 3.0×10^7 m):

$$W_g = Gm_p m\left(\frac{1}{R} - \frac{1}{R_0}\right) \approx \left(\frac{2}{3} \times 10^{-10}\right)(9 \times 10^{24})(8 \times 10^3)\left(\frac{1}{3 \times 10^7} - \frac{1}{4 \times 10^7}\right)$$

If not using a calculator, it is convenient to reorganize and factor out the 10^7:

$$W_g \approx \frac{(2)(9)(8)}{(3)} \times \frac{10^{-10}10^{24}10^3}{10^7}\left(\frac{1}{3} - \frac{1}{4}\right) = 48 \times \frac{10^{17}}{10^7}\left(\frac{4}{12} - \frac{3}{12}\right) = 48 \times 10^{10}\left(\frac{1}{12}\right)$$

$$W_g \approx 4.0 \times 10^{10} \text{ J}$$

The work done by gravity came out positive because the satellite came closer to the planet.

Example: As illustrated below, a 60-kg box of bananas slides down a curved hill shaped like a quarter circle with a radius of 25 m. Find the work done by normal force.

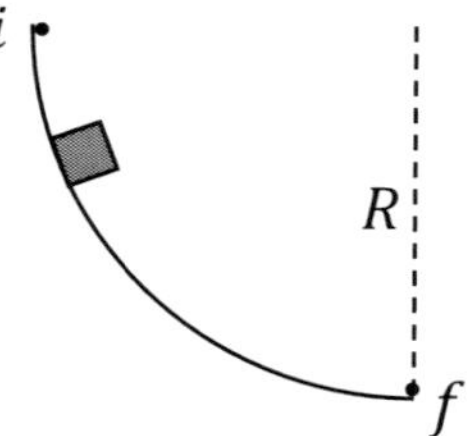

Normal force is perpendicular to the surface at all times, such that the angle between $\vec{\mathbf{F}}$ and $\vec{\mathbf{s}}$ is 90°. Since $\cos 90° = 0$, normal force doesn't do any work: $W_N = 0$.

Example: A 40-kg box of bananas slides 5.0 m along horizontal ground. The coefficient of friction between the box and ground is $\frac{1}{4}$. How much work is done by friction?

First, we need a FBD (Chapter 14) so that we can find normal force.

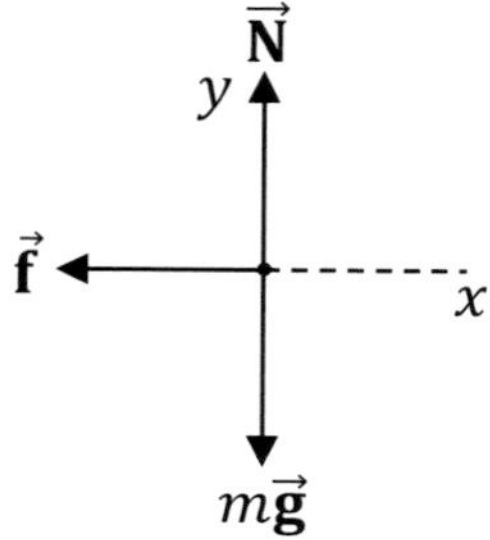

Sum the y-components of the forces to find normal force:

$$\sum F_y = ma_y$$
$$N - mg = 0$$
$$N = mg$$

We set $a_y = 0$ because the box is accelerating horizontally, not vertically. Substitute this expression for normal force into the equation for the work done by friction:

$$W_{nc} = -\mu\, N\, s = -\mu\, mg\, s = -\frac{1}{4}(40)(9.81)(5) \approx -\frac{1}{4}(40)(10)(5) = -500 \text{ J}$$

The work done by friction is negative because it subtracts mechanical energy from the system.

Example: A force $\vec{\mathbf{F}} = 3\,\hat{\mathbf{x}} - 2\,\hat{\mathbf{y}}$ causes the displacement $\vec{\mathbf{s}} = 4\,\hat{\mathbf{x}} - 5\,\hat{\mathbf{y}}$, where SI units have been suppressed. Find the work done.

For a constant force (there are no variables in $3\,\hat{\mathbf{x}} - 2\,\hat{\mathbf{y}}$, since $\hat{\mathbf{x}}$ are $\hat{\mathbf{y}}$ unit vectors), work is the scalar product (Chapter 20) between force and displacement:

$$W = \vec{\mathbf{F}} \cdot \vec{\mathbf{s}} = F_x \Delta x + F_y \Delta y + F_z \Delta z = (3)(4) + (-2)(-5) = 12 + 10 = 22 \text{ J}$$

Example: As illustrated below, a monkey pulls a 6.0-kg box of bananas with a force of 80 N in a direction that is parallel to the incline. The box travels 5.0 m up the incline. Find the work done by the monkey's pull.

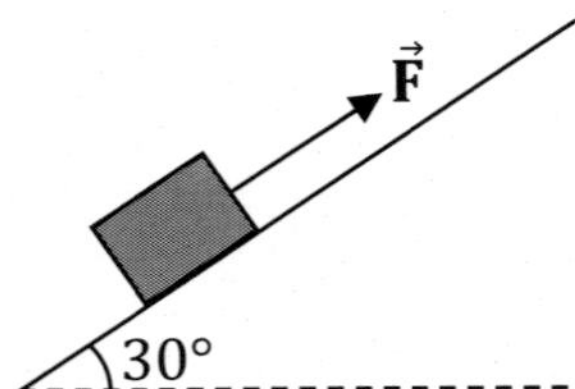

Since none of the other equations apply, and since the force is constant, use the scalar product equation for work:

$$W = \vec{\mathbf{F}} \cdot \vec{\mathbf{s}} = F\,s\cos\theta = (80)(5)\cos 0° = 400 \text{ J}$$

Note that the correct angle to use in this example is 0° because we want the work done by the monkey's pull and the pull is parallel to the displacement (along the incline). The angle θ in the work equation is the angle between the force and the displacement.

Tip: Always take a moment to draw the specified force and the displacement, as this makes it easier to work out the correct angle to use.

117. A monkey drops a 50-kg box of bananas. The box falls 1.5 m. How much work is done by gravity?

Answer: 750 J

118. A monkey stretches a spring 4.0 m from equilibrium. The spring constant is 8.0 N/m. How much work is done by the spring?

Answer: −64 J

119. Planet Coconut has a mass of 5.0×10^{24} kg and a radius of 4.0×10^{6} m. Find the work done by gravity when a 300-kg rocket climbs from Coconut's surface up to an altitude of 4.0×10^{6} m above Coconut's surface.

Answer: -1.25×10^{10} J

120. A 200-kg bananamobile uniformly accelerates from rest to 60 m/s in 5.0 s. Find the net work done.

Want help? Check the hints section at the back of the book.

Answer: 360 kJ

121. A 20-kg monkey slides 8.0 m down a 30° incline. The coefficient of friction between the monkey and the incline is $\frac{\sqrt{3}}{4}$.

(A) Find the work done by gravity.

(B) Find the work done by the friction force.

(C) Find the work done by the normal force.

(D) Find the net work.

Want help? Check the hints section at the back of the book.

Answers: 800 J, −600 J, 0, 200 J

122. A force $\vec{\mathbf{F}} = 2\,\hat{\mathbf{x}} + 6\,\hat{\mathbf{y}}$ causes the displacement $\vec{\mathbf{s}} = 3\,\hat{\mathbf{x}} - 8\,\hat{\mathbf{y}}$, where SI units have been suppressed. Find the work done.

Answer: –42 J

123. A force $\vec{\mathbf{F}} = 7\,\hat{\mathbf{x}} - 9\,\hat{\mathbf{y}} + 5\,\hat{\mathbf{z}}$ causes the displacement $\vec{\mathbf{s}} = 6\,\hat{\mathbf{x}} - \hat{\mathbf{z}}$, where SI units have been suppressed. Find the work done.

Answer: 37 J

124. A monkey pulls on a box of bananas with a force of 300 N. The box of bananas travels 12 m along a straight line. The angle between the monkey's pull and the displacement is 60°. How much work does the monkey do?

Want help? Check the hints section at the back of the book.

Answer: 1.8 kJ

125. As illustrated below, a monkey pulls a 20-kg box of bananas with a force of 160 N at an angle of 30° above the horizontal. The coefficient of friction between the box and the ground is $\frac{\sqrt{3}}{6}$. The box travels 7.0 m.

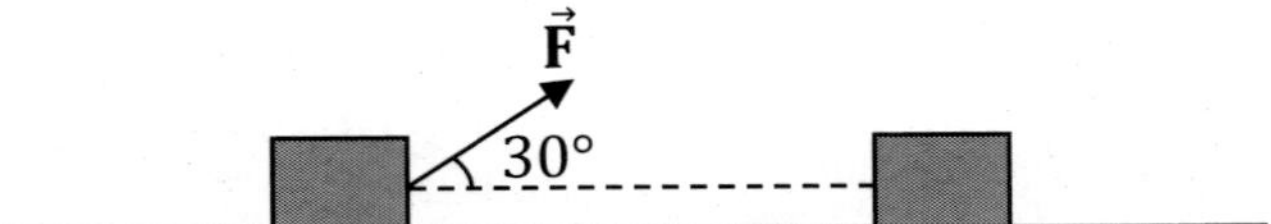

(A) Find the work done by the pull.

(B) Find the work done by the friction force.

(C) Find the work done by the normal force.

Want help? Check the hints section at the back of the book.

Answers: $560\sqrt{3}$ J, $-140\sqrt{3}$ J, 0

Strategy for Performing the Work Integral

If a problem gives you force as a function of the Cartesian coordinates x, y, and z, follow these steps to perform the work integral:

1. Identify the x-, y-, and z-components of the force. If the force is expressed in terms of Cartesian unit vectors as $\vec{\mathbf{F}} = F_x\hat{\mathbf{x}} + F_y\hat{\mathbf{y}} + F_z\hat{\mathbf{z}}$, the components of the force are F_x, F_y, and F_z (the coefficients of $\hat{\mathbf{x}}$, $\hat{\mathbf{y}}$, and $\hat{\mathbf{z}}$).
2. Write the equation for the work integral.
$$W = \int_{x=x_0}^{x} F_x\,dx + \int_{y=y_0}^{y} F_y\,dy + \int_{z=z_0}^{z} F_z\,dz$$
3. Substitute the expressions for F_x, F_y, and F_z from Step 1 into the integrals in Step 2.
4. Express F_x strictly as a function of x. If F_x is a function of both x and y, use the equation of the path taken from the initial position to the final position to express y in terms of x.
5. Express F_y strictly as a function of y. If F_y is a function of both x and y, use the equation of the path taken from the initial position to the final position to express x in terms of y.
6. Similar to Steps 4 and 5, express F_z strictly as a function of z.
7. Perform the integrals over F_x, F_y, and F_z using methods from Chapters 2 and 9.

This strategy is illustrated in the examples that follow.

Example: A monkey exerts a force $\vec{\mathbf{F}} = 6x^2\,\hat{\mathbf{x}}$ (where SI units have been suppressed) as a box of bananas is displaced in a straight line from (2.0 m, 0) to (3.0 m, 0). Find the work done by the monkey.

Compare the force $\vec{\mathbf{F}} = 6x^2\,\hat{\mathbf{x}}$ to the general equation $\vec{\mathbf{F}} = F_x\hat{\mathbf{x}} + F_y\hat{\mathbf{y}} + F_z\hat{\mathbf{z}}$ to see that $F_x = 6x^2$, $F_y = 0$, and $F_z = 0$. Substitute F_x, F_y, and F_z into the work integral. Since the box travels from (2.0 m, 0) to (3.0 m, 0), the limits of the x-integration are from $x_0 = 2$ to $x = 3$.
$$W = \int_{x=x_0}^{x} F_x\,dx + \int_{y=y_0}^{y} F_y\,dy + \int_{z=z_0}^{z} F_z\,dz = \int_{x=2}^{3} 6x^2\,dx + 0 + 0$$
This integral can be performed using the method from Chapter 2:
$$W = \int_{x=2}^{3} 6x^2\,dx = \left[6\frac{x^{2+1}}{2+1}\right]_{x=2}^{3} = \left[6\frac{x^3}{3}\right]_{x=2}^{3} = [2x^3]_{x=2}^{3}$$
Evaluate $2x^3$ over the limits:
$$W = 2(3)^3 - 2(2)^3 = 2(27) - 2(8) = 54 - 16 = 38\text{ J}$$

Example: A monkey exerts a force $\vec{\mathbf{F}} = y^2\,\hat{\mathbf{x}} - xy\,\hat{\mathbf{y}}$ (where SI units have been suppressed) as a box of bananas is displaced in a straight line from (2.0 m, 1.0 m) to (2.0 m, 4.0 m). Find the work done by the monkey.

Compare the force $\vec{\mathbf{F}} = y^2\,\hat{\mathbf{x}} - xy\,\hat{\mathbf{y}}$ to the general equation $\vec{\mathbf{F}} = F_x\hat{\mathbf{x}} + F_y\hat{\mathbf{y}} + F_z\hat{\mathbf{z}}$ to see that $F_x = y^2$, $F_y = -xy$, and $F_z = 0$. Substitute F_x, F_y, and F_z into the work integral. Since the box travels from (2.0 m, 1.0 m) to (2.0 m, 4.0 m), the limits of the x-integration are from $x_0 = 2$ to $x = 2$ and the limits of the y-integration are from $y_0 = 1$ to $y = 4$.

$$W = \int_{x=x_0}^{x} F_x\,dx + \int_{y=y_0}^{y} F_y\,dy + \int_{z=z_0}^{z} F_z\,dz = \int_{x=2}^{2} y^2\,dx - \int_{y=1}^{4} xy\,dy + 0 = 0 - \int_{y=1}^{4} xy\,dy + 0$$

The x-integral is zero since the lower and upper limits are identical (both equal 2). We can't integrate over y yet because the integrand contains x. What we can do is realize that x is a constant and equal to 2 because the box travels in a straight line from (2.0 m, 1.0 m) to (2.0 m, 4.0 m). This means that we can substitute $x = 2$ into the y-integral.

$$W = -\int_{y=1}^{4} xy\,dy = -\int_{y=1}^{4} 2y\,dy = -\left[2\frac{y^{1+1}}{1+1}\right]_{y=1}^{4} = -\left[2\frac{y^2}{2}\right]_{y=1}^{4} = -[y^2]_{y=1}^{4}$$

Evaluate y^2 over the limits:

$$W = -(4^2 - 1^2) = -(16 - 1) = -15 \text{ J}$$

Example: A monkey exerts a force $\vec{\mathbf{F}} = y\,\hat{\mathbf{x}} + xy\,\hat{\mathbf{y}}$ (where SI units have been suppressed) as a box of bananas is displaced along the line $y = 2x$ from (0, 0) to (3.0 m, 6.0 m). Find the work done by the monkey.

Compare the force $\vec{\mathbf{F}} = y\,\hat{\mathbf{x}} + xy\,\hat{\mathbf{y}}$ to the general equation $\vec{\mathbf{F}} = F_x\hat{\mathbf{x}} + F_y\hat{\mathbf{y}} + F_z\hat{\mathbf{z}}$ to see that $F_x = y$, $F_y = xy$, and $F_z = 0$. Substitute F_x, F_y, and F_z into the work integral. Since the box travels from (0, 0) to (3.0 m, 6.0 m), the limits of the x-integration are from $x_0 = 0$ to $x = 3$ and the limits of the y-integration are from $y_0 = 0$ to $y = 6$.

$$W = \int_{x=x_0}^{x} F_x\,dx + \int_{y=y_0}^{y} F_y\,dy + \int_{z=z_0}^{z} F_z\,dz = \int_{x=0}^{3} y\,dx + \int_{y=0}^{6} xy\,dy + 0$$

We can't integrate yet because the x-integrand contains y and vice-versa. Since the box travels along the line $y = 2x$, we can substitute this into the x-integral. Solving for x in the equation $y = 2x$, we get $x = \frac{y}{2}$, which we can substitute into the y-integral.

$$W = \int_{x=0}^{3} 2x\,dx + \int_{y=0}^{6} \frac{y}{2}y\,dy = \int_{x=0}^{3} 2x\,dx + \int_{y=0}^{6} \frac{y^2}{2}dy = [x^2]_{x=0}^{3} + \left[\frac{y^3}{6}\right]_{y=0}^{6}$$

$$W = 3^2 - 0^2 + \frac{6^3}{6} - \frac{0^3}{6} = 9 + 36 = 45 \text{ J}$$

126. A chimpanzee exerts a force $\vec{\mathbf{F}} = 15y^4\,\hat{\mathbf{y}}$ (where SI units have been suppressed) as a box of pineapples is displaced along a straight line from $(0, -1.0 \text{ m})$ to $(0, 2.0 \text{ m})$. Find the work done by the chimpanzee.

Want help? Check the hints section at the back of the book.

Answer: 99 J

127. A lemur exerts a force $\vec{\mathbf{F}} = x\,\hat{\mathbf{x}} + y^2\,\hat{\mathbf{y}}$ (where SI units have been suppressed) as a box of grapes is displaced along a straight line from $(2.0 \text{ m}, 3.0 \text{ m})$ to $(4.0 \text{ m}, 6.0 \text{ m})$. Find the work done by the lemur.

Want help? Check the hints section at the back of the book.

Answer: 69 J

128. A gorilla exerts a force $\vec{\mathbf{F}} = xy\,\hat{\mathbf{x}} + x^2\,\hat{\mathbf{y}}$ (where SI units have been suppressed) as a box of coconuts is displaced along the parabola $y = x^2$ from $(0, 0)$ to $(4.0 \text{ m}, 16.0 \text{ m})$. Find the work done by the gorilla.

Want help? Check the hints section at the back of the book.

Answer: 192 J

Power Equations

Instantaneous power (P) is a derivative of work (W) done with respect to time (t):

$$P = \frac{dW}{dt}$$

The equation for **average** power ($\bar{P}$) is work (W) done divided by the elapsed time (t):

$$\bar{P} = \frac{W}{t}$$

If the force ($\vec{\mathbf{F}}$) is **constant**, the average power can be found from the scalar product:

$$\bar{P} = \vec{\mathbf{F}} \cdot \vec{\bar{\boldsymbol{v}}} = F_x \bar{v}_x + F_y \bar{v}_y + F_z \bar{v}_z = F\, \bar{v} \cos\theta$$

The average velocity ($\vec{\bar{\boldsymbol{v}}}$) arises from the displacement ($\vec{\mathbf{s}}$) divided by the time (Chapter 8):

$$\vec{\bar{\boldsymbol{v}}} = \frac{\vec{\mathbf{s}}}{t}$$

Important Distinction

Instantaneous power involves a derivative, whereas **average** power does not. Pay careful attention to the wording.

Strategy for Finding Power

How you find power depends on whether the problem asks you to find instantaneous power or average power:

- Instantaneous power (P) is a derivative of work (W) done with respect to time (t).

  $$P = \frac{dW}{dt}$$

- How you calculate **average** power ($\bar{P}$) depends on what you know:
  - One way is to first find the work (W) done and then divide by the elapsed time (t):

    $$\bar{P} = \frac{W}{t}$$

    The problem-solving strategies from pages 187 and 196 help you find work. You may need to use the equations from Chapters 3-4 to find time.
  - If the force is constant, average power equals the scalar product between force ($\vec{\mathbf{F}}$) and average velocity ($\vec{\bar{\boldsymbol{v}}}$).

    $\bar{P} = \vec{\mathbf{F}} \cdot \vec{\bar{\boldsymbol{v}}}$ or $\bar{P} = F_x \bar{v}_x + F_y \bar{v}_y + F_z \bar{v}_z$ or $\bar{P} = F\, \bar{v} \cos\theta$

    In this context, θ is the angle between $\vec{\mathbf{F}}$ and $\vec{\bar{\boldsymbol{v}}}$.

Symbols and Units

Symbol	Name	Units
$\vec{\mathbf{F}}$	force	N
$\vec{\bar{v}}$	average velocity	m/s
θ	angle between $\vec{\mathbf{F}}$ and $\vec{\bar{v}}$	° or rad
W	work	J
P	instantaneous power	W
$\bar{P}$	average power	W
t	time	s

Notes Regarding Units

The SI unit of power is the Watt (W). It follows from the power equation,

$$P = \frac{dW}{dt}$$

that a Watt is equivalent to a Joule per second:

$$1\ \text{W} = 1\frac{\text{J}}{\text{s}}$$

In terms of the base SI units, a Watt can be expressed as:

$$1\ \text{W} = 1\frac{\text{kg} \cdot \text{m}^2}{\text{s}^3}$$

One horsepower (hp) is equivalent to 746 Watts: $1\ \text{hp} = 746\ \text{W}$.
Notice that the W's don't match:

- Work (W) is measured in Joules (J).
- Power (P) is measured in Watts (W).

Example: A monkey designs a motor that does 300 J of work in 4.0 s. What average power does the motor deliver?

Use the equation for average power that involves work and time:

$$\bar{P} = \frac{W}{t} = \frac{300}{4} = 75\ \text{W}$$

Example: A monkey pulls a box of bananas with a force of 40 N at an angle of 60° above the horizontal, while the box of bananas travels with a constant speed of 3.0 m/s horizontally. What average power does the monkey deliver to the box of bananas?

Use the equation for average power that involves average speed:

$$\bar{P} = F\,\bar{v}\cos\theta = (40)(3)\cos 60° = (40)(3)\left(\frac{1}{2}\right) = 60 \text{ W}$$

The angle θ is 60° because it is defined as the angle between the force and the velocity.

Example: The work done by a monkey is given by the expression $W = 4t^3$ (where SI units have been suppressed). (A) What is the instantaneous power at $t = 5.0$ s? (B) What is the average power from $t = 0$ to $t = 5.0$ s?

(A) Instantaneous power is a derivative of work with respect to time. This derivative can be found using the technique from Chapter 2:

$$P = \frac{dW}{dt} = \frac{d}{dt}(4t^3) = (3)(4)t^{3-1} = 12t^2$$

Evaluate the derivative at the specified time:

$$P(\text{at } t = 5 \text{ s}) = 12(5)^2 = 12(25) = 300 \text{ W}$$

(B) Average power equals the work done divided by the elapsed time.

$$\bar{P} = \frac{W}{t} = \frac{4t^3}{t} = 4t^2$$

The elapsed time is 5.0 s:

$$\bar{P} = 4(5)^2 = 4(25) = 100 \text{ W}$$

Example: A monkey exerts a force $\vec{\mathbf{F}} = 3\,\hat{\mathbf{x}} + 5\,\hat{\mathbf{y}}$ on a box that has an average velocity of $\vec{\bar{v}} = -2\,\hat{\mathbf{x}} + 4\,\hat{\mathbf{y}}$, where SI units have been suppressed. Find the average power delivered by the monkey.

The average power equals the scalar product (Chapter 20) between force and average velocity. Compare $\vec{\mathbf{F}} = 3\,\hat{\mathbf{x}} + 5\,\hat{\mathbf{y}}$ to $\vec{\mathbf{F}} = F_x\hat{\mathbf{x}} + F_y\hat{\mathbf{y}} + F_z\hat{\mathbf{z}}$ to see that $F_x = 3$, $F_y = 5$, and $F_z = 0$. Similarly, $\bar{v}_x = -2$, $\bar{v}_y = 4$, and $\bar{v}_z = 0$.

$$\bar{P} = \vec{\mathbf{F}} \cdot \vec{\bar{v}} = F_x\bar{v}_x + F_y\bar{v}_y + F_z\bar{v}_z$$

$$\bar{P} = (3)(-2) + (5)(4) + (0)(0) = -6 + 20 + 0$$

$$\bar{P} = 14 \text{ W}$$

129. The work done by a monkey is given by the expression $W = 72\sqrt{t}$ (where SI units have been suppressed). (A) What is the instantaneous power at $t = 9.0$ s? (B) What is the average power from $t = 0$ to $t = 9.0$ s?

Answer: 12 W, 24 W

130. A monkey exerts a force $\vec{\mathbf{F}} = 8\,\hat{\mathbf{x}} - 2\,\hat{\mathbf{y}}$ on a box that has an average velocity of $\vec{\bar{v}} = 6\,\hat{\mathbf{x}} - \hat{\mathbf{y}}$, where SI units have been suppressed. Find the average power delivered by the monkey.

Answer: 50 W

131. A 200-kg bananamobile uniformly accelerates from rest to 60 m/s in 5.0 s. Find the average power delivered.

Want help? Check the hints section at the back of the book.

Answer: 72 kW

22 CONSERVATION OF ENERGY

Relevant Terminology

Energy – the ability to do work, meaning that a force is available to contribute towards the displacement of an object.
Potential energy – work that can be done by changing position. All forms of potential energy are stored energy. For example, gravity stores energy: If you let go of an object, gravity will do work to change the object's height. Similarly, a spring stores energy: A compressed spring can be used to do work, displacing an object toward (and eventually beyond) its equilibrium position (for example, to launch an object up into the air).
Kinetic energy – work that can be done by changing speed. Moving objects have kinetic energy. Hence, kinetic energy is considered to be energy of motion.
Nonconservative work – energy that the system exchanges with the surroundings (which includes changes to the system's internal energy). In many kinematics problems, nonconservative work is work done by resistive forces such as friction.
Reference height – the origin of your coordinate system from which you are measuring all of your heights.
Escape speed – the initial speed that a rocket needs to escape the pull of an astronomical body (like a planet or moon).

Conservation of Energy

Energy is conserved in nature. The total energy of any completely isolated system remains constant. If the system isn't completely isolated, energy is still conserved, but you must account for exchanges of energy between the system and the surroundings. While energy can neither be created nor destroyed, it can be **transformed** from one kind into another. For example, potential energy can be converted into kinetic energy (or vice-versa), or mechanical energy can be transformed into heat energy.

Conservation of energy can be expressed mathematically as follows, where PE stands for potential energy, KE stands for kinetic energy, and W_{nc} represents nonconservative work:

$$PE_0 + KE_0 + W_{nc} = PE + KE$$

On the left-hand side, $PE_0 + KE_0$ represents the total initial energy of the system, while on the right-hand side, $PE + KE$ represents the total final energy of the system. The term W_{nc} represents energy exchanged between the system and surroundings (such as work done by friction). If W_{nc} equals zero (meaning that no nonconservative work is done), then the total initial energy of the system equals the total final energy of the system.

Work-Energy Theorem

An alternate, yet equivalent, way to express the idea of conservation of energy is via the work-energy theorem. According to the work-energy theorem, the **net work** done on the system equals the **change in the system's kinetic energy**:

$$W_{net} = \Delta KE = KE - KE_0$$

Here, the net work (W_{net}) includes both nonconservative work (W_{nc}) and work done by conservative forces (W_c): $W_{net} = W_c + W_{nc}$. The work done by conservative forces (W_c) equals the negative of the change in the system's potential energy (ΔPE): $W_c = -\Delta PE = -(PE - PE_0) = PE_0 - PE$. The equation for the work-energy theorem becomes the equation for conservation of energy if you make these two substitutions:

$$W_c + W_{nc} = KE - KE_0$$
$$-(PE - PE_0) + W_{nc} = KE - KE_0$$
$$PE_0 + KE_0 + W_{nc} = PE + KE$$

Although the work-energy is equivalent to conservation of energy, students tend to make more mistakes trying to apply the work-energy theorem than they do trying to apply conservation of energy to solve a problem. So unless a problem specifically asks you to find the net work or the change in kinetic energy, applying **conservation of energy** is the **recommended** problem-solving strategy for problems that involve changes in position and changes in speed.

Forms of Energy

Energy can be classified as follows:

- **Potential energy** (PE) is work that can be done by changing position.
- **Kinetic energy** (KE) is work that can be done by changing speed.
- **Nonconservative work** (W_{nc}) involves exchanges of energy between the system and its surroundings (including changes in the system's internal energy). A common example is work done by resistive forces such as **friction**.

Here are two common examples of how potential energy relates to position:

- **Gravitational potential energy** depends on **height**. The greater an object's height, the more gravitational potential energy it has. Simply remove whatever forces may be supporting an object and gravity will immediately begin doing work on the object, causing the object's height to change.
- **Spring potential energy** depends on how much the spring is compressed or stretched from its equilibrium position. The further a spring is from equilibrium, the more spring potential energy it has. If you stretch a spring and release it, for example, the spring will do work, compressing itself toward its equilibrium position.

Escape Speed

As we will see in one of the examples, you can determine an object's escape speed by applying the law of conservation of energy. Every astronomical body has an escape speed, which represents the minimum initial speed that an object would need in order to escape the astronomical body's gravitational pull.

If you jump upward, you fall back down. If you throw a rock upward, it eventually falls back down to earth. But if you could launch a rocket with a great enough speed, it could leave earth's gravitational pull forever. Earth's escape speed tells us exactly how fast a rocket would need to be traveling to escape from earth's gravitational field.

The way to calculate escape speed is to **set the final kinetic energy equal to zero** when the rocket gets **infinitely** far away. If it has enough initial kinetic energy to eventually reach the edge of the universe, then it will escape the influence of the astronomical body. We'll explore the details of this calculation in one of the examples.

Symbols and SI Units

Symbol	Name	SI Units
W_{net}	net work	J
PE_0	initial potential energy	J
PE	final potential energy	J
KE_0	initial kinetic energy	J
KE	final kinetic energy	J
W_{nc}	nonconservative work	J
m	mass	kg
g	gravitational acceleration	m/s^2
h_0	initial height (relative to the reference height)	m
h	final height (relative to the reference height)	m
v_0	initial speed	m/s
v	final speed	m/s

G	gravitational constant	$\frac{\text{N·m}^2}{\text{kg}^2}$ or $\frac{\text{m}^3}{\text{kg·s}^2}$
m_p	mass of large astronomical body	kg
R	distance from the center of the planet	m
k	spring constant	N/m (or kg/s^2)
x	displacement of a spring from equilibrium	m

Energy Equations

Conservation of energy can be expressed as:

$$PE_0 + KE_0 + W_{nc} = PE + KE$$

Since energy is the ability to do work, the equations for potential energy are very similar to the equations for work (Chapter 21). In fact, the equation for potential energy can be found from the corresponding equation for work by using the following relationship: $W_c = -\Delta PE = -(PE - PE_0) = PE_0 - PE$.

- **Gravitational potential energy** is proportional to **height** (h). For a non-astronomical change in altitude:

  $$PE_{g0} = mgh_0 \quad , \quad PE_g = mgh$$

  Here, height is measured relative a reference height, which is a point that serves as the origin of your coordinate system. For an astronomical change in altitude, use the following equations instead, where R is measured from the center of the **astronomical** body (like a planet or moon):

  $$PE_{g0} = -G\frac{m_p m}{R_0} \quad , \quad PE_g = -G\frac{m_p m}{R}$$

- **Spring potential energy** is proportional to the square of its displacement from equilibrium (x), where k is the spring constant:

  $$PE_{s0} = \frac{1}{2}kx_0^2 \quad , \quad PE_s = \frac{1}{2}kx^2$$

Kinetic energy is proportional to **speed** (v) squared:

$$KE_0 = \frac{1}{2}mv_0^2 \quad , \quad KE = \frac{1}{2}mv^2$$

Nonconservative work is often the work done by friction (see Chapter 21), where μ is the coefficient of friction, N is normal force (found by drawing a FBD and setting $a_y = 0$ in $\sum F_y = ma_y$), and s is the total distance traveled for which there is **friction**:

$$W_{nc} = -\mu N s$$

The **work-energy theorem** states that the net work equals the change in kinetic energy:

$$W_{net} = \Delta KE = KE - KE_0$$

Conservation of Energy Strategy

It is useful to apply the law of conservation of energy to problems that involve a change in position and a change in speed. For problems that give you acceleration or ask you to find acceleration, it may be more efficient to apply a different strategy (like Newton's second law, depending on the nature of the problem). To apply the law of conservation of energy to a problem, follow these steps:

1. Draw a diagram of the path. Label the initial position (i), final position (f), and the **reference height** (RH). Note: The reference height is not itself an actual value to use. Rather, the reference height is like the origin of your coordinate system, indicating from where you are measuring the initial height (h_0) and final height (h).
2. Is there a **spring** involved in the problem? If so, also mark these positions in your diagram: equilibrium (EQ), fully compressed (FC), and fully stretched (FS).
3. Write out the law of conservation of energy in symbols:
$$PE_0 + KE_0 + W_{nc} = PE + KE$$
4. Rewrite each term of the conservation of energy equation in symbols as follows:
   - At the **initial** position (i), is the object's **height** different from the reference height (RH)? If so, then $PE_{g0} = mgh_0$ (except for a large change in altitude, where $PE_{g0} = -G\frac{m_p m}{R_0}$). Otherwise, if at the initial position (i) the object is at the reference height (RH), then PE_{g0} equals zero.
   - At the **initial** position (i), is the object connected to a **spring**, and if so, is the spring either compressed or stretched from its equilibrium position (EQ)? If so, then $PE_{s0} = \frac{1}{2}kx_0^2$. Note that if the spring is in its equilibrium position (EQ) at the initial position (i), then PE_{s0} equals zero.
   - Initial potential energy may include both gravitational and spring potential energy: $PE_0 = PE_{g0} + PE_{s0}$.
   - At the **initial** position (i), is the object **moving**? If so, then $KE_0 = \frac{1}{2}mv_0^2$. Otherwise, if at the initial position (i) the object is at rest, then KE_0 is zero.
   - Is any nonconservative work being done between the initial position (i) and the final position (f)? A common example is work done by a resistive force such as **friction**. If there is friction in the problem, write $W_{nc} = -\mu Ns$. Draw a FBD and set $a_y = 0$ in $\sum F_y = ma_y$ to solve for normal force (N). If non-conservative work isn't done, then W_{nc} equals zero.
   - At the **final** position (f), is the object's **height** different from the reference height (RH)? If so, then $PE_g = mgh$ (except for a large change in altitude, where $PE_g = -G\frac{m_p m}{R}$). Otherwise, if at the final position (f) the object is at the reference height (RH), then PE_g equals zero.

- At the **final** position (f), is the object connected to a **spring**, and if so, is the spring either compressed or stretched from its equilibrium position (EQ)? If so, then $PE_s = \frac{1}{2}kx^2$. Note that if the spring is in its equilibrium position (EQ) at the final position (f), then PE_s equals zero.
- Final potential energy may include both gravitational and spring potential energy: $PE = PE_g + PE_s$.
- At the **final** position (f), is the object **moving**? If so, then $KE = \frac{1}{2}mv^2$. Otherwise, if at the final position (f) the object is at rest, then KE is zero.

5. It may help to think of the conservation of energy equation as looking like this:
$$mgh_0 + \frac{1}{2}kx_0^2 + \frac{1}{2}mv_0^2 - \mu N s = mgh + \frac{1}{2}kx^2 + \frac{1}{2}mv^2$$
In practice, one or more of these terms will often be zero, in each of the cases noted in Step 4.
6. Use algebra to solve for the desired unknown.

Note Regarding Subscripts

It's very important to distinguish **initial** from **final**:

- h_0 is the initial height.
- h is the final height.
- x_0 is the initial displacement of a spring from equilibrium.
- x is the final displacement of a spring from equilibrium.
- v_0 is the initial speed.
- v is the final speed.

When students forget the subscript zero (called "nought"), they get cancellations in their algebra that shouldn't occur. For example, if you write mgh on the left-hand side instead of mgh_0, and if you also have mgh on the right-hand side, then your mgh's will cancel out in the algebra in a situation where they don't really cancel because the two h's are different.

Reference Height

Label a reference height (RH) in your diagram. This special point serves as the origin of your coordinate system from which all heights are measured. For example, you could measure all of your heights from the ground, or you could measure them all from the top of a building. Either way, the change in height will come out the same. The reference height makes it clear **where you are choosing to measure your heights from**. (When using the equation $PE = -G\frac{m_p m}{R}$, the reference height is at infinity, as we will discuss in an example.)

Example: As shown below, a box of bananas slides from rest down a curved frictionless hill, descending 20 m in height. Determine the speed of the box when it travels along the horizontal surface at the bottom of the hill.

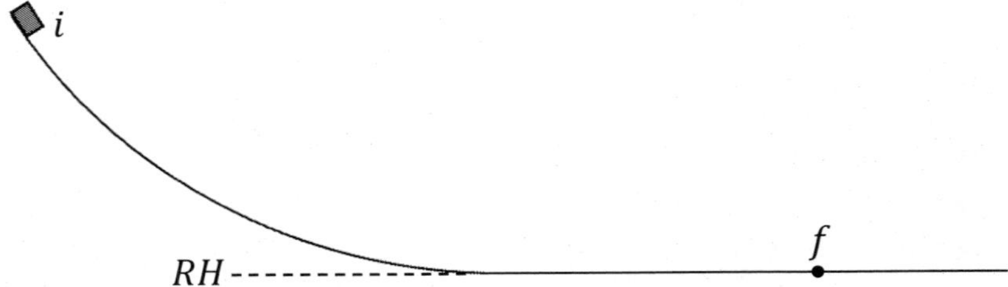

Begin by drawing the path and labeling the initial position (i), final position (f), and reference height (RH). The initial position (i) is where it starts, while the final position (f) is anywhere along the horizontal. (The speed will be the same everywhere along the horizontal since it is frictionless and the object has inertia – see Chapter 13 – but the speed will change along the curved portion where the height changes). We choose the reference height (RH) to be at the bottom of the hill. Write out conservation of energy:

$$PE_0 + KE_0 + W_{nc} = PE + KE$$

Let's analyze this term by term:

- $PE_{g0} = mgh_0$ since i is **not** at the same height as RH.
- $PE_{s0} = 0$ since at i there isn't a spring compressed or stretched from equilibrium.
- $PE_0 = PE_{g0} + PE_{s0} = mgh_0$.
- $KE_0 = 0$ since the box is at rest at i.
- $W_{nc} = 0$ since there are no frictional forces.
- $PE_g = 0$ since f is at the same height as RH.
- $PE_s = 0$ since at f there isn't a spring compressed or stretched from equilibrium.
- $PE = PE_g + PE_s = 0$.
- $KE = \frac{1}{2}mv^2$ since the box is moving at f.

Make the above substitutions into the conservation of energy equation:

$$mgh_0 = \frac{1}{2}mv^2$$

Solve for the final speed:

$$2gh_0 = v^2$$

$$v = \sqrt{2gh_0} = \sqrt{2(9.81)(20)} \approx \sqrt{2(10)(20)} = \sqrt{400} = 20 \text{ m/s}$$

Example: As shown below, a box of bananas slides horizontally with an initial speed of 20 m/s. The coefficient of friction between the box and the ground is $\frac{1}{2}$. How far does the box of bananas travel?

$RH\ i$ •————————► f

Begin by drawing the path and labeling the initial position (i), final position (f), and reference height (RH). The initial position (i) is where it starts, while the final position (f) is where the box comes to rest. We choose the reference height (RH) to be on the ground. Write out conservation of energy:

$$PE_0 + KE_0 + W_{nc} = PE + KE$$

Let's analyze this term by term:

- $PE_{g0} = 0$ since i is at the same height as RH.
- $PE_{s0} = 0$ since at i there isn't a spring compressed or stretched from equilibrium.
- $PE_0 = PE_{g0} + PE_{s0} = 0$.
- $KE_0 = \frac{1}{2}mv_0^2$ since the box is moving at i.
- $W_{nc} = -\mu Ns$ since there is work done by friction.
- $PE_g = 0$ since f is at the same height as RH.
- $PE_s = 0$ since at f there isn't a spring compressed or stretched from equilibrium.
- $PE = PE_g + PE_s = 0$.
- $KE = 0$ since the box comes to rest at f.

Make the above substitutions into the conservation of energy equation:

$$\frac{1}{2}mv_0^2 - \mu Ns = 0$$

We need to find normal force. Draw a FBD: Weight pulls down, normal force pushes up, and friction pulls backwards (decelerating the box).

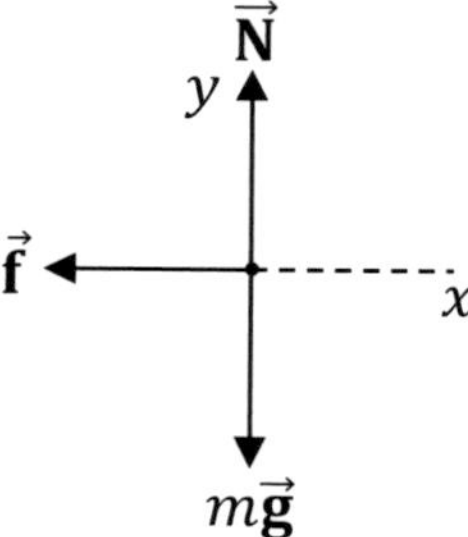

Apply Newton's second law (Chapter 14). Set $a_y = 0$ since the box doesn't accelerate up or down (just horizontally):

$$\sum F_y = ma_y$$
$$N - mg = 0$$
$$N = mg$$

Plug the expression for normal force into the equation we obtained from conservation of energy:

$$\frac{1}{2}mv_0^2 - \mu N s = 0$$

$$\frac{1}{2}mv_0^2 - \mu m g s = 0$$

Solve for the distance that the box travels (s):

$$\frac{1}{2}v_0^2 = \mu g s$$

$$s = \frac{v_0^2}{2\mu g}$$

$$s = \frac{20^2}{2\left(\frac{1}{2}\right)(9.81)} \approx \frac{20^2}{2\left(\frac{1}{2}\right)(10)} = \frac{400}{10} = 40 \text{ m}$$

Example: As shown below, a pendulum swings back and forth, beginning from rest at the leftmost position. The length of the pendulum is 40 cm. Find the speed of the pendulum bob when it passes through the bottom of the arc.

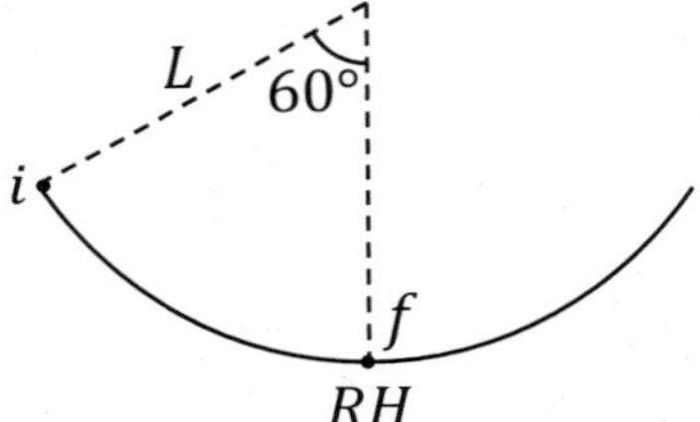

Begin by drawing the path and labeling the initial position (i), final position (f), and reference height (RH). The initial position (i) is where it starts, while the final position (f) is at the bottom of the arc (where we're trying to find the speed). We choose the reference height (RH) to be at the bottom of the arc. Write out conservation of energy:

$$PE_0 + KE_0 + W_{nc} = PE + KE$$

Let's analyze this term by term:

- $PE_{g0} = mgh_0$ since i is **<u>not</u>** at the same height as RH.
- $PE_{s0} = 0$ since at i there isn't a spring compressed or stretched from equilibrium.
- $PE_0 = PE_{g0} + PE_{s0} = mgh_0$.
- $KE_0 = 0$ since the pendulum at rest at i.
- $W_{nc} = 0$ neglecting any frictional forces.
- $PE_g = 0$ since f is at the same height as RH.
- $PE_s = 0$ since at f there isn't a spring compressed or stretched from equilibrium.
- $PE = PE_g + PE_s = 0$.
- $KE = \frac{1}{2}mv^2$ since the pendulum is moving at f.

Make the above substitutions into the conservation of energy equation:

$$mgh_0 = \frac{1}{2}mv^2$$

We know the length of the pendulum (L), not the initial height (h_0). We can solve for h_0 using geometry. Study the diagram below.

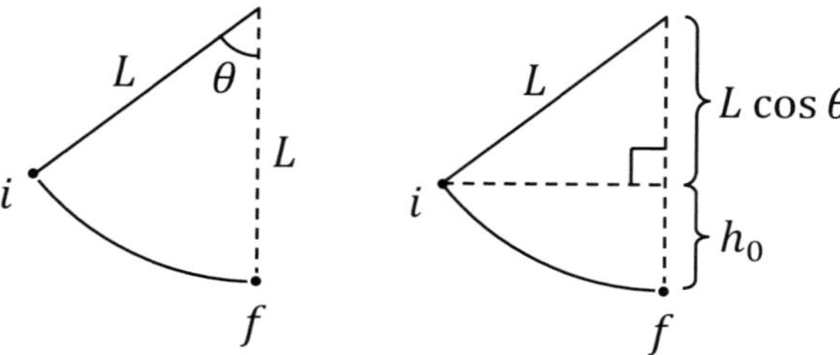

In the figure above, the two distances at the right add up to the length L:

$$L = h_0 + L\cos\theta$$
$$h_0 = L - L\cos\theta$$

Convert L from 40 cm to meters: $L = 40 \text{ cm} = 0.40 \text{ m} = \frac{2}{5}$ m.

$$h_0 = \frac{2}{5} - \frac{2}{5}\cos 60° = \frac{2}{5} - \frac{2}{5}\frac{1}{2} = \frac{2}{5} - \frac{1}{5} = \frac{1}{5} \text{ m}$$

Now we can return to the equation that we reached from conservation of energy:

$$mgh_0 = \frac{1}{2}mv^2$$
$$2gh_0 = v^2$$
$$v = \sqrt{2gh_0} = \sqrt{2(9.81)\left(\frac{1}{5}\right)} \approx \sqrt{2(10)\left(\frac{1}{5}\right)} = \sqrt{4} = 2.0 \text{ m/s}$$

Example: As illustrated below, a box of bananas slides from rest down a curved hill, passes once through the circular loop, and then continues along the horizontal. Neglecting friction, what is the minimum initial height needed to ensure that the box of bananas makes it safely through the loop?

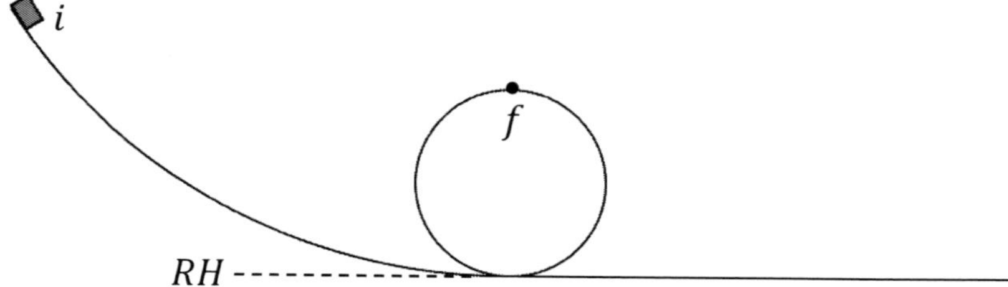

The trick to this problem is to realize that the wording "makes it safely through the loop" demands that the normal force be positive. As long as there is normal force, the box of bananas will be in contact with the surface. If the box of bananas loses contact with the surface, the normal force will be zero. Therefore, we will solve this problem by demanding that normal force be positive at the top of the loop: If the box of bananas reaches the top of the loop safely, its inertia will carry it on through. As always, the way to find normal force is to draw a FBD. Normal force pushes down (perpendicular to the surface) because the box reaches the top on the inside of the loop. Since the box is traveling in a circle, we must follow the prescription from Chapter 17 and work with inward (in) and tangential (tan) directions (not x and y).

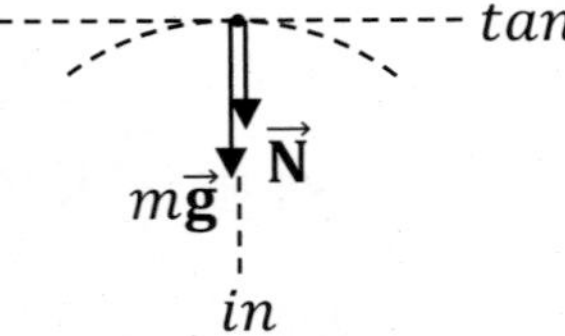

Applying Newton's second law, where the acceleration is centripetal (a_c), we obtain:

$$\sum F_{in} = ma_c$$
$$N + mg = ma_c$$
$$N = m(a_c - g)$$

If the centripetal acceleration is greater than g, the normal force will be positive.

$$N > 0 \;\; \Rightarrow \; a_c > g$$

(The symbol $\Rightarrow$ stands for "implies that.") Recall the equation for centripetal acceleration $\left(a_c = \frac{v^2}{R}\right)$. Substitute this into the above inequality.

$$a_c > g$$
$$\frac{v^2}{R} > g$$
$$v^2 > Rg$$

Now we use conservation of energy to relate the initial height to the speed at the top of the loop. We'll save the above inequality for later.

Begin by drawing the path and labeling the initial position (i), final position (f), and reference height (RH). (This is drawn on the previous page.) The initial position (i) is where it starts, while the final position (f) is at the top of the loop. We choose the reference height (RH) to be at the horizontal. Write out conservation of energy:

$$PE_0 + KE_0 + W_{nc} = PE + KE$$

Let's analyze this term by term:

- $PE_{g0} = mgh_0$ since i is **not** at the same height as RH.
- $PE_{s0} = 0$ since at i there isn't a spring compressed or stretched from equilibrium.
- $PE_0 = PE_{g0} + PE_{s0} = mgh_0$.
- $KE_0 = 0$ since the box is at rest at i.
- $W_{nc} = 0$ since there are no frictional forces.
- $PE_g = mgh$ since f is **not** at the same height as RH.
- $PE_s = 0$ since at f there isn't a spring compressed or stretched from equilibrium.

- $PE = PE_g + PE_s = mgh$.
- $KE = \frac{1}{2}mv^2$ since the box is moving at f.

Make the above substitutions into the conservation of energy equation:

$$mgh_0 = mgh + \frac{1}{2}mv^2$$

$$gh_0 = gh + \frac{v^2}{2}$$

Note that $h = 2R$ since f is one diameter above the reference height:

$$gh_0 = g2R + \frac{v^2}{2}$$

Recall from our normal force constraint that $v^2 > Rg$. Using this, we get:

$$gh_0 > g2R + \frac{Rg}{2}$$

$$h_0 > 2R + \frac{R}{2}$$

$$h_0 > \frac{4R}{2} + \frac{R}{2}$$

$$h_0 > \frac{5R}{2}$$

As long as the initial height is at least 2.5 times the radius, the box of bananas will make it safely through the loop, meaning that it won't lose contact with the surface at any time.

Example: As shown below, one end of a horizontal 96 N/m spring is fixed to a vertical wall, while a 6.0-kg box of banana-shaped chocolates is connected to its free end. There is no friction between the box and the horizontal. The spring is compressed 2.0-m from the equilibrium position and released from rest. What is the speed of the box when the system passes through the equilibrium position?

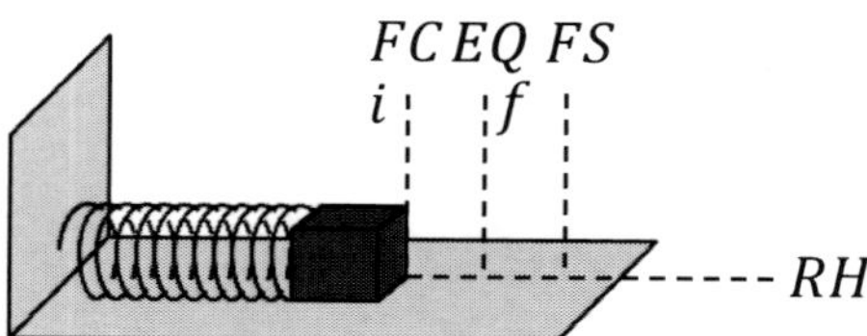

Begin by drawing the path and labeling the initial position (i), final position (f), and reference height (RH). For a spring, we also label the equilibrium (EQ), fully compressed (FC), and fully stretched (FS) positions. We choose the initial position (i) to be where the spring is fully compressed (FC) and the final position (f) to be at the equilibrium (EQ) position. We choose the reference height (RH) to be along the horizontal. Write out conservation of energy:

$$PE_0 + KE_0 + W_{nc} = PE + KE$$

Let's analyze this term by term:

- $PE_{g0} = 0$ since i is at the same height as RH.
- $PE_{s0} = \frac{1}{2}kx_0^2$ since a spring is compressed from equilibrium at i.
- $PE_0 = PE_{g0} + PE_{s0} = \frac{1}{2}kx_0^2$.
- $KE_0 = 0$ since the box is at rest at i.
- $W_{nc} = 0$ since there are no frictional forces.
- $PE_g = 0$ since f is at the same height as RH.
- $PE_s = 0$ since the spring is in its equilibrium position at f (a spring needs to be compressed or stretched from equilibrium in order to have potential energy).
- $PE = PE_g + PE_s = 0$.
- $KE = \frac{1}{2}mv^2$ since the box is moving at f.

Make the above substitutions into the conservation of energy equation:

$$\frac{1}{2}kx_0^2 = \frac{1}{2}mv^2$$

Solve for the final speed:

$$kx_0^2 = mv^2$$

$$\frac{kx_0^2}{m} = v^2$$

$$v = \sqrt{\frac{kx_0^2}{m}} = x_0\sqrt{\frac{k}{m}} = 2\sqrt{\frac{96}{6}} = 2\sqrt{16} = (2)(4) = 8.0 \text{ m/s}$$

Example: Planet Watermelon has a mass of 6.0×10^{26} kg and a radius of 2.0×10^6 m. What is the escape speed of a rocket launched from the surface of Watermelon?

Begin by drawing the path and labeling the initial position (i), final position (f), and reference height (RH). The initial position (i) is on the surface of Watermelon, while the final position (f) is infinitely far away. Why infinitely far away? That's how far away the rocket would eventually need to get in order to completely escape the influence of Watermelon's gravitational pull. The reference height (RH) is also infinitely far away from Watermelon. The reference height is the place where gravitational potential energy equals zero. For a small change in altitude, when we work with the expression $PE_g = mgh$, the reference height is the place where h would be zero, since that would make mgh zero. However, for a significant change in altitude (which is the case for the rocket), we work with the expression $PE_g = -\frac{Gm_pm}{R}$, so the reference height is the place where R is infinite (since $\frac{1}{R}$ approaches zero as R approaches infinity). Write out conservation of energy:

$$PE_0 + KE_0 + W_{nc} = PE + KE$$

Let's analyze this term by term:

- $PE_{g0} = -\frac{Gm_p m}{R_0}$ since i is not at the same height as RH.
- $PE_{s0} = 0$ since at i there isn't a spring compressed or stretched from equilibrium.
- $PE_0 = PE_{g0} + PE_{s0} = -\frac{Gm_p m}{R_0}$.
- $KE_0 = \frac{1}{2}mv_0^2$ since the rocket must be moving at i in order to leave the planet.
- $W_{nc} = 0$ since there are no resistive forces like friction. (There is air friction while the rocket travels through the atmosphere, but we always neglect air resistance unless the problem states otherwise.)
- $PE_g = 0$ since f is at the same height as RH (and since $-\frac{Gm_p m}{R}$ approaches zero as R approaches infinity).
- $PE_s = 0$ since at f there isn't a spring compressed or stretched from equilibrium.
- $PE = PE_g + PE_s = 0$.
- $KE = 0$ since the escape speed is the minimum initial speed needed for the rocket to reach f. (If the rocket has any extra speed when it reaches the final position, it would have been able to get there with a smaller initial speed.)

Make the above substitutions into the conservation of energy equation:

$$-\frac{Gm_p m}{R_0} + \frac{1}{2}mv_0^2 = 0$$

Solve for the initial speed:

$$\frac{1}{2}mv_0^2 = \frac{Gm_p m}{R_0}$$

$$v_0^2 = \frac{2Gm_p}{R_0}$$

$$v_0 = \sqrt{\frac{2Gm_p}{R_0}} \approx \sqrt{\frac{2\left(\frac{2}{3} \times 10^{-10}\right)(6 \times 10^{26})}{2 \times 10^6}} = \sqrt{\frac{4 \times 10^{16}}{10^6}} = \sqrt{4 \times 10^{10}} = 2.0 \times 10^5 \text{ m/s}$$

A rocket needs an initial speed of at least 2.0×10^5 m/s in order to escape Watermelon's gravitational pull. This is the escape speed for any object leaving planet Watermelon.

132. From the edge of an 80-m tall cliff, a monkey throws your textbook with an initial speed of 30 m/s at an angle of 30° above the horizontal. Your textbook lands on horizontal ground below.

Draw/label a diagram with the path, initial (i), final (f), and the reference height (RH).

Use conservation of energy to determine the speed of your textbook just before impact.

Want help? Check the hints section at the back of the book.

Answer: 50 m/s

133. As shown below, a box of bananas has an initial speed of 40 m/s. The box of bananas travels along the frictionless surface, which curves up a hill.

Draw/label a diagram with the path, initial (i), final (f), and the reference height (RH).

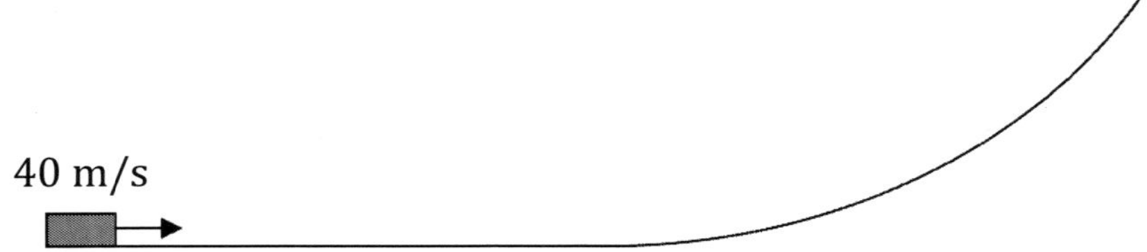

What is the maximum height of the box of bananas?

Want help? Check the hints section at the back of the book.

Answer: 80 m

134. A monkey makes a pendulum by tying a 10-m long cord to a banana. The pendulum swings back and forth, beginning from rest at the leftmost position in the diagram below.

Draw/label a diagram with the path, the initial position (i), the final position for each part (f_A and f_B), and the reference height (RH).

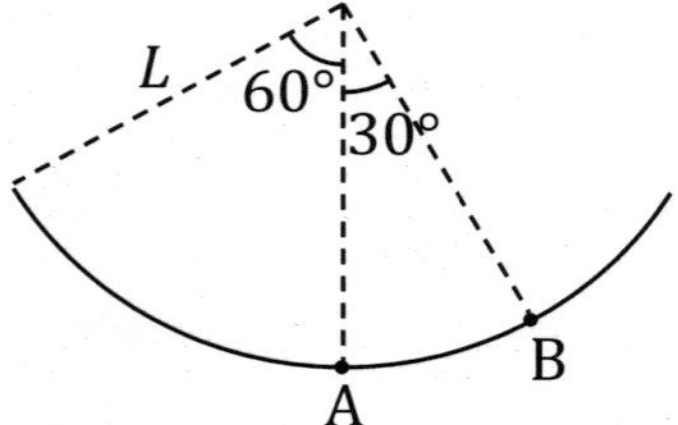

(A) Determine the speed of the banana at point A.

(B) Determine the speed of the banana at point B.

Want help? Check the hints section at the back of the book.

Answers: 10 m/s, $10\sqrt{\sqrt{3}-1}$ m/s (or 8.47 m/s in decimal form)

135. A box of bananas begins from rest at the top of the circular hill illustrated below (the hill is exactly one quarter of a circle). The radius of the circular hill is 20 m. The hill is frictionless, but the horizontal surface is not. The coefficient of friction between the box of bananas and the horizontal surface is $\frac{1}{5}$. The box of bananas comes to rest at point B.

Draw/label a diagram with the path, the initial position (i), the final position for each part (f_A and f_B), and the reference height (RH).

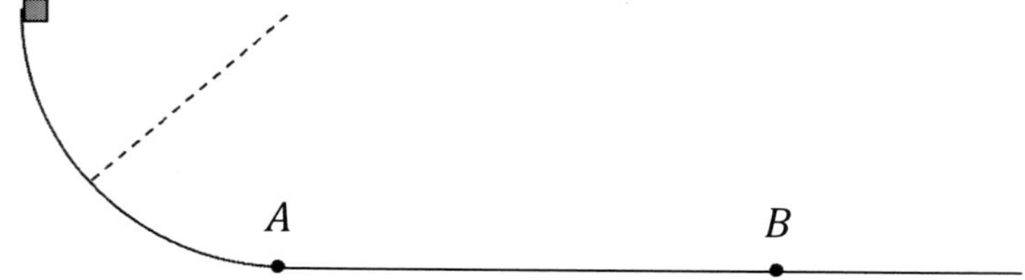

(A) Determine the speed of the box of bananas at point A.

(B) Determine the distance between points A and B.

Want help? Check the hints section at the back of the book.

Answers: 20 m/s, 100 m

136. A box of bananas slides $240\sqrt{2}$ m down a $45°$ incline from rest. The coefficient of friction between the box of bananas and the incline is $\frac{1}{4}$.

Draw/label a diagram with the path, initial (i), final (f), and the reference height (RH).

Use energy methods to determine the speed of the box of bananas as it reaches the bottom of the incline.

Want help? Check the hints section at the back of the book.

Answer: 60 m/s

137. A monkey rides the roller coaster illustrated below. The initial speed of the roller coaster is 50 m/s. The total mass of the roller coaster is 210 kg. The diameter of the loop is 35 m. Neglect friction.

Draw/label a diagram with the path, initial (i), final (f), and the reference height (RH).

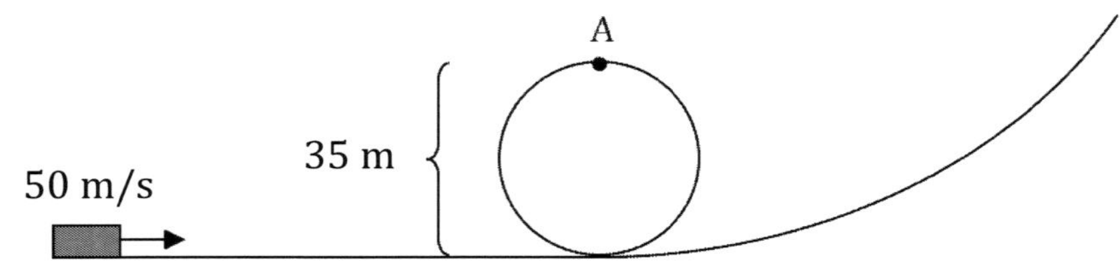

(A) How fast is the roller coaster moving at point A?

(B) What normal force is exerted on the roller coaster at point A?

Want help? Check the hints section at the back of the book.

Answers: $30\sqrt{2}$ m/s, 19.5 kN

138. A monkey uses a spring to launch a pellet with a mass of $\frac{1}{200}$ kg straight upward. The spring is compressed $\frac{1}{8}$ m from equilibrium and released from rest. The pellet rises to a maximum height of 45 m above its initial position (the initial position is where the spring is fully compressed).

Draw/label a diagram with the path, initial (i), final (f), and the reference height (RH). Also label the equilibrium (EQ) and fully compressed (FC) positions for the spring.

What is the spring constant?

Want help? Check the hints section at the back of the book.

Answer: 288 N/m

139. One end of a horizontal 75 N/m spring is fixed to a vertical wall, while a 3.0-kg box of banana-shaped chocolates is connected to its free end. There is no friction between the box and the horizontal. The spring is compressed 5.0-m from the equilibrium position and released from rest.

Draw/label a diagram with the path, the initial position (i), the final position for each part (f_A and f_B), and the reference height (RH). Also label the equilibrium (EQ), fully compressed (FC), and fully stretched (FS) positions for the spring.

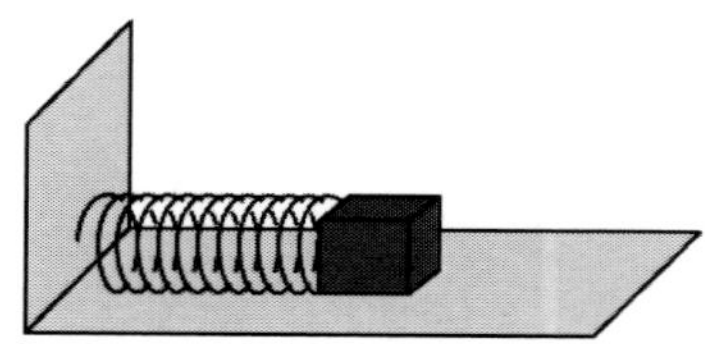

(A) What is the speed of the box when the system passes through equilibrium?

(B) What is the speed of the box when the spring is stretched 4.0 m from equilibrium?

Want help? Check the hints section at the back of the book.

Answers: 25 m/s, 15 m/s

140. Planet FurryTail has a mass of 1.5×10^{26} kg and a radius of 2.0×10^{6} m.

Draw/label a diagram with the path, initial (i), final (f), and the reference height (RH).

What is the escape speed for a projectile leaving the surface of FurryTail?

Want help? Check the hints section at the back of the book.

Answer: 100 km/s

141. Planet SillyMonk has a mass of 3.0×10^{24} kg. A 2000-kg rocket traveling in a free fall trajectory is moving $2000\sqrt{19}$ m/s when it is 8.0×10^{7} m from the center of SillyMonk.

Draw/label a diagram with the path, initial (i), final (f), and the reference height (RH).

What is the speed of the rocket when it is 4.0×10^{7} m from the center of SillyMonk?

Want help? Check the hints section at the back of the book.

Answer: 9000 m/s

23 ONE-DIMENSIONAL COLLISIONS

Relevant Terminology

Momentum – the product of mass and velocity.
Impulse – the change in an object's momentum.
Elastic collision – a collision where both momentum and mechanical energy are conserved.
Inelastic collision – a collision where only momentum is conserved. Mechanical energy is either lost or gained.
Perfectly inelastic collision – a collision where the objects stick together afterward. Only momentum is conserved. Mechanical energy is either lost or gained.
Inverse perfectly inelastic collision – a process where two objects are together initially and then separate. If the process is run in reverse, it would look like a perfectly inelastic collision. Only momentum is conserved. Mechanical energy is either lost or gained.
Ballistic pendulum – a pendulum where a perfectly inelastic collision at the bottom of the arc causes the pendulum to swing upward.

Essential Concepts

Momentum equals mass times velocity. Moving objects have **momentum**.

Impulse equals change in momentum. Impulse also equals average force times the time interval. The average force exerted during the collision causes the change in momentum. The concept of impulse can be useful conceptually. For example, if you jump off the roof of a building, would you rather flex your knees when you land or lock your legs straight? You probably know the answer from experience, but you can reason out the explanation using the concept of impulse. Before your collision with the ground, either way your initial momentum is fixed: It equals your mass times your initial velocity. Your final momentum is also fixed: It will be zero after the collision. Therefore, your change in momentum and your impulse are also fixed. What you can control is the duration of the collision: the time interval. By flexing your knees and bending your legs with the collision, you extend the time interval. Since impulse (which is the same either way) equals average force times the time interval, by extending the time interval, you effectively reduce the average collision force that the ground exerts on your legs. On the other hand, if you lock your legs stiff, you reduce the time interval, resulting in a greater average force, which may cause a broken leg.

Note that **elastic** and **inelastic** aren't distinguished by whether or not the objects stick together, but by whether or not mechanical energy is conserved in addition to momentum. The wording that tells you if objects stick together is "**perfectly inelastic**."

Collision Equations

Momentum ($\vec{\boldsymbol{p}}$) equals mass (m) times velocity ($\vec{\boldsymbol{v}}$):

$$\vec{\boldsymbol{p}} = m\vec{\boldsymbol{v}}$$

Impulse ($\vec{\boldsymbol{J}}$) equals change in momentum ($\Delta\vec{\boldsymbol{p}} = \vec{\boldsymbol{p}} - \vec{\boldsymbol{p}}_0$). Impulse also equals the average collision force ($\vec{\boldsymbol{F}}_c$) times the time interval (Δt) of the collision.

$$\vec{\boldsymbol{J}} = \Delta\vec{\boldsymbol{p}} = \vec{\boldsymbol{p}} - \vec{\boldsymbol{p}}_0 = \vec{\boldsymbol{F}}_c\Delta t$$

The law of conservation of momentum generally applies to collisions. Strictly speaking, the total momentum of a system is only conserved if the net external force acting on the system equals zero. However, in practice, most collisions last for only a very short duration, in which case the total momentum of the system may be approximately conserved even if the net force acting on the system isn't zero. In problems involving collisions, it is almost always necessary to express **conservation of momentum** for the system:

$$m_1\vec{\boldsymbol{v}}_{10} + m_2\vec{\boldsymbol{v}}_{20} = m_1\vec{\boldsymbol{v}}_1 + m_2\vec{\boldsymbol{v}}_2$$

Mechanical energy is often **not** conserved for collisions. Mechanical energy is only conserved for **elastic** collisions. Even then, it's not practical to write out conservation of mechanical energy, since speed is squared in the equation for kinetic energy ($KE = \frac{1}{2}mv^2$). For **elastic** collisions, it's more efficient to write out the following equation in place of the conservation of energy equation, as it makes the algebra more efficient. Don't use the following equation by itself: Combine it with the conservation of momentum equation (above) and solve the system of two equations in order to solve a problem with an elastic collision. (The following equation follows from the combination of both conservation of momentum and mechanical energy for an elastic collision.) Note that the following equation separates object 1 from object 2: It does **not** separate initial from final.

$$\vec{\boldsymbol{v}}_{10} + \vec{\boldsymbol{v}}_1 = \vec{\boldsymbol{v}}_{20} + \vec{\boldsymbol{v}}_2$$

There are two special cases of the conservation of momentum equation worth knowing:

- For a **perfectly inelastic collision**, the final velocities are equal (since the objects stick together in a perfectly inelastic collision):

$$m_1\vec{\boldsymbol{v}}_{10} + m_2\vec{\boldsymbol{v}}_{20} = (m_1 + m_2)\vec{\boldsymbol{v}}$$

- For an **inverse perfectly inelastic collision**, the initial velocities are equal (in this case the objects are together initially and then separate):

$$(m_1 + m_2)\vec{\boldsymbol{v}}_0 = m_1\vec{\boldsymbol{v}}_1 + m_2\vec{\boldsymbol{v}}_2$$

To calculate how much **kinetic energy** the system gains or loses, compare the total final kinetic energy ($KE = \frac{1}{2}m_1v_1^2 + \frac{1}{2}m_2v_2^2$) to the total initial kinetic energy ($KE_0 = \frac{1}{2}m_1v_{10}^2 + \frac{1}{2}m_2v_{20}^2$). Note that the numerator is in absolute values.

$$\%\text{ change} = \frac{|KE - KE_0|}{KE_0} \times 100\%$$

The 'Tricky' Collision

There is one type of collision that might not seem like a "collision" when you read the problem. This is the **inverse perfectly inelastic collision**. This is a process where two objects have the same initial velocity and then separate.

Sign Note

An arrow above a symbol, as in $\vec{v}$, reminds you that a quantity is a **vector**: It includes **direction**. In one-dimension, if two objects are heading in opposite directions initially, a minus **sign** is crucial toward obtaining the correct answers.

Symbols and SI Units

Symbol	Name	SI Units
$\vec{p}$	momentum	kg·m/s
m	mass	kg
$\vec{v}$	velocity	m/s
v	speed	m/s
$\vec{J}$	impulse	Ns
$\Delta\vec{p}$	change in momentum	kg·m/s
F_c	average collision force	N
Δt	duration of the collision (time interval)	s
m_1	mass of object 1	kg
m_2	mass of object 2	kg
$\vec{v}_{10}$	initial velocity of object 1	m/s
$\vec{v}_{20}$	initial velocity of object 2	m/s
$\vec{v}_1$	final velocity of object 1	m/s
$\vec{v}_2$	final velocity of object 2	m/s

Notes Regarding Units

The SI units of momentum are kg·m/s. This follows from the momentum equation:

$$\vec{\boldsymbol{p}} = m\vec{\boldsymbol{v}}$$

Since impulse equals change in momentum ($\vec{\boldsymbol{J}} = \Delta\vec{\boldsymbol{p}}$), it follows that impulse shares the same SI units as momentum. However, since impulse also equals $\vec{\boldsymbol{J}} = \vec{\boldsymbol{F}}_c\Delta t$, the SI units of impulse can also be expressed as N·s, which is more commonly used for impulse.

Strategy for One-dimensional Collisions

To solve a problem with a one-dimensional collision, follow these steps:

1. Declare your choice of the $+x$ (or y) axis. If an object is heading in the $-x$ (or $-y$) direction at any time, its velocity will be negative.
2. How you solve a collision problem depends on the nature of the collision. Following are the different types of one-dimensional collisions. (For a two-dimensional collision, see Chapter 24.)
   - If the problem makes it clear that the objects stick together after the collision, or if the problem calls the collision "**perfectly inelastic**," use this equation:
     $$m_1\vec{\boldsymbol{v}}_{10} + m_2\vec{\boldsymbol{v}}_{20} = (m_1 + m_2)\vec{\boldsymbol{v}}$$
   - If two objects are together initially and then separate, it's an **inverse** perfectly inelastic collision. One example involves a monkey walking across a canoe in water (or walking on a wooden plank on top of ice), where initially either the system was at rest or the monkey was simply riding in the canoe.
     $$(m_1 + m_2)\vec{\boldsymbol{v}}_0 = m_1\vec{\boldsymbol{v}}_1 + m_2\vec{\boldsymbol{v}}_2$$
   - If the problem calls the collision "**elastic**," or if the problem tells you that mechanical energy is conserved for the collision, use the two equations below. You will need to solve the system with a substitution (Chapter 1):
     $$m_1\vec{\boldsymbol{v}}_{10} + m_2\vec{\boldsymbol{v}}_{20} = m_1\vec{\boldsymbol{v}}_1 + m_2\vec{\boldsymbol{v}}_2$$
     $$\vec{\boldsymbol{v}}_{10} + \vec{\boldsymbol{v}}_1 = \vec{\boldsymbol{v}}_{20} + \vec{\boldsymbol{v}}_2$$
   - A rare problem could involve an **inelastic** collision where the objects are neither stuck together after nor before the collision, in which case only the following equation applies. In such a rare problem, you would know three of the four velocities.
     $$m_1\vec{\boldsymbol{v}}_{10} + m_2\vec{\boldsymbol{v}}_{20} = m_1\vec{\boldsymbol{v}}_1 + m_2\vec{\boldsymbol{v}}_2$$
3. If the problem asks you to determine the percentage of **kinetic energy** lost or gained in the collision, first find $KE = \frac{1}{2}m_1v_1^2 + \frac{1}{2}m_2v_2^2$ and $KE_0 = \frac{1}{2}m_1v_{10}^2 + \frac{1}{2}m_2v_{20}^2$:
   $$\%\text{ change} = \frac{|KE - KE_0|}{KE_0} \times 100\%$$
   (For an **elastic** collision, the % change will be zero: It doesn't change at all.)

Example: A 50-kg sumo wrestling orangutan traveling 4.0 m/s to the north collides head-on with a 70-kg sumo wrestling orangutan traveling 5.0 m/s to the south. This wrestling match occurs on horizontal frictionless ice. The two orangutans stick together after the collision. Determine the final velocity.

First setup a coordinate system. We choose $+x$ to point north. With this choice, $\vec{\boldsymbol{v}}_{10}$ will be along $\hat{\mathbf{x}}$ and $\vec{\boldsymbol{v}}_{20}$ will be along $-\hat{\mathbf{x}}$ (see Chapter 10). Since the orangutans stick together, use the equation for a perfectly inelastic collision. They have the same final velocity:

$$m_1\vec{\boldsymbol{v}}_{10} + m_2\vec{\boldsymbol{v}}_{20} = (m_1 + m_2)\vec{\boldsymbol{v}}$$
$$(50)(4\hat{\mathbf{x}}) + (70)(-5\hat{\mathbf{x}}) = (50 + 70)\vec{\boldsymbol{v}}$$
$$200\hat{\mathbf{x}} - 350\hat{\mathbf{x}} = -150\hat{\mathbf{x}} = 120\vec{\boldsymbol{v}}$$
$$\vec{\boldsymbol{v}} = -\frac{150}{120}\hat{\mathbf{x}} = -\frac{5}{4}\hat{\mathbf{x}}\text{ m/s}$$

The orangutans travel $\frac{5}{4}$ m/s to the south after the collision, since the $\hat{\mathbf{x}}$ unit vector points to the north. (See the end of Chapter 10 if you need to review **unit vectors**.)

Example: A 40-kg monkey is at rest on horizontal frictionless ice. The monkey throws a 10-kg watermelon to the east with a speed of 4.0 m/s relative to the ice (it's a little trickier if the problem says "relative to the monkey," but that case just requires vector subtraction to express the velocity relative to the ice). Determine the final velocity of the monkey.

We can see conceptually what will happen from Newton's third law (Chapter 13): The watermelon will exert a force on the monkey to the west, equal in magnitude to the force that the monkey exerts on the watermelon. Therefore, we know that the monkey will travel to the west. However, Newton's third law doesn't tell us how fast the monkey will travel. We must apply the law of conservation of momentum to determine that.

First setup a coordinate system. We choose $+x$ to point east. With this choice, the final velocity of the watermelon, $\vec{\boldsymbol{v}}_2$, will be along $\hat{\mathbf{x}}$ while the final velocity of the monkey, $\vec{\boldsymbol{v}}_2$, will be along $-\hat{\mathbf{x}}$. The monkey and watermelon are together initially, and then separate. Thus, this is an inverse perfectly inelastic collision. They have the same initial velocity (zero), since the monkey and watermelon are initially both at rest:

$$(m_1 + m_2)\vec{\boldsymbol{v}}_0 = m_1\vec{\boldsymbol{v}}_1 + m_2\vec{\boldsymbol{v}}_2$$
$$0 = (10)(4\hat{\mathbf{x}}) + (40)\vec{\boldsymbol{v}}_2$$
$$0 = 40\hat{\mathbf{x}} + 40\vec{\boldsymbol{v}}_2$$
$$-40\hat{\mathbf{x}} = 40\vec{\boldsymbol{v}}_2$$
$$\vec{\boldsymbol{v}}_2 = -\frac{40}{40}\hat{\mathbf{x}} = -1.0\hat{\mathbf{x}}\text{ m/s}$$

The significance of the minus sign is that the monkey travels 1.0 m/s to the west after throwing the watermelon, since the $\hat{\mathbf{x}}$ unit vector points to the east.

Example: A 15-kg box of lemons traveling 8.0 m/s to the north collides head-on with a 25-kg box of oranges traveling 4.0 m/s to the south on horizontal frictionless ice. The collision is elastic. Determine the final velocities.

First setup a coordinate system. We choose $+x$ to point north. With this choice, $\vec{\boldsymbol{v}}_{10}$ will be along $\hat{\mathbf{x}}$ and $\vec{\boldsymbol{v}}_{20}$ will be along $-\hat{\mathbf{x}}$. Since the collision is elastic, there are two different equations to use. One equation expresses conservation of momentum:

$$m_1\vec{\boldsymbol{v}}_{10} + m_2\vec{\boldsymbol{v}}_{20} = m_1\vec{\boldsymbol{v}}_1 + m_2\vec{\boldsymbol{v}}_2$$
$$(15)(8\hat{\mathbf{x}}) + (25)(-4\hat{\mathbf{x}}) = 15\vec{\boldsymbol{v}}_1 + 25\vec{\boldsymbol{v}}_2$$
$$120\hat{\mathbf{x}} - 100\hat{\mathbf{x}} = 20\hat{\mathbf{x}} = 15\vec{\boldsymbol{v}}_1 + 25\vec{\boldsymbol{v}}_2$$

The second equation derives from a combination of the law of conservation of momentum and the law of conservation of energy, and only applies to elastic collisions:

$$\vec{\boldsymbol{v}}_{10} + \vec{\boldsymbol{v}}_1 = \vec{\boldsymbol{v}}_{20} + \vec{\boldsymbol{v}}_2$$
$$8\hat{\mathbf{x}} + \vec{\boldsymbol{v}}_1 = -4\hat{\mathbf{x}} + \vec{\boldsymbol{v}}_2$$

We must use both equations to solve for the two unknown final velocities, $\vec{\boldsymbol{v}}_1$ and $\vec{\boldsymbol{v}}_2$. It is simplest to first isolate one of the final velocities from the previous equation:

$$\vec{\boldsymbol{v}}_2 = 8\hat{\mathbf{x}} + 4\hat{\mathbf{x}} + \vec{\boldsymbol{v}}_1 = 12\hat{\mathbf{x}} + \vec{\boldsymbol{v}}_1$$

Substitute this expression into the equation that came from conservation of momentum:

$$20\hat{\mathbf{x}} = 15\vec{\boldsymbol{v}}_1 + 25\vec{\boldsymbol{v}}_2$$
$$20\hat{\mathbf{x}} = 15\vec{\boldsymbol{v}}_1 + 25(12\hat{\mathbf{x}} + \vec{\boldsymbol{v}}_1)$$
$$20\hat{\mathbf{x}} = 15\vec{\boldsymbol{v}}_1 + 25(12\hat{\mathbf{x}}) + 25\vec{\boldsymbol{v}}_1$$
$$20\hat{\mathbf{x}} = 40\vec{\boldsymbol{v}}_1 + 300\hat{\mathbf{x}}$$
$$\vec{\boldsymbol{v}}_1 = -\frac{280}{40}\hat{\mathbf{x}} = -7.0\hat{\mathbf{x}}\text{ m/s}$$

Now plug this into any of the previous equations to solve for $\vec{\boldsymbol{v}}_2$:

$$8\hat{\mathbf{x}} + \vec{\boldsymbol{v}}_1 = -4\hat{\mathbf{x}} + \vec{\boldsymbol{v}}_2$$
$$8\hat{\mathbf{x}} + (-7\hat{\mathbf{x}}) = -4\hat{\mathbf{x}} + \vec{\boldsymbol{v}}_2$$
$$\hat{\mathbf{x}} = -4\hat{\mathbf{x}} + \vec{\boldsymbol{v}}_2$$
$$\vec{\boldsymbol{v}}_2 = \hat{\mathbf{x}} + 4\hat{\mathbf{x}} = 5.0\hat{\mathbf{x}}\text{ m/s}$$

Since we chose $+x$ to point north, after the collision the box of lemons travels 7.0 m/s to the south while the box of oranges travels 5.0 m/s to the north.

Example: A 250-kg car initially traveling 30 m/s along $\hat{\mathbf{x}}$ is bumped from behind. The collision lasts for a duration of 500 ms. Immediately after the collision, the car is traveling 40 m/s along $\hat{\mathbf{x}}$. Determine the average collision force.

Combine the two equations for impulse. Also, convert 500 milliseconds to 0.500 s.

$$\vec{\boldsymbol{J}} = \vec{\boldsymbol{p}} - \vec{\boldsymbol{p}}_0 = \vec{\boldsymbol{F}}_c\Delta t$$
$$\vec{\boldsymbol{F}}_c = \frac{\vec{\boldsymbol{p}} - \vec{\boldsymbol{p}}_0}{\Delta t} = \frac{m\vec{\boldsymbol{v}} - m\vec{\boldsymbol{v}}_0}{\Delta t} = \frac{m(\vec{\boldsymbol{v}} - \vec{\boldsymbol{v}}_0)}{\Delta t} = \frac{250(40\hat{\mathbf{x}} - 30\hat{\mathbf{x}})}{0.5} = 5000\hat{\mathbf{x}}\text{ N} = 5.0\hat{\mathbf{x}}\text{ kN}$$

Example: As illustrated below, a monkey shoots a 250-g bullet into a 2.0-kg block of wood. The bullet is traveling horizontally just before impact. The block of wood is initially at rest. The bullet sticks inside the block of wood and the system rises upward to a maximum angle with the vertical of 60°. The length of the pendulum from the pivot to the center of mass of the system is 250 cm. This problem is called a **ballistic pendulum**. Determine the initial speed of the bullet, just before impact.

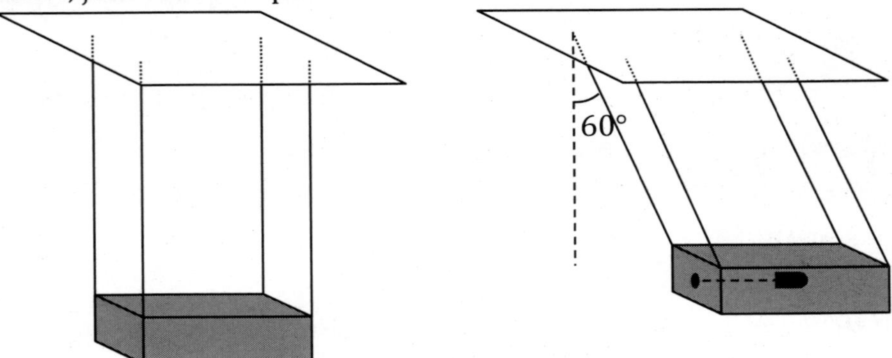

The ballistic pendulum involves more than just a collision:

- It begins with a perfectly inelastic collision at the bottom.
- After the collision, the pendulum and bullet swing together upward in an arc.

We must treat the two processes separately:

- Only momentum is conserved for the perfectly inelastic collision. (As the collision is inelastic, mechanical energy isn't conserved for this process.)
- Only mechanical energy is conserved for the swing upward. (The momentum of the system is clearly lost during the swing upward, since the system has momentum at the bottom and comes to rest at the top.)

This example gives us information about the final position ($\theta = 60°$), and asks for the initial speed of the bullet prior to the collision. We must therefore solve this problem in reverse, beginning with the swing upward and treating the collision last.

Apply conservation of energy (Chapter 22) to the swing upward. Begin by drawing the path and labeling the initial position (i), final position (f), and reference height (RH). The initial position (i) is just **after** the collision (because mechanical energy is not conserved for an inelastic collision), while the final position (f) is at the top of the arc. We choose the reference height (RH) to be at the bottom of the arc. Write out conservation of energy:

$$PE_{ac} + KE_{ac} + W_{nc} = PE + KE$$

The subscript ac stands for "after collision." The reason for this is that the initial position of the swing upward will be the final position for the collision. It would be confusing to call the initial speed v_0 for the swing upward and then proceed to use the same symbol v_0 for the initial speed for the collision (since the two "initial" speed differ). The hope is that using v_{ac} for the **initial** speed of the swing upward and the **final** speed for the collision may make the notation less confusing. Let's analyze this term by term:

- $PE_{gac} = 0$ since i is at the same height as RH. Remember that ac stands for "after collision." What we would usually call nought (0), like PE_{g0}, we are instead calling ac, as in PE_{gac}, since what is "initial" for the swing upward will be "final" for the collision. Hopefully, if you can remember that ac stands for "after collision," it will help to avoid other possible confusion.
- $PE_{sac} = 0$ since at i there isn't a spring compressed or stretched from equilibrium.
- $PE_{ac} = PE_{gac} + PE_{sac} = 0$.
- $KE_{ac} = \frac{1}{2}(m_b + m_p)v_{ac}^2$ since the bullet and pendulum are moving at i.
- $W_{nc} = 0$ neglecting any frictional forces.
- $PE_g = (m_b + m_p)gh$ since f is **not** at the same height as RH.
- $PE_s = 0$ since at f there isn't a spring compressed or stretched from equilibrium.
- $PE = PE_g + PE_s = (m_b + m_p)gh$.
- $KE = 0$ since the system is at rest at f (otherwise, it would rise higher).

Make the above substitutions into the conservation of energy equation:

$$\frac{1}{2}(m_b + m_p)v_{ac}^2 = (m_b + m_p)gh$$

We need to relate the length of the pendulum (L) to the final height (h). We can do this using geometry. Study the diagram below.

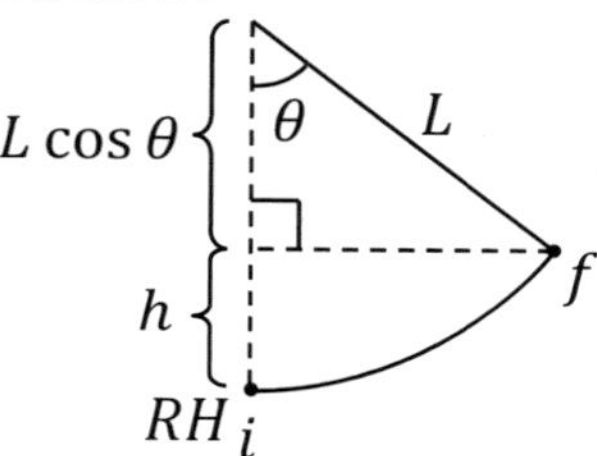

In the figure above, the two distances at the left add up to the length L. Convert L from 250 cm to $\frac{5}{2}$ m.

$$L = h + L\cos\theta$$

$$h = L - L\cos\theta = L(1 - \cos\theta) = \frac{5}{2}(1 - \cos 60°) = \frac{5}{2}\left(1 - \frac{1}{2}\right) = \frac{5}{2}\left(\frac{1}{2}\right) = \frac{5}{4} \text{ m}$$

Now we can return to the equation that we reached from conservation of energy:

$$\frac{1}{2}(m_b + m_p)v_{ac}^2 = (m_b + m_p)gh$$

Mass cancels out:

$$v_{ac}^2 = 2gh$$

$$v_{ac} = \sqrt{2(9.81)\left(\frac{1}{4}\right)} \approx \sqrt{2(10)\left(\frac{5}{4}\right)} = \sqrt{25} = 5.0 \text{ m/s}$$

Now we are prepared to treat the collision. We choose $+x$ to be the forward direction of the bullet. Only momentum is conserved for the perfectly inelastic collision at the bottom:

$$m_b \vec{\boldsymbol{v}}_{b0} + m_p \vec{\boldsymbol{v}}_{p0} = (m_b + m_p)\vec{\boldsymbol{v}}_{ac}$$

We choose to work with the subscripts b for bullet and p for pendulum rather than the usual 1 and 2 to help keep track of which object is which. The final velocity of the collision, $\vec{\boldsymbol{v}}_{ac}$, is the same as the initial velocity for the swing upward. We already found v_{ac}. The pendulum is initially at rest: $\vec{\boldsymbol{v}}_{p0} = 0$. Convert the mass of the bullet (m_b) from 250 g to $\frac{1}{4}$ kg. Solve for the initial velocity of the bullet:

$$m_b \vec{\boldsymbol{v}}_{b0} + m_p \vec{\boldsymbol{v}}_{p0} = (m_b + m_p)\vec{\boldsymbol{v}}_{ac}$$

$$\frac{1}{4}\vec{\boldsymbol{v}}_{b0} + 2(0) = \left(\frac{1}{4} + 2\right)(5\hat{\mathbf{x}})$$

$$\frac{1}{4}\vec{\boldsymbol{v}}_{b0} = \left(\frac{1}{4} + \frac{8}{4}\right)(5\hat{\mathbf{x}})$$

$$\frac{1}{4}\vec{\boldsymbol{v}}_{b0} = \left(\frac{9}{4}\right)(5\hat{\mathbf{x}}) = \frac{45}{4}\hat{\mathbf{x}}$$

$$\vec{\boldsymbol{v}}_{b0} = 45\hat{\mathbf{x}} \text{ m/s}$$

The bullet is initially moving 45 m/s along $\hat{\mathbf{x}}$ (where the unit vector $\hat{\mathbf{x}}$ points horizontally to the right in the diagram on page 235). (**Unit vectors** were introduced in Chapter 10.)

142. A monkey places a 500-g banana on top of his head. Another monkey shoots the banana with a 250-g arrow. The arrow sticks in the banana. The arrow is traveling 45 m/s horizontally just before impact. How fast do the banana and arrow travel just after impact?

Want help? Check the hints section at the back of the book.

Answer: 15 m/s

143. A 50-kg monkey is initially sitting in a 250-kg canoe. The canoe is initially at rest relative to the lake. The monkey begins to walk 3.0 m/s (relative to the water) to the south from one end of the canoe to the other. What is the velocity of the canoe while the monkey walks to the south?

Want help? Check the hints section at the back of the book.

Answers: $\frac{3}{5}$ m/s to the north

144. A 6.0-kg box of bananas traveling 5.0 m/s to the east collides head-on with a 12.0-kg box of apples traveling 4.0 m/s to the west on horizontal frictionless ice.

(A) Determine the final speed of each box if the collision is perfectly inelastic.

(B) Determine the final speed of each box if the collision is elastic.

Want help? Check the hints section at the back of the book.

Answers: 1.0 m/s to the west; 7.0 m/s to the west and 2.0 m/s to the east

145. A 3.0-kg box of coconuts traveling 16.0 m/s to the north collides head-on with a 9.0-kg box of grapefruit traveling 6.0 m/s to the south on horizontal frictionless ice. The collision is elastic. The collision lasts for a duration of 250 ms.

(A) Determine the final velocity of each box.

(B) What is the magnitude of the average collision force?

Want help? Check the hints section at the back of the book.

Answers: 17.0 m/s to the south, 5.0 m/s to the north; 396 N

146. A 20-kg box of bananas begins from rest at the top of the circular hill illustrated below (the hill is exactly one quarter of a circle). The radius of the circular hill is 5.0 m. The hill and horizontal are both frictionless. When the box of bananas reaches point A, it collides with a 30-kg box of pineapples which is initially at rest. The two boxes stick together after the collision. Determine the final speed of the boxes.

Want help? Check the hints section at the back of the book.

Answer: 4.0 m/s

24 TWO-DIMENSIONAL COLLISIONS

Essential Concepts

Momentum ($\vec{\boldsymbol{p}}$) and velocity ($\vec{\boldsymbol{v}}$) are vectors. The equation for conservation of momentum involves vector addition.

$$m_1\vec{\boldsymbol{v}}_{10} + m_2\vec{\boldsymbol{v}}_{20} = m_1\vec{\boldsymbol{v}}_1 + m_2\vec{\boldsymbol{v}}_2$$

The way to add vectors is to work with components (Chapter 10). This means that, for two-dimensional collisions, we will express the law of conservation of momentum using the x- and y-components of velocity.

$$m_1 v_{10x} + m_2 v_{20x} = m_1 v_{1x} + m_2 v_{2x}$$

$$m_1 v_{10y} + m_2 v_{20y} = m_1 v_{1y} + m_2 v_{2y}$$

Momentum is separately conserved along both the x- and y-components of the motion.

Two-Dimensional Collision Equations

The initial velocities can be resolved into components using trig. These equations are just like the first step of vector addition (Chapter 10). Below, v_{10} and v_{20} are the initial speeds of the objects, and θ_{10} and θ_{20} are the directions of the initial velocities.

$$v_{10x} = v_{10}\cos\theta_{10} \quad , \quad v_{20x} = v_{20}\cos\theta_{20}$$

$$v_{10y} = v_{10}\sin\theta_{10} \quad , \quad v_{20y} = v_{20}\sin\theta_{20}$$

Conservation of momentum involves two components in two-dimensional collisions:

$$m_1 v_{10x} + m_2 v_{20x} = m_1 v_{1x} + m_2 v_{2x}$$

$$m_1 v_{10y} + m_2 v_{20y} = m_1 v_{1y} + m_2 v_{2y}$$

For an **elastic** collision, mechanical energy is conserved in addition to momentum.

$$\frac{1}{2}m_1 v_{10}^2 + \frac{1}{2}m_2 v_{20}^2 = \frac{1}{2}m_1 v_1^2 + \frac{1}{2}m_2 v_2^2$$

A special case worth noting is an **elastic** two-dimensional collision with **equal masses** ($m_1 = m_2 = m$). In this special case, the final velocities are perpendicular. If we work with the reference angles (Chapter 9) corresponding to θ_1 and θ_2, we can express this as:

$$\theta_{1ref} + \theta_{2ref} = 90°$$

The final speed of each object can be found from the Pythagorean theorem, and the direction of each final velocity can be found with an inverse tangent. These equations are just like the third step of vector addition (Chapter 10).

$$v_1 = \sqrt{v_{1x}^2 + v_{1y}^2} \quad , \quad v_2 = \sqrt{v_{2x}^2 + v_{2y}^2}$$

$$\theta_1 = \tan^{-1}\left(\frac{v_{1y}}{v_{1x}}\right) \quad , \quad \theta_2 = \tan^{-1}\left(\frac{v_{2y}}{v_{2x}}\right)$$

Symbols and Units

Symbol	Name	Units
m_1	mass of object 1	kg
m_2	mass of object 2	kg
$\vec{\boldsymbol{v}}_{10}$	initial velocity of object 1	m/s
$\vec{\boldsymbol{v}}_{20}$	initial velocity of object 2	m/s
v_{10}	initial speed of object 1	m/s
v_{20}	initial speed of object 2	m/s
θ_{10}	initial direction of object 1	°
θ_{20}	initial direction of object 2	°
v_{10x}	initial x-component of velocity of object 1	m/s
v_{10y}	initial y-component of velocity of object 1	m/s
v_{20x}	initial x-component of velocity of object 2	m/s
v_{20y}	initial y-component of velocity of object 2	m/s
$\vec{\boldsymbol{v}}_1$	final velocity of object 1	m/s
$\vec{\boldsymbol{v}}_2$	final velocity of object 2	m/s
v_1	final speed of object 1	m/s
v_2	final speed of object 2	m/s
θ_1	final direction of object 1	°
θ_2	final direction of object 2	°
v_{1x}	final x-component of velocity of object 1	m/s
v_{1y}	final y-component of velocity of object 1	m/s
v_{2x}	final x-component of velocity of object 2	m/s
v_{2y}	final y-component of velocity of object 2	m/s

Strategy for Two-dimensional Collisions

To solve a problem with a two-dimensional collision, follow these steps:

1. Setup your (x, y) coordinate system (if it isn't already established in the problem).
2. Relate the initial speeds (v_{10} and v_{20}) and the directions of the initial velocities (θ_{10} and θ_{20}) to the components of the initial velocities (v_{10x}, v_{10y}, v_{20x}, and v_{20y}):
$$v_{10x} = v_{10}\cos\theta_{10} \quad , \quad v_{20x} = v_{20}\cos\theta_{20}$$
$$v_{10y} = v_{10}\sin\theta_{10} \quad , \quad v_{20y} = v_{20}\sin\theta_{20}$$
The angles θ_{10} and θ_{20} are measured counterclockwise from the $+x$-axis.
3. This step depends on the nature of the collision. Following are the types of two-dimensional collisions commonly encountered in physics problems.
   - If the problem makes it clear that the objects stick together after the collision, or if the problem calls the collision "**perfectly inelastic**," use these equations:
   $$m_1 v_{10x} + m_2 v_{20x} = (m_1 + m_2)v_x$$
   $$m_1 v_{10y} + m_2 v_{20y} = (m_1 + m_2)v_y$$
   - If the problem calls the collision "**elastic**," or if the problem tells you that mechanical energy is conserved for the collision, the following equations apply:
   $$m_1 v_{10x} + m_2 v_{20x} = m_1 v_{1x} + m_2 v_{2x}$$
   $$m_1 v_{10y} + m_2 v_{20y} = m_1 v_{1y} + m_2 v_{2y}$$
   $$\frac{1}{2}m_1 v_{10}^2 + \frac{1}{2}m_2 v_{20}^2 = \frac{1}{2}m_1 v_1^2 + \frac{1}{2}m_2 v_2^2$$
   - For an **elastic** collision with **equal masses** ($m_1 = m_2$), use the following equations. (The first two are the same as conservation of momentum, except that mass cancels out since $m_1 = m_2$.)
   $$v_{10x} + v_{20x} = v_{1x} + v_{2x}$$
   $$v_{10y} + v_{20y} = v_{1y} + v_{2y}$$
   $$\theta_{1ref} + \theta_{2ref} = 90°$$
4. Relate the components of the final velocities (v_{1x}, v_{1y}, v_{2x}, and v_{2y}) to the final speeds (v_1 and v_2) and the directions of the final velocities (θ_1 and θ_2):
   - For a perfectly inelastic collision, use these equations:
   $$v = \sqrt{v_x^2 + v_y^2}$$
   $$\theta = \tan^{-1}\left(\frac{v_y}{v_x}\right)$$
   - For an elastic collision, use these equations:
   $$v_1 = \sqrt{v_{1x}^2 + v_{1y}^2} \quad , \quad v_2 = \sqrt{v_{2x}^2 + v_{2y}^2}$$
   $$\theta_1 = \tan^{-1}\left(\frac{v_{1y}}{v_{1x}}\right) \quad , \quad \theta_2 = \tan^{-1}\left(\frac{v_{2y}}{v_{2x}}\right)$$

Example: A 200-kg bananamobile traveling 30 m/s to the north collides with a 100-kg bananamobile traveling $60\sqrt{3}$ m/s to the west. The two bananamobiles stick together after the collision. Determine the speed of the bananamobiles just after the collision and the direction of the final velocity.

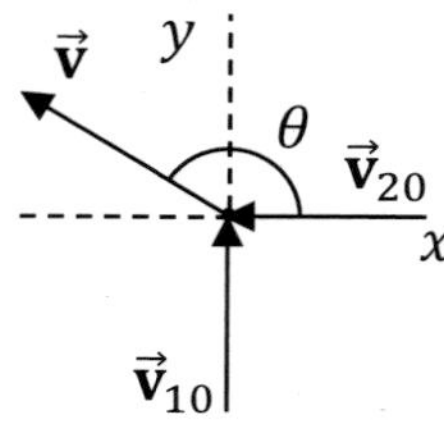

First setup a coordinate system. We choose $+x$ to point east and $+y$ to point north. With this choice, $\theta_{10} = 90°$ (along $+y$) and $\theta_{10} = 180°$ (along $-x$). Resolve the initial velocities into components:

$$v_{10x} = v_{10}\cos\theta_{10} = 30\cos 90° = 0 \quad , \quad v_{20x} = v_{20}\cos\theta_{20} = 60\sqrt{3}\cos 180° = -60\sqrt{3}\text{ m/s}$$

$$v_{10y} = v_{10}\sin\theta_{10} = 30\sin 90° = 30\text{ m/s} \quad , \quad v_{20y} = v_{20}\sin\theta_{20} = 60\sqrt{3}\sin 180° = 0$$

Since the bananamobiles stick together, use the equations for a perfectly inelastic collision. They have the same final velocity:

$$m_1 v_{10x} + m_2 v_{20x} = (m_1 + m_2)v_x$$

$$200(0) + 100(-60\sqrt{3}) = (200 + 100)v_x$$

$$-6000\sqrt{3} = 300 v_x$$

$$v_x = \frac{-6000\sqrt{3}}{300} = -20\sqrt{3}\text{ m/s}$$

$$m_1 v_{10y} + m_2 v_{20y} = (m_1 + m_2)v_y$$

$$200(30) + 100(0) = (200 + 100)v_y$$

$$6000 = 300 v_y$$

$$v_y = \frac{6000}{300} = 20\text{ m/s}$$

Use the components of the final velocity to determine the final speed and the direction of the final velocity.

$$v = \sqrt{v_x^2 + v_y^2} = \sqrt{\left(-20\sqrt{3}\right)^2 + (20)^2} = \sqrt{1200 + 400} = \sqrt{1600} = 40\text{ m/s}$$

$$\theta = \tan^{-1}\left(\frac{v_y}{v_x}\right) = \tan^{-1}\left(\frac{20}{-20\sqrt{3}}\right) = \tan^{-1}\left(-\frac{1}{\sqrt{3}}\right) = \tan^{-1}\left(-\frac{1}{\sqrt{3}}\frac{\sqrt{3}}{\sqrt{3}}\right) = \tan^{-1}\left(-\frac{\sqrt{3}}{3}\right)$$

The reference angle is 30°, but the reference angle isn't the answer. The direction of the final velocity lies in Quadrant II because v_x is negative and v_y is positive. The Quadrant II angle corresponding to a reference angel of 30° is:

$$\theta_{III} = 180° - \theta_{ref} = 180° - 30° = 150°$$

Example: A billiard ball traveling 4.0 m/s collides with another billiard ball of equal mass that is initially at rest. After the collision, the second billiard ball travels 2.0 m/s along a line that is tilted 60° relative to the initial velocity of the first billiard ball. The collision is <u>elastic</u>. Determine the magnitude and direction of the final velocity of the first billiard ball.

First setup a coordinate system. We choose $+x$ to point along the initial velocity of the first billiard ball and $+y$ to be perpendicular to that. With this choice, $\theta_{10} = 0°$ (along $+x$). Observe that $v_{20} = 0$ since the second billiard ball is initially at rest. Resolve the initial velocities into components:

$$v_{10x} = v_{10}\cos\theta_{10} = 4\cos 0° = 4 \text{ m/s} \quad , \quad v_{20x} = v_{20}\cos\theta_{20} = 0$$
$$v_{10y} = v_{10}\sin\theta_{10} = 4\sin 0° = 0 \quad , \quad v_{20y} = v_{20}\sin\theta_{20} = 0$$

According to the problem, the collision is elastic and the billiard balls have equal mass ($m_1 = m_2 = m$). Use the appropriate equations:

$$v_{10x} + v_{20x} = v_{1x} + v_{2x}$$
$$4 + 0 = 4 = v_{1x} + v_{2x}$$
$$v_{10y} + v_{20y} = v_{1y} + v_{2y}$$
$$0 + 0 = 0 = v_{1y} + v_{2y}$$
$$v_{2y} = -v_{1y}$$
$$\theta_{1ref} + \theta_{2ref} = 90°$$
$$60° + \theta_{2ref} = 90°$$
$$\theta_{2ref} = 30°$$

Since this problem gives us the final speed and the direction of the final velocity of the first billiard ball, we will resolve these into the components of that ball's final velocity:

$$v_{1x} = v_1\cos\theta_1 = 2\cos 60° = 1.0 \text{ m/s}$$
$$v_{1y} = v_1\sin\theta_1 = 2\sin 60° = \sqrt{3} \text{ m/s}$$

Plug these values into the equations above:

$$4 = v_{1x} + v_{2x}$$
$$4 = 1 + v_{2x}$$
$$v_{2x} = 4 - 1 = 3.0 \text{ m/s}$$
$$v_{2y} = -v_{1y} = -\sqrt{3} \text{ m/s}$$

Use the components of the final velocity to determine the final speed and the direction of the final velocity for the second billiard ball.

$$v_2 = \sqrt{v_{2x}^2 + v_{2y}^2} = \sqrt{(3)^2 + \left(-\sqrt{3}\right)^2} = \sqrt{9+3} = \sqrt{12} = \sqrt{(4)(3)} = 2\sqrt{3} \text{ m/s}$$

$$\theta_2 = \tan^{-1}\left(\frac{v_{2y}}{v_{2x}}\right) = \tan^{-1}\left(\frac{-\sqrt{3}}{3}\right)$$

The reference angle is 30°, but θ_2 lies in Quadrant IV since v_{2x} is positive and v_{2y} is negative: $\theta_2 = \theta_{IV} = 360° - \theta_{ref} = 360° - 30° = 330°$.

147. A 300-kg bananamobile traveling 20 m/s to the north collides with a 200-kg bananamobile traveling 40 m/s to the east. The two bananamobiles stick together after the collision. Determine the speed of the bananamobiles just after the collision.

Want help? Check the hints section at the back of the book.

Answer: 20 m/s

148. Two bananamobiles of equal mass collide at an intersection and stick together. One bananamobile is traveling 20 m/s to the south prior to the collision. After the collision, the two bananamobiles travel $10\sqrt{2}$ m/s to the southwest. Determine the initial speed and the direction of the initial velocity of the other bananamobile prior to the collision.

Want help? Check the hints section at the back of the book.

Answers: 20 m/s to the west

149. A billiard ball traveling $6\sqrt{3}$ m/s collides with another billiard ball of equal mass that is initially at rest. After the collision, the first billiard ball travels $3\sqrt{3}$ m/s along a line that is deflected 60° relative to its original direction. The collision is elastic. Determine the magnitude and direction of the final velocity of the second billiard ball.

Want help? Check the hints section at the back of the book.

Answers: 9.0 m/s, 330°

25 ROCKET PROPULSION

Relevant Terminology

Propulsion – the act of being propelled by a force. Rocket propulsion refers to the thrust created from Newton's third law by ejecting gases. (See the essential concepts below.)
Inertial reference frame – a coordinate system that travels with constant velocity. (If a rocket accelerates, a coordinate system attached to the rocket would **not** serve as an inertial reference frame.)
Thrust – the force exerted on the rocket as a reaction to ejecting the gases.
Free fall – the state of a rocket that isn't ejecting gases (to apply thrust). A rocket is in a free fall orbit when the only forces acting on the rocket are gravitational.
Payload – the mass of the rocket without any fuel.
Burn time – the time the rocket spends burning its fuel.
Burnout – the point when all of the rocket's fuel has been burned.

Essential Concepts

Rocket propulsion works according to **Newton's third law** (Chapter 13): When the rocket ejects gases, the gases exert a force on the rocket which is equal in magnitude and opposite in direction to the force that the rocket exerts on the gases. At its most basic level, this is the same principle with which you walk or drive a car. When you walk, your shoe exerts a force on the ground backward, and the ground pushes your shoe forward through an equal force in the opposite direction.

Newton's second law takes on a different form in the context of rocket propulsion. In Chapters 13-14, we set the net force ($\sum \vec{\mathbf{F}}$) equal to mass (m) times acceleration ($\vec{\mathbf{a}}$). However, that's only true for an object with constant mass. A rocket continually changes mass when it ejects gases, so it would be incorrect to set net force equal to mass times acceleration when a rocket ejects gases. In this case, we must apply the more general form of **Newton's second law**:

$$\sum \vec{\mathbf{F}} = \frac{d\vec{\mathbf{p}}_{tot}}{dt}$$

Here, $\vec{\mathbf{p}}_{tot}$ is the total **momentum** of the system. Recall from Chapters 23-24 that momentum ($\vec{\mathbf{p}}$) equals mass (m) times velocity ($\vec{\boldsymbol{v}}$): $\vec{\mathbf{p}} = m\vec{\boldsymbol{v}}$.

We can see that the general form of Newton's second law reduces to the more common form of Newton's second law in the special case of an object with constant mass. When mass is constant, $\frac{d\vec{\mathbf{p}}}{dt} = m\frac{d\vec{\boldsymbol{v}}}{dt} = m\vec{\mathbf{a}}$. However, m isn't constant in rocket propulsion.

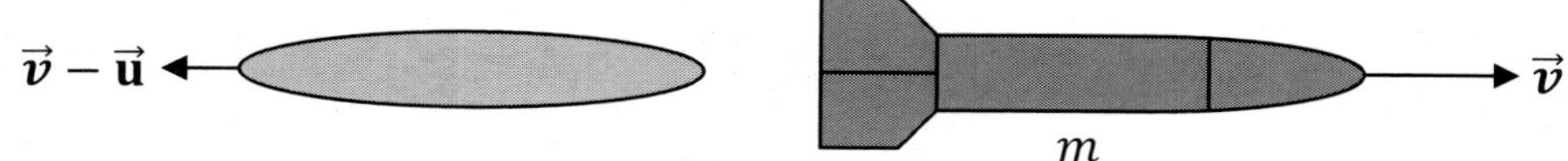

Reference Frames

Rocket problems involve two different reference frames:
- An inertial reference frame is a coordinate system with constant velocity.
- The reference frame of the rocket is non-inertial for a rocket that either changes speed or direction.

Measurements of velocity depend on not only which object we're talking about (the rocket or the gases), but also which reference frame it is relative to:
- $\vec{v}$ is the instantaneous velocity of the rocket relative to the inertial reference frame.
- $\vec{\mathbf{u}}$ is the instantaneous velocity of the ejected gases relative to the rocket.
- $\vec{v} - \vec{\mathbf{u}}$ is the instantaneous velocity of the ejected gases relative to the inertial reference frame. See the diagram above.

Newton's Second Law for Rocket Propulsion

We define the system to include the rocket plus the unburned fuel plus the ejected gases. Newton's second law relates the net external force acting on the rocket to the total momentum of the system:

$$\sum \vec{\mathbf{F}} = \frac{d\vec{\mathbf{p}}_{tot}}{dt}$$

Note that the momentum of the system isn't conserved in general. Rockets are under the influence of external gravitational forces created by planets, moons, stars, etc. In terms of the instantaneous velocity of the rocket and the instantaneous velocity of the ejected gases (see the previous discussion of reference frames), Newton's second law is:

$$\sum \vec{\mathbf{F}} = m\frac{d\vec{v}}{dt} + \vec{\mathbf{u}}\frac{dm}{dt}$$

The **thrust** ($\vec{\mathbf{F}}_t$) of the rocket equals:

$$\vec{\mathbf{F}}_t = -\vec{\mathbf{u}}\frac{dm}{dt}$$

Using the equation for thrust, Newton's second law can be written in the following form:

$$\vec{\mathbf{F}}_t + \sum \vec{\mathbf{F}} = m\frac{d\vec{v}}{dt}$$

Burn Rate

The **burn rate** (R_b) of a rocket is the instantaneous rate at which fuel is burned:

$$R_b = \left|\frac{dm}{dt}\right|$$

Payload Equations

The **payload** (m_p) refers to the mass of the rocket without any fuel. The payload is often expressed as a ratio compared to the initial mass m_0 of the rocket including all fuel.

$$r_p = \frac{m_p}{m_0}$$

The **initial mass** m_0 equals the payload (m_p) plus the initial mass of the fuel (m_{f0}):

$$m_0 = m_p + m_{f0}$$

The **final mass** m equals the payload (m_p) plus the mass of any remaining fuel (m_f):

$$m = m_p + m_f$$

The ratio of the final mass to the initial mass is:

$$\frac{m}{m_0} = \frac{m_p + m_f}{m_p + m_{f0}}$$

The above ratio becomes r_p in the limit that all of the fuel is spent (when $m_f \to 0$).

Special Case: Uniform Gravity and Constant Burn Rate

In an approximately uniform gravitational field with a constant burn rate, Newton's second law becomes:

$$F_t - mg = m\frac{dv_y}{dt}$$

Using the definition of thrust $\left(F_t = -u\frac{dm}{dt}\right)$, this can be expressed as:

$$-u\frac{dm}{dt} - mg = m\frac{dv_y}{dt}$$

The solution to this differential equation is:*

$$v_y = v_{y0} - gt_b + u\ln\left(\frac{m_0}{m}\right)$$

The burn time (t_b) is related to the burn rate (R_b) by:

$$m - m_0 = R_b t_b$$

Special Case: Zero-gravity

In an approximately zero-gravity region, Newton's second law simplifies to:

$$-u\frac{dm}{dt} = m\frac{dv_y}{dt}$$

The solution to this differential equation is:*

$$v_y = v_{y0} + u\ln\left(\frac{m_0}{m}\right)$$

* This book is focused on how to solve the problems. If you would like to see the calculus involved in solving these differential equations, consult a sufficiently advanced calculus-based physics or mechanics textbook.

Symbols and SI Units

Symbol	Name	SI Units
m_0	initial mass of the rocket (including all of the fuel)	kg
m_{f0}	mass of the initial fuel of the rocket	kg
m	mass of rocket plus the mass of any unburned fuel	kg
m_f	mass of the fuel remaining in the rocket	kg
m_p	the payload (the mass of the rocket without any fuel)	kg
r_p	the ratio of the payload (m_p) to the initial mass (m_0)	unitless
y	position coordinate	m
v_{y0}	initial y-component of the rocket's velocity	m/s
v_y	y-component of the rocket's velocity	m/s
$\vec{v}$	instantaneous velocity of the rocket relative to the inertial reference frame	m/s
$\vec{\mathbf{u}}$	instantaneous velocity of the ejected gases relative to the rocket	m/s
$\vec{v} - \vec{\mathbf{u}}$	instantaneous velocity of the ejected gases relative to the inertial reference frame	m/s
$\vec{\mathbf{p}}_{tot}$	the total momentum of the system	kg·m/s
g	gravitational acceleration	m/s^2
a_y	y-component of acceleration	m/s^2
$\vec{\mathbf{F}}$	force	N
$\vec{\mathbf{F}}_t$	the rocket's thrust	N
R_b	the instantaneous rate at which the rocket burns fuel	kg/s
t	time	s
t_b	burn time (the time spent burning fuel)	s

Strategy for Solving a Rocket Propulsion Problem

Most rocket propulsion problems in first-year physics courses fall into one of the two classes listed under Step 2 below. To solve a rocket propulsion system like those, follow these steps:

1. Make a list of the known quantities, and identify the desired unknown(s). Consult the table of symbols on the previous page (also pay attention to the SI units). Read the question carefully to help distinguish among similar quantities (like the five different symbols for mass).
2. Find and use the relevant equations. Think about the unknown(s) you are solving for and which quantities you know. We will first list equations that apply to all rocket propulsion problems and then we will consider two common special cases. The following equations relate the five different kinds of masses.
$$r_p = \frac{m_p}{m_0} \quad , \quad m_0 = m_p + m_{f0} \quad , \quad m = m_p + m_f \quad , \quad \frac{m}{m_0} = \frac{m_p + m_f}{m_p + m_{f0}}$$
The **burn rate** (R_b) in kg/s (**not** to be confused with the mass ratio r_p) is defined as:
$$R_b = \left|\frac{dm}{dt}\right|$$
If the burn rate (R_b) is constant, the burn rate is related to the burn time (t_b) by:
$$m - m_0 = -R_b t_b$$
The **thrust** of the rocket ($\vec{\mathbf{F}}_t$) is defined in terms of the velocity ($\vec{\mathbf{u}}$) of the **gases** relative to the rocket (**not** relative to the inertial reference frame).
$$\vec{\mathbf{F}}_t = -\vec{\mathbf{u}}\frac{dm}{dt}$$
Following are equations that apply to two special cases:
   - For one-dimensional rocket propulsion in an approximately uniform gravitational field with a constant burn rate, the final velocity (v_y) is related to the burn time (t_b) via the velocity of the ejected gases (u) relative to the rocket according to the following equation:
   $$v_y = v_{y0} - g t_b + u \ln\left(\frac{m_0}{m}\right)$$
   In this case, Newton's second law takes on the form:
   $$F_{thrust} - mg = m a_y$$
   - For one-dimensional rocket propulsion in an approximately zero-gravity region of space, the final velocity (v_y) is related to the velocity of the gases (u) relative to the rocket according to the following equation:
   $$v_y = v_{y0} + u \ln\left(\frac{m_0}{m}\right)$$
   For the zero-gravity problem, Newton's second law is:
   $$F_{thrust} = m a_y$$

Example: A monkey launches a homemade rocket straight upward from rest. The initial mass of the rocket is 80% water, which is ejected in the form of steam with an exhaust speed of 200 m/s relative to the rocket with a constant burn rate of 50 kg/s. Burnout occurs 10 s after launch. Determine the velocity and acceleration of the rocket when burnout occurs, assuming that $g \approx$ const.

Choosing $+y$ to point upward, the y-component of the velocity when burnout occurs is

$$v_y(t_b) = v_{y0} - gt_b + u \ln\left(\frac{m_0}{m}\right) = 0 - (9.81)(10) + (200) \ln\left(\frac{1}{r_p}\right)$$

The payload is 20%, so $r_p = \frac{1}{5}$. As usual, we approximate $g \approx 10 \text{ m/s}^2$. At burnout,

$$v_y(t_{burnout}) \approx -(10)(10) + (200) \ln\left(\frac{1}{1/5}\right) = -100 + (200)\ln(5) = 222 \text{ m/s}$$

(You need a calculator to find the natural log of 5.) The thrust of the rocket is

$$F_t = -u\frac{dm}{dt} = -(200)(-50) = 10 \text{ kN}$$

Note that $\frac{dm}{dt}$ is **negative** because the rocket loses mass (so m is a decreasing function of time). The acceleration of the rocket can be found from Newton's second law:

$$F_{thrust} - mg = ma_y$$

At burnout, the mass of the rocket equals the payload:

$$a_y(t_{burnout}) = \frac{F_{thrust}}{m_p} - g$$

We must determine the mass of the payload before we can find the acceleration. Since the burn rate is constant, $m - m_0 = -R_b t_b$ and $m_p - m_0 = -R_b t_{burnout}$. The initial mass of the fuel was $m_{f0} = m_0 - m_p = R_b t_{burnout} = 50(10) = 500$ kg. The payload is 20% or $\frac{1}{5}$.

$$r_p = \frac{m_p}{m_0} = \frac{m_0 - m_{f0}}{m_0} = \frac{1}{5}$$

$$5m_0 - 5m_{f0} = m_0$$

$$4m_0 = 5m_{f0}$$

$$m_0 = \frac{5m_{f0}}{4} = \frac{5(500)}{4} = 625 \text{ kg}$$

The payload is

$$m_p = m_0 - m_{f0} = 625 - 500 = 125 \text{ kg}$$

The acceleration of the rocket at burnout is thus found to be

$$a_y(t_{burnout}) = \frac{F_{thrust}}{m_p} - g = \frac{10{,}000}{125} - 9.81 \approx \frac{10{,}000}{125} - 10 = 70 \text{ m/s}^2$$

Just as the water is completely exhausted, the rocket is accelerating upward at approximately seven gravities.

150. A 2,000-kg rocket (not including the fuel) in deep space – i.e. where gravity is negligible compared to the thrust of the rocket – has a payload of 25% and an initial speed of 200 m/s relative to the nearest (yet quite distant) star. Fuel is ejected with an exhaust speed of 1500 m/s relative to the rocket at a constant burn rate of 50 kg/s in order to accelerate the rocket to a final speed of 600 m/s.

Note: This problem requires a calculator in order to evaluate the natural logarithm.

(A) What percentage of the mass of the fuel is spent to accomplish this?

(B) How much time does this take?

Want help? Check the hints section at the back of the book.

Answers: 31%, 37 s

151. A 3,000-kg rocket (not including the fuel) has 6,000 kg of fuel and an initial speed of 100 m/s as it accelerates upward away from earth. With a constant burn rate, the rocket accelerates to 500 m/s in 12 s. The exhaust speed is 2000 m/s relative to the rocket. Assume that the rocket's initial altitude is low enough that any variation in gravitational acceleration from 9.81 m/s^2 may be neglected during this time interval.

Note: This problem requires a calculator in order to evaluate the natural logarithm.

(A) What percentage of the mass of the fuel is spent to accomplish this?

(B) What is the burn rate?

(C) What is the thrust?

(D) What is the initial acceleration?

(E) What is the final acceleration?

Want help? Check the hints section at the back of the book.

Answers: 34%, 171 kg/s, 342 kN, 28 m/s^2, 39 m/s^2

26 TECHNIQUES OF INTEGRATION AND COORDINATE SYSTEMS

Strategy for Integrating by Substitution

To integrate via a substitution, follow these steps:

1. Visualize a substitution of the form $x = x(u)$, meaning that the old variable x is a function of the new variable u, which will transform the integral from $\int_{x=x_0}^{x} f(x)\,dx$ to the form $\int_{u=u_0}^{u} g(u)\,du$ in such a way that you can perform the new integral (in terms of the variable u) with a known technique (see Chapters 2 and 9). Two common substitutions are:
   - **Polynomial** substitutions. Example: $\int (6x+4)^3\,dx$. Try $u = 6x + 4$.
   - **Trig** substitutions. This is common with quadratic functions in squareroots. The idea is to make a substitution that collapses two terms down to a single term through the trig identities $\sin^2 u + \cos^2 u = 1$ or $\tan^2 u + 1 = \sec^2 u$.
     - Example: $\int \sqrt{a^2 - x^2}\,dx$. Try $x = a \sin u$ because $1 - \sin^2 u = \cos^2 u$.
     - Example: $\int \sqrt{x^2 + a^2}\,dx$. Try $x = a \tan u$ because $\tan^2 u + 1 = \sec^2 u$.
     - Example: $\int \sqrt{x^2 - a^2}\,dx$. Try $x = a \sec u$ because $\sec^2 u - 1 = \tan^2 u$.
2. Implicitly differentiate the function $x = x(u)$ in order to write du in terms of dx. On one side, take a derivative with respect to u and multiply by du, and on the other side, take a derivative with respect to x and multiply by dx. See the examples.
3. Solve for dx from the equation in Step 2.
4. Determine the new lower and upper limits of integration for the variable u which correspond to the old limits of integration for the variable x. Use the equation from Step 1 to determine the new limits from the old ones.
5. Make **three** substitutions in the original integral:
   - First replace x with the function of u from Step 1.
   - Next replace dx with the equation from Step 3. Don't forget the du.
   - Replace the old limits of integration with the new ones for u.
6. You should now be able to do the new integral in terms of u. Occasionally, the substitution that you try in Step 1 doesn't pan out as you expect. When this happens, you just need to start over with a different substitution. If you think about what went wrong, it may help you revise your original substitution in order to make one that works better. Study the examples and practice working out exercises and thinking through the logic in order to become more adept at the integration.
7. Carry out the new integral over the variable u. Review Chapters 2 and 9, if needed.

Example: Perform the following definite integral: $\int_{x=1}^{3}(8x-4)^3\,dx$.

We make the substitution $u = 8x - 4$. Take an implicit derivative:

- On the left-hand side: $\frac{d}{du}(u) = 1$. Multiply by du to get $1\,du = du$.
- On the right-hand side: $\frac{d}{dx}(8x-4) = 8$. Multiply by dx to get $8\,dx$.

The implicit derivative of $u = 8x - 4$ is therefore $du = 8\,dx$. Solve for dx to get $dx = \frac{du}{8}$.

$$u = 8x - 4 \quad , \quad dx = \frac{du}{8}$$

Now we must adjust the limits. Plug each limit into $u = 8x - 4$:

$$u(x = 1) = 8(1) - 4 = 8 - 4 = 4$$
$$u(x = 3) = 8(3) - 4 = 24 - 4 = 20$$

The integral becomes:

$$\int_{x=1}^{3}(8x-4)^3\,dx = \int_{u=4}^{20} u^3\frac{du}{8} = \frac{1}{8}\int_{u=4}^{20} u^3\,du$$

We can integrate this following the method from Chapter 2:

$$\frac{1}{8}\int_{u=4}^{20} u^3\,du = \frac{1}{8}\left[\frac{u^{3+1}}{3+1}\right]_{u=4}^{20} = \frac{1}{8}\left[\frac{u^4}{4}\right]_{u=4}^{20} = \frac{1}{8}\frac{(20)^4}{4} - \frac{1}{8}\frac{(4)^4}{4} = 5000 - 8 = 4992$$

Example: Perform the following definite integral: $\int_{x=0}^{\pi/3}\sin\left(2x + \frac{\pi}{6}\right)dx$.

We make the substitution $u = 2x + \frac{\pi}{6}$. Take an implicit derivative:

- On the left-hand side: $\frac{d}{du}(u) = 1$. Multiply by du to get $1\,du = du$.
- On the right-hand side: $\frac{d}{dx}\left(2x + \frac{\pi}{6}\right) = 2$. Multiply by dx to get $2\,dx$.

The implicit derivative of $u = 2x + \frac{\pi}{6}$ is therefore $du = 2\,dx$. Solve for dx to get $dx = \frac{du}{2}$.

$$u = 2x + \frac{\pi}{6} \quad , \quad dx = \frac{du}{2}$$

Now we must adjust the limits. Plug each limit into $u = 2x + \pi/6$:

$$u(x = 0) = 2(0) + \frac{\pi}{6} = \frac{\pi}{6}$$

$$u\left(x = \frac{\pi}{3}\right) = 2\left(\frac{\pi}{3}\right) + \frac{\pi}{6} = \frac{2\pi}{3} + \frac{\pi}{6} = \frac{4\pi}{6} + \frac{\pi}{6} = \frac{4\pi + \pi}{6} = \frac{5\pi}{6}$$

The integral becomes:

$$\int_{x=0}^{\pi/3}\sin\left(2x + \frac{\pi}{6}\right)dx = \int_{u=\pi/6}^{5\pi/6}\sin(u)\frac{du}{2} = \frac{1}{2}\int_{u=\pi/6}^{5\pi/6}\sin(u)\,du$$

We can integrate this following the method from Chapter 9. These angles are in **radians**.

$$\frac{1}{2}\int_{u=\pi/6}^{5\pi/6}\sin(u)\,du = \frac{1}{2}[-\cos u]_{u=\pi/6}^{5\pi/6} = \frac{1}{2}\left(-\cos\frac{5\pi}{6}\right) - \frac{1}{2}\left(-\cos\frac{\pi}{6}\right) = \frac{1}{2}\frac{\sqrt{3}}{2} + \frac{1}{2}\frac{\sqrt{3}}{2} = \frac{\sqrt{3}}{2}$$

Example: Perform the following definite integral:

$$\int_{x=\sqrt{3}}^{3} \frac{dx}{\sqrt{x^2+9}}$$

Write $x^2 + 9$ as $x^2 + 3^2$. Now it looks like $x^2 + a^2$ with $a = 3$. Following the suggestion in the strategy, we make the substitution $x = 3\tan u$. Take an implicit derivative:

- On the left-hand side: $\frac{d}{dx}(x) = 1$. Multiply by dx to get $1\,dx = dx$.
- On the right-hand side: $\frac{d}{du}(3\tan u) = 3\sec^2 u$. Multiply by du to get $3\sec^2 u\ du$.

The implicit derivative of $x = 3\tan u$ is therefore $dx = 3\sec^2 u\ du$. (If you need to review your trig derivatives, see Chapter 9.) We will make the following pair of substitutions in the original integral:

$$x = 3\tan u \quad , \quad dx = 3\sec^2 u\ du$$

Now we must adjust the limits. Solve for u to obtain $u = \tan^{-1}\left(\frac{x}{3}\right)$:

$$u\left(x = \sqrt{3}\right) = \tan^{-1}\left(\frac{\sqrt{3}}{3}\right) = 30°$$

$$u(x = 3) = \tan^{-1}\left(\frac{3}{3}\right) = \tan^{-1}(1) = 45°$$

The integral becomes:

$$\int_{x=\sqrt{3}}^{3} \frac{dx}{\sqrt{x^2+9}} = \int_{u=30°}^{45°} \frac{3\sec^2 u\ du}{\sqrt{(3\tan u)^2+9}} = \int_{u=30°}^{45°} \frac{3\sec^2 u\ du}{\sqrt{9\tan^2 u+9}}$$

Factor out the 9 to write $\sqrt{9\tan^2 u + 9} = \sqrt{9(\tan^2 u + 1)} = 3\sqrt{\tan^2 u + 1}$

$$\int_{x=\sqrt{3}}^{3} \frac{dx}{\sqrt{x^2+9}} = \int_{u=30°}^{45°} \frac{3\sec^2 u\ du}{3\sqrt{\tan^2 u+1}}$$

Use the trig identity $\tan^2 u + 1 = \sec^2 u$ to replace $\sqrt{\tan^2 u + 1}$ with $\sqrt{\sec^2 u}$.

$$\int_{x=\sqrt{3}}^{3} \frac{dx}{\sqrt{x^2+9}} = \int_{u=30°}^{45°} \frac{\sec^2 u\ du}{\sqrt{\sec^2 u}} = \int_{u=30°}^{45°} \frac{\sec^2 u\ du}{\sec u} = \int_{u=30°}^{45°} \sec u\, du$$

We can integrate this following the method from Chapter 9:

$$\int_{u=30°}^{45°} \sec u\, du = [\ln|\sec u + \tan u|]_{u=30°}^{45°} = \ln|\sec 45° + \tan 45°| - \ln|\sec 30° + \tan 30°|$$

$$\int_{u=30°}^{45°} \sec u\, du = \ln\left|\sqrt{2}+1\right| - \ln\left|\frac{2\sqrt{3}}{3} + \frac{\sqrt{3}}{3}\right| = \ln\left(\frac{\sqrt{2}+1}{\frac{2\sqrt{3}}{3}+\frac{\sqrt{3}}{3}}\right) = \ln\left(\frac{\sqrt{2}+1}{\sqrt{3}}\right) = \ln\left(\frac{\sqrt{6}+\sqrt{3}}{3}\right)$$

We used the logarithm identity $\ln\left(\frac{x}{y}\right) = \ln x - \ln y$ and we rationalized the denominator.

152. Perform the following definite integral.

$$\int_{x=0}^{15} \left(\frac{x}{5} - 1\right)^4 dx =$$

Answer: 33

153. Perform the following definite integral (where the angles are in **radians**).

$$\int_{x=0}^{2\pi} \cos\left(\frac{\pi}{2} - \frac{x}{6}\right) dx =$$

Want help? Check the hints section at the back of the book.

Answer: 3

154. Perform the following definite integral.

$$\int_{x=0}^{\sqrt{3}} \frac{dx}{\sqrt{1+x^2}} =$$

Answer: $\ln(2+\sqrt{3})$

155. Perform the following definite integral.

$$\int_{x=0}^{2} \sqrt{4-x^2}\, dx =$$

Want help? Check the hints section at the back of the book.

Answer: π

Strategy for Performing Multiple Integrals

To perform a **double integral** or a **triple integral**, follow these steps:

1. Feel free to reverse the order of the differentials. For example, it doesn't really matter whether you write $\iint f(x,y)\,dxdy$ or $\iint f(x,y)\,dydx$. What matters (if there is a variable limit) is which integral you do first. The order of the differentials does not tell you which integral to do first. Look at the limits for this (see Step 2).
2. If any of the integrals has a variable limit, you must perform that integration first. The following examples will help you decide which integral to do first.
   - In the example below, the upper y-limit is a function of x. Since the y-limit has a variable, you must integrate over y before integrating over x.
   $$\int_{x=0}^{1}\int_{y=0}^{x^2} f(x,y)\,dxdy = \int_{x=0}^{1}\left(\int_{y=0}^{x^2} f(x,y)\,dy\right)dx$$
   - In the example below, the upper x-limit is a function of y. Since the x-limit has a variable, you must integrate over x before integrating over y.
   $$\int_{x=0}^{4y}\int_{y=-8}^{8} f(x,y)\,dxdy = \int_{y=-8}^{8}\left(\int_{x=0}^{4y} f(x,y)\,dx\right)dy$$
   - In the example below, all of the limits are constants. In this case, you can do the integrals in any order.
   $$\int_{x=0}^{2}\int_{y=-1}^{1} f(x,y)\,dxdy = \int_{x=0}^{2}\left(\int_{y=-1}^{1} f(x,y)\,dy\right)dx = \int_{y=-1}^{1}\left(\int_{x=0}^{2} f(x,y)\,dx\right)dy$$

   Note: In the above equations, on the right-hand side the order of the differentials does matter, as the integrals have been separated (that is, they were reorganized to show which integral will be performed first).
3. When integrating over one variable, treat the other independent variables as constants. For example:
   - When integrating over x, treat y and z the same way as you would treat any other constants.
   - When integrating over y, treat x and z the same way as you would treat any other constants.
   - When integrating over z, treat x and y the same way as you would treat any other constants.
4. Evaluate the first integral over its limits before performing the second integral.
5. A triple integral works the same way as a double integral, except that you do it in three stages, one integral at a time.

The following examples illustrate how to perform a double or triple integral.

Example: Perform the following double integral:

$$\int_{x=0}^{2}\int_{y=0}^{\sqrt{x}} x^2 y^3\, dxdy$$

Since the upper y-limit is a function of x, we must carry out the y-integration first.

$$\int_{x=0}^{2}\int_{y=0}^{\sqrt{x}} x^2 y^3\, dxdy = \int_{x=0}^{2}\left(\int_{y=0}^{\sqrt{x}} x^2 y^3\, dy\right)dx$$

When integrating over y, we treat x like any other constant. Therefore, we can factor x^2 out of the y integral (but be careful not to pull x^2 out of the x integral). We're applying the same concept as $\int cf(y)\,dy = c\int f(y)\,dy$.

$$\int_{x=0}^{2}\left(\int_{y=0}^{\sqrt{x}} x^2 y^3\, dy\right)dx = \int_{x=0}^{2} x^2\left(\int_{y=0}^{\sqrt{x}} y^3\, dy\right)dx$$

To help make this clear, we will carry out the complete definite integral over y in parentheses before proceeding. Students who do this tend to make fewer mistakes when they're learning how to perform multiple integrals.

$$\int_{x=0}^{2} x^2\left(\int_{y=0}^{\sqrt{x}} y^3\, dy\right)dx = \int_{x=0}^{2} x^2\left(\left[\frac{y^{3+1}}{3+1}\right]_{y=0}^{y=\sqrt{x}}\right)dx = \int_{x=0}^{2} x^2\left(\left[\frac{y^4}{4}\right]_{y=0}^{y=\sqrt{x}}\right)dx$$

$$\int_{x=0}^{2} x^2\left(\left[\frac{y^4}{4}\right]_{y=0}^{y=\sqrt{x}}\right)dx = \int_{x=0}^{2} x^2\left(\frac{\left(\sqrt{x}\right)^4}{4} - \frac{(0)^4}{4}\right)dx = \int_{x=0}^{2} x^2\left(\frac{x^2}{4}\right)dx = \int_{x=0}^{2} \frac{x^4}{4}dx$$

Now we have a single integral which we can integrate using the method from Chapter 2.

$$\int_{x=0}^{2} \frac{x^4}{4}dx = \frac{1}{4}\left[\frac{x^{4+1}}{4+1}\right]_{x=0}^{x=2} = \frac{1}{4}\left[\frac{x^5}{5}\right]_{x=0}^{x=2} = \frac{2^5}{20} - \frac{0^5}{20} = \frac{32}{20} - 0 = \frac{32}{20} = \frac{8}{5}$$

Example: Perform the following double integral:

$$\int_{x=0}^{3}\int_{y=-1}^{1} 2xy^2\, dxdy$$

Since all of the limits are constants, in this example we are free to integrate in any order.

$$\int_{x=0}^{3}\int_{y=-1}^{1} 2xy^2\, dxdy = \int_{x=0}^{3}\left(\int_{y=-1}^{1} 2xy^2\, dy\right)dx$$

When integrating over y, we treat x like a constant. We can factor $2x$ out of the y integral.

$$\int_{x=0}^{3} 2x\left(\int_{y=-1}^{1} y^2\, dy\right)dx = \int_{x=0}^{3} 2x\left(\left[\frac{y^3}{3}\right]_{y=-1}^{y=1}\right)dx = \int_{x=0}^{3} 2x\left(\frac{1^3}{3} - \frac{(-1)^3}{3}\right)dx$$

$$\int_{x=0}^{3} 2x\left(\frac{1}{3}+\frac{1}{3}\right)dx = \int_{x=0}^{3} 2x\frac{2}{3}dx = \frac{4}{3}\int_{x=0}^{3} x\, dx = \frac{4}{3}\left[\frac{x^2}{2}\right]_{x=0}^{x=3} = \frac{4}{3}\frac{(3)^2}{2} - \frac{4}{3}\frac{(0)^2}{2} = 6$$

Example: Perform the following double integral:

$$\int_{x=y^2}^{4y^2}\int_{y=0}^{6}\frac{y}{\sqrt{x}}dxdy$$

Since the x-limits are functions of y, we must carry out the x-integration first.

$$\int_{x=y^2}^{4y^2}\int_{y=0}^{6}\frac{y\,dxdy}{\sqrt{x}} = \int_{y=0}^{6}\left(\int_{x=y^2}^{4y^2}\frac{ydx}{\sqrt{x}}\right)dy$$

When integrating over x, we treat y like a constant. We can factor y out of the x integral.

$$\int_{y=0}^{6}\left(\int_{x=y^2}^{4y^2}\frac{y}{\sqrt{x}}dx\right)dy = \int_{y=0}^{6}y\left(\int_{x=y^2}^{4y^2}\frac{dx}{\sqrt{x}}\right)dy$$

As we learned in Chapter 2, we write $\frac{1}{\sqrt{x}}$ as $x^{-1/2}$. We then compare $x^{-1/2}$ to ax^b to see that $a = 1$ and $b = -\frac{1}{2}$. Since the anti-derivative of ax^b is $\frac{ax^{b+1}}{b+1}$, it follows that the anti-derivative of $x^{-1/2}$ is $\frac{(1)x^{-1/2+1}}{-1/2+1} = \frac{x^{1/2}}{1/2} = \frac{\sqrt{x}}{1/2} = \sqrt{x} \div \frac{1}{2} = \sqrt{x}\left(\frac{2}{1}\right) = 2\sqrt{x}$. What this means is that $\int \frac{dx}{\sqrt{x}} = 2\sqrt{x}$.

$$\int_{y=0}^{6}y\left(\int_{x=y^2}^{4y^2}\frac{dx}{\sqrt{x}}\right)dy = \int_{y=0}^{6}y\left(\left[2\sqrt{x}\right]_{x=y^2}^{x=4y^2}\right)dy$$

Evaluate the expression $2\sqrt{x}$ from $x = y^2$ to $x = 4y^2$.

$$\int_{y=0}^{6}y\left(\left[2\sqrt{x}\right]_{x=y^2}^{x=4y^2}\right)dy = \int_{y=0}^{6}y\left(2\sqrt{4y^2} - 2\sqrt{y^2}\right)dy$$

Factor the 4 out of the first squareroot: $2\sqrt{4y^2} = 2\sqrt{4}\sqrt{y^2} = 2(2)\sqrt{y^2} = 4\sqrt{y^2}$.

$$\int_{y=0}^{6}y\left(2\sqrt{4y^2} - 2\sqrt{y^2}\right)dy = \int_{y=0}^{6}y\left(4\sqrt{y^2} - 2\sqrt{y^2}\right)dy$$

Recognize that $\sqrt{y^2} = y$.

$$\int_{y=0}^{6}y\left(4\sqrt{y^2} - 2\sqrt{y^2}\right)dy = \int_{y=0}^{6}y(4y - 2y)\,dy = \int_{y=0}^{6}y(2y)\,dy = \int_{y=0}^{6}2y^2\,dy$$

Now we have a single integral which we can integrate using the method from Chapter 2.

$$\int_{y=0}^{6}2y^2\,dy = 2\left[\frac{y^{2+1}}{2+1}\right]_{y=0}^{y=6} = 2\left[\frac{y^3}{3}\right]_{y=0}^{y=6} = \frac{2(6)^3}{3} - \frac{2(0)^3}{3} = 144$$

Example: Perform the following triple integral:

$$\int_{x=-1}^{2}\int_{y=0}^{x^2}\int_{z=0}^{y} xy\,dx\,dy\,dz$$

Since the y- and z-limits both involve variables, we must perform the y and z integrals first. Furthermore, in this example we must perform the z integral **before** performing the y integral. Why? Because there is a y in the upper limit of the z integral, we won't be able to integrate over y until we perform the z-integration. Thus we begin with the z-integration.

$$\int_{x=-1}^{2}\int_{y=0}^{x^2}\left(\int_{z=0}^{y} xy\,dz\right)dy\,dx$$

When integrating over z, we treat the independent variables x and y like constants. We can factor xy out of the z integral. We're left with a trivial integral: $\int dz = z$.

$$\int_{x=-1}^{2}\int_{y=0}^{x^2} xy\left(\int_{z=0}^{y} dz\right)dy\,dx = \int_{x=-1}^{2}\int_{y=0}^{x^2} xy\left([z]_{z=0}^{z=y}\right)dy\,dx = \int_{x=-1}^{2}\int_{y=0}^{x^2} xy(y-0)\,dy\,dx$$

This simplifies to:

$$\int_{x=-1}^{2}\int_{y=0}^{x^2} xy(y)\,dy\,dx = \int_{x=-1}^{2}\int_{y=0}^{x^2} xy^2\,dy\,dx$$

Now we have a double integral similar to the previous examples. We perform the y-integration next because there is a variable in the upper limit of the y integral.

$$\int_{x=-1}^{2}\int_{y=0}^{x^2} xy^2\,dy\,dx = \int_{x=-1}^{2}\left(\int_{y=0}^{x^2} xy^2\,dy\right)dx$$

We factor out the x since we treat the independent variable x as a constant when integrating over y.

$$\int_{x=-1}^{2} x\left(\int_{y=0}^{x^2} y^2\,dy\right)dx = \int_{x=-1}^{2} x\left(\left[\frac{y^{2+1}}{2+1}\right]_{y=0}^{x^2}\right)dx = \int_{x=-1}^{2} x\left(\left[\frac{y^3}{3}\right]_{y=0}^{x^2}\right)dx$$

$$\int_{x=-1}^{2} x\left(\left[\frac{y^3}{3}\right]_{y=0}^{x^2}\right)dx = \int_{x=-1}^{2} x\left(\frac{(x^2)^3}{3}-\frac{(0)^3}{3}\right)dx = \int_{x=-1}^{2} x\left(\frac{x^6}{3}\right)dx = \int_{x=-1}^{2}\frac{x^7}{3}dx$$

In the last two steps, we applied the rules $(x^m)^n = x^{mn}$ and $x^m x^n = x^{m+n}$.

$$\int_{x=-1}^{2}\frac{x^7}{3}dx = \left[\frac{x^{7+1}}{(3)(7+1)}\right]_{x=-1}^{2} = \left[\frac{x^8}{24}\right]_{x=-1}^{2} = \frac{2^8}{24}-\frac{(-1)^8}{24} = \frac{256}{24}-\frac{1}{24} = \frac{255}{24} = \frac{85}{8}$$

156. Perform the following double integral.

$$\int_{x=0}^{2}\int_{y=0}^{x^3} xy\,dxdy =$$

Answer: 16

157. Perform the following double integral.

$$\int_{x=0}^{5}\int_{y=0}^{4} x^3y^4\,dxdy =$$

Want help? Check the hints section at the back of the book.

Answer: 32,000

158. Perform the following double integral.

$$\int_{x=0}^{4}\int_{y=0}^{9}\sqrt{xy}\,dxdy =$$

Answer: 96

159. Perform the following double integral.

$$\int_{x=0}^{3}\int_{y=0}^{x}6(x-y)\,dxdy =$$

Want help? Check the hints section at the back of the book.

Answer: 27

160. Perform the following triple integral.

$$\int_{x=0}^{2}\int_{y=0}^{x^2}\int_{z=0}^{y} 27z^2\,dx\,dy\,dz =$$

Answer: 128

161. Perform the following triple integral.

$$\int_{x=0}^{z}\int_{y=0}^{\sqrt{x}}\int_{z=0}^{9} 14xy^2\,dx\,dy\,dz =$$

Want help? Check the hints section at the back of the book.

Answer: 5832

Essential Concepts

Physics problems are often easier to solve if you work with a coordinate system well-suited to the geometry. We will see examples of this in Chapters 27 and 32.

- **Cartesian coordinates**. A problem featuring straight edges and flat sides (such as a triangle or cube) is often easier to solve using Cartesian coordinates (also called rectangular coordinates).

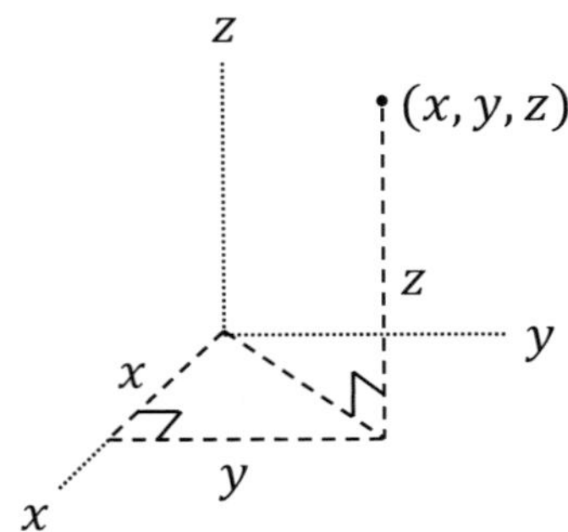

  - x is measured along the x-axis. In the diagram above, x is out of the page.
  - y is measured along the y-axis. In the diagram above, y is to the right.
  - z is measured along the z-axis. In the diagram above, z is up.
- **Polar coordinates**. A problem featuring a circular shape (such as a disc, ring, or circular arc) is often easier to solve using 2D polar coordinates:

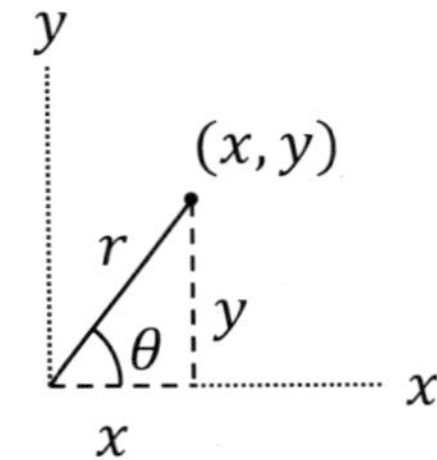

  - r extends radially outward from the origin.
  - θ is measured counterclockwise from the $+x$-axis.
- **Cylindrical coordinates**. A problem featuring cylindrical symmetry (such as a cylinder, cone, or helix) is often easier to solve using cylindrical coordinates:

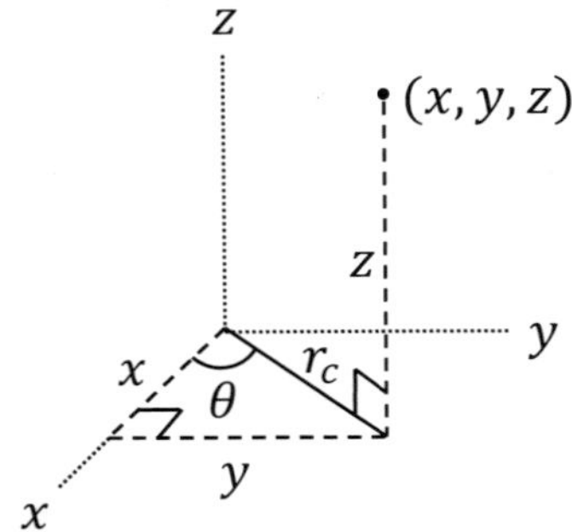

- r_c extends horizontally outward from the z-axis (**not** from the origin).
- θ is measured counterclockwise from the $+x$-axis.
- z is measured along the z-axis. In the previous diagram, z is up.

Cylindrical coordinates are merely a combination of 2D polar coordinates and the z-coordinate.

- **Spherical coordinates.** A problem featuring a spherical shape (such as a solid sphere or a hemispherical shell) is often easier to solve using spherical coordinates.

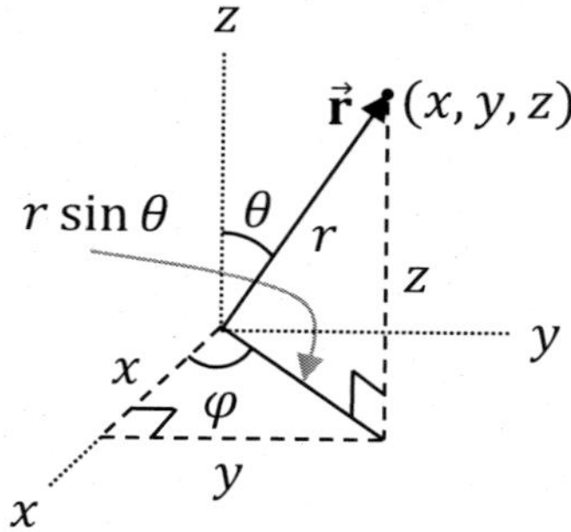

- r extends radially outward from the origin.
- θ is measured down from the $+z$-axis.
- φ is measured counterclockwise from the $+x$-axis after projecting down onto the xy plane. The symbol φ is the lowercase Greek letter phi.

Important Distinctions

Note that the r of spherical coordinates is different from the r_c of cylindrical coordinates: r extends outward from the origin, whereas r_c extends horizontally from the z-axis.

The roles of the angles θ and φ are often swapped in physics textbooks compared to math textbooks. Therefore, if you study physics and vector calculus simultaneously, for example, you may have an additional challenge of trying to keep these two angles straight.

In physics, setting φ constant creates a great vertical circle called a longitude, whereas setting θ constant creates a horizontal circle called a latitude. As shown below, except for the equator, all of the latitudes are smaller than the longitudes.

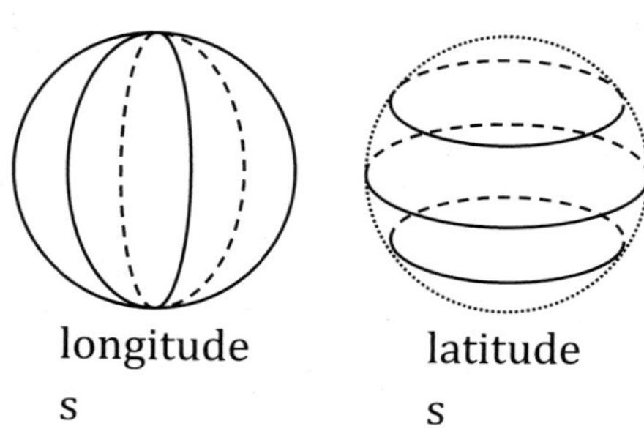

longitude
s

latitude
s

Relations between Different Coordinate Systems

The 2D polar coordinates r and θ are related to the Cartesian coordinates x and y by:

$$x = r\cos\theta \quad , \quad y = r\sin\theta$$

$$r = \sqrt{x^2 + y^2} \quad , \quad \theta = \tan^{-1}\left(\frac{y}{x}\right)$$

The cylindrical coordinates r_c, θ, and z are related to the Cartesian coordinates x, y, and z by (where z is the same in both coordinate systems):

$$x = r_c\cos\theta \quad , \quad y = r_c\sin\theta$$

$$r_c = \sqrt{x^2 + y^2} \quad , \quad \theta = \tan^{-1}\left(\frac{y}{x}\right)$$

The spherical coordinates r, θ, and φ are related to the Cartesian coordinates x, y, and z by:

$$x = r\sin\theta\cos\varphi \quad , \quad y = r\sin\theta\sin\varphi \quad , \quad z = r\cos\theta$$

$$r = \sqrt{x^2 + y^2 + z^2} \quad , \quad \theta = \cos^{-1}\left(\frac{z}{r}\right) \quad , \quad \varphi = \tan^{-1}\left(\frac{y}{x}\right)$$

Differential Elements

The differential arc length is represented in general by ds. If you integrate over ds, you get a finite length of arc: $s = \int ds$. The arc length s is finite, whereas ds is infinitesimal.

- For a straight line parallel to the x-axis, use $ds = dx$. For a straight line parallel to the y-axis, use $ds = dy$.
- For a circular arc of radius R, in 2D polar coordinates use $ds = Rd\theta$. **Notation**: Uppercase R is a constant (the radius), whereas lowercase r is a variable.

The differential area element is represented in general by dA. If you integrate over dA, you get surface area: $A = \int dA$. This will be done as a double integral in the examples.

- For a flat surface bounded entirely by straight lines (like a triangle or rectangle) lying in the xy plane, use $dA = dxdy$.
- If one of the sides of a flat surface is circular (like a solid semicircle or like a slice of pizza), in 2D polar coordinates use $dA = rdrd\theta$.
- If the surface is spherical (like a thin hemispherical shell) with radius R, in spherical coordinates use $dA = R^2\sin\theta\, d\theta d\varphi$. Recall that R is a constant (unlike r).

The differential volume element is represented in general by dV. If you integrate over dV, you get volume: $V = \int dV$. This will be done as a triple integral in the examples.

- For a solid consisting of flat sides and straight edges (like a cube), use $dV = dxdydz$.
- For a cylindrical solid (like a cylinder or cone), use $dV = r_c dr_c d\theta dz$.
- For a spherical solid (like a solid hemisphere), use $dV = r^2\sin\theta\, drd\theta d\varphi$.

Strategy for Integrating over Arc Length, Surface Area, or Volume

To integrate over arc length, surface area, or volume, follow these steps:

1. First make the appropriate substitution for the differential element:
   - For a straight line parallel to the x-axis, $ds = dx$.
   - For a straight line parallel to the y-axis, $ds = dy$.
   - For a circular arc of radius R, $ds = Rd\theta$.
   - For a solid polygon like a rectangle or triangle, $dA = dxdy$.
   - For a solid semicircle or pie slice, $dA = rdrd\theta$.
   - For a very thin spherical shell of radius R, $dA = R^2 \sin\theta \, d\theta d\varphi$.
   - For a solid polyhedron like a cube $dV = dxdydz$.
   - For a solid cylinder or cone, $dV = r_c dr_c d\theta dz$.
   - For a portion of a solid sphere like a hemisphere, $dV = r^2 \sin\theta \, drd\theta d\varphi$.
2. An integral over ds is a single integral, an integral over dA is a double integral, and an integral over dV is a volume integral.
3. Set the limits of each integration variable that map out the region of integration, as illustrated in the examples.
4. Perform the integral using techniques from this chapter and also from Chapters 2 and 9.

Example: Find the circumference of a circle by integrating over arc length.

For a circular arc, $ds = Rd\theta$. This entails a single integral over θ. For a full circle, θ varies form 0 to 2π radians.

$$C = \int ds = \int_{\theta=0}^{2\pi} Rd\theta$$

When integrating over a circular arc length, the radius R is constant. (In contrast, when integrating over a surface area, with $dA = rdrd\theta$, lowercase r is a variable.)

$$C = R \int_{\theta=0}^{2\pi} d\theta = R[\theta]_{\theta=0}^{2\pi} = R(2\pi - 0) = 2\pi R$$

The result is the usual equation for the circumference of a circle (Chapter 5).

Example: Find the area of the triangle illustrated below using a double integral.

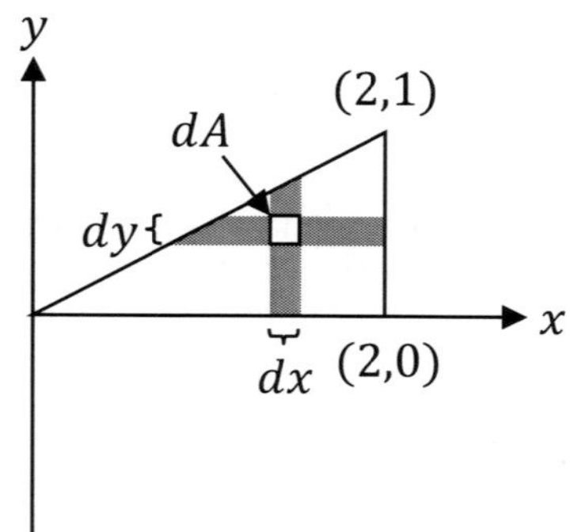

For a triangle, use $dA = dxdy$. This requires integrating over both x and y. We **don't** want to let x vary from 0 to 2 and let y vary from 0 to 1 because that would give us a rectangle instead of a triangle. If we let x vary from 0 to 2, observe that for a given value of x, y varies from 0 to the hypotenuse of the triangle, which is less than 1 (except when x reaches its upper limit).

To find the proper upper limit for y, we need the equation of the line that serves as the triangle's hypotenuse. That line has a slope equal to $\frac{y_2-y_1}{x_2-x_1} = \frac{1-0}{2-0} = \frac{1}{2}$ and a y-intercept of 0. Since the general equation for a line is $y = mx + b$, the equation for this line is $y = \frac{1}{2}x + 0$. For a given value of x, y will vary from 0 to $y = \frac{1}{2}x$, where the upper limit came from the equation for the line (of the hypotenuse). See the vertical gray band in the previous diagram. Now we have the integration limits.

$$A = \int dA = \int_{x=0}^{2} \int_{y=0}^{\frac{x}{2}} dydx$$

We must perform the y-integration first because it has x in its upper limit.

$$A = \int_{x=0}^{2} \left(\int_{y=0}^{\frac{x}{2}} dy \right) dx = \int_{x=0}^{2} \left([y]_{y=0}^{y=\frac{x}{2}} \right) dx = \int_{x=0}^{2} \left(\frac{x}{2} - 0 \right) dx = \int_{x=0}^{2} \frac{x}{2} dx$$

$$A = \int_{x=0}^{2} \frac{x}{2} dx = \left[\frac{x^2}{4} \right]_{x=0}^{x=2} = \frac{2^2}{4} - \frac{0^2}{4} = \frac{4}{4} = 1$$

Of course, we don't need calculus to find the area of a triangle. We could just use the formula one-half base times height (Chapter 5): $A = \frac{1}{2}bh = \frac{1}{2}(2)(1) = 1$. However, there are some integrals that can only be done with calculus. (On a similar note, while area can be calculated as a single integral, there are some physics integrals that can only be done as double or triple integrals, so multi-integration is a necessary skill.)

Example: Find the area of a circle using a double integral.

For a solid circle, use $dA = rdrd\theta$. This requires integrating over both r and θ.

Unlike in the previous example, r and θ each have constant limits:
- $0 \le r \le R$
- $0 \le \theta \le 2\pi$

Note that R is a constant (the radius of the circle), whereas r is a variable of integration. For any value of θ, the variable r ranges between 0 and R. Contrast this with the previous example where the upper limit of y depended on the value of x.

Substitute $dA = rdrd\theta$ into the area integral.

$$A = \int dA = \int_{r=0}^{R} \int_{\theta=0}^{2\pi} rdrd\theta$$

Since all of the limits are constant, we can do these integrals in any order.

$$A = \int_{r=0}^{R} \left(\int_{\theta=0}^{2\pi} rd\theta \right) dr$$

When integrating over θ, treat the independent variable r as a constant. This means that you can pull r out of the θ integral (but be careful not to pull r out of the r integral).

$$A = \int_{r=0}^{R} r \left(\int_{\theta=0}^{2\pi} d\theta \right) dr = \int_{r=0}^{R} r[\theta]_{\theta=0}^{2\pi}\, dr = \int_{r=0}^{R} r(2\pi - 0)\, dr = \int_{r=0}^{R} 2\pi r\, dr$$

$$A = \int_{r=0}^{R} 2\pi r\, dr = 2\pi \int_{r=0}^{R} r\, dr = 2\pi \left[\frac{r^2}{2}\right]_{r=0}^{R} = 2\pi \left(\frac{R^2}{2} - \frac{0^2}{2}\right) = 2\pi \frac{R^2}{2} = \pi R^2$$

As expected, the area of a circle is $A = \pi R^2$ (Chapter 5).

162. Find the area of the triangle illustrated below using a double integral.

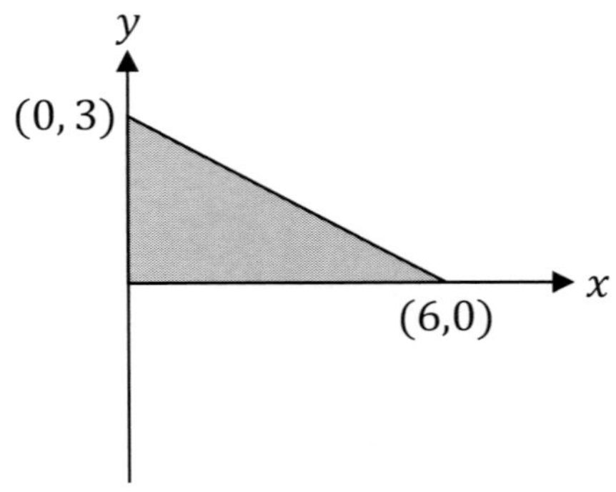

Want help? Check the hints section at the back of the book.

Answer: 9

163. A thick circular ring has inner radius R_1 and outer radius R_2. Find the area of this ring using a double integral.

Want help? Check the hints section at the back of the book.

Answer: $\pi(R_2^2 - R_1^2)$

27 CENTER OF MASS

Strategy for Finding the Center of Mass for a Discrete System

To find the center of mass of a system of discrete objects, follow these steps. (If instead you have a continuous object like a rod or disc, see pages 287-293.)

1. If they're not already given, determine the (x, y) coordinates of the center of each object in the system. For example, if an object has the shape of a circle or a square and the object is uniform, its center will lie at the geometric center of the circle or square. For other kinds of shapes, see the strategy on page 287.
2. Plug the mass and (x, y) coordinates of the center of each object into the formulas below:
$$x_{cm} = \frac{m_1 x_1 + m_2 x_2 + \cdots + m_N x_N}{m_1 + m_2 + \cdots + m_N}$$
$$y_{cm} = \frac{m_1 y_1 + m_2 y_2 + \cdots + m_N y_N}{m_1 + m_2 + \cdots + m_N}$$
The symbol $\cdots$ means "and so on" and the subscript N represents the number of objects in the system.
3. The **center of mass** of the system lies at (x_{cm}, y_{cm}).
4. If there is a z-coordinate, do the same thing to find z_{cm}. The coordinates of the center of mass will be (x_{cm}, y_{cm}, z_{cm}).

Symbols and SI Units

Symbol	Name	SI Units
m_i	mass of object i	kg
x_i	x-coordinate of object i	m
y_i	y-coordinate of object i	m
N	number of objects in the system	unitless
x_{cm}	x-coordinate of the center of mass	m
y_{cm}	y-coordinate of the center of mass	m
(x_{cm}, y_{cm})	coordinates of the center of mass	m

Example: A 30-kg monkey stands 12 m away from a 60-kg monkey. Where is the center of mass of the system?

Setup a coordinate system for this problem. We choose to put the origin on the 30-kg monkey and orient the $+x$-axis toward the 60-kg monkey. With this choice, $(x_1, y_1) = (0,0)$ and $(x_2, y_2) = (12 \text{ m}, 0)$. Use the formula for the center of mass with $N = 2$ objects:

$$x_{cm} = \frac{m_1 x_1 + m_2 x_2}{m_1 + m_2} = \frac{(30)(0) + (60)(12)}{30 + 60} = \frac{720}{90} = 8.0 \text{ m} \quad \text{and} \quad y_{cm} = 0$$

Example: Find the center of mass of the system illustrated below, where each square has uniform density and an edge length of 6.0 m.

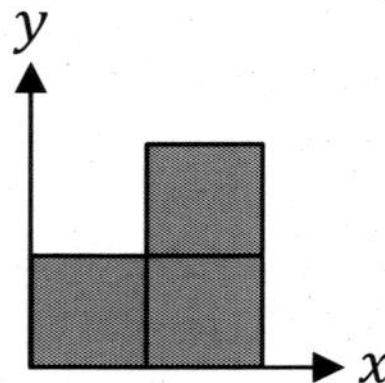

First determine the coordinates of the center of each square, given that each square has dimensions of 6.0 m × 6.0 m. It may help to visualize the center of each square (or draw a dot in the center of each square and label tick marks in 6.0-m increments along each axis).

- left square: (3.0 m, 3.0 m)
- right bottom square: (9.0 m, 3.0 m)
- top square: (9.0 m, 9.0 m)

Each square has the same mass, which we choose to call m_s. Use the formulas for center of mass with $N = 3$ objects:

$$x_{cm} = \frac{m_1 x_1 + m_2 x_2 + m_3 x_3}{m_1 + m_2 + m_3} = \frac{3m_s + 9m_s + 9m_s}{m_s + m_s + m_s} = \frac{21m_s}{3m_s} = 7.0 \text{ m}$$

$$y_{cm} = \frac{m_1 y_1 + m_2 y_2 + m_3 y_3}{m_1 + m_2 + m_3} = \frac{3m_s + 3m_s + 9m_s}{m_s + m_s + m_s} = \frac{15m_s}{3m_s} = 5.0 \text{ m}$$

The center of mass of the composite object is located at (7.0 m, 5.0 m).

Example: A 25-kg monkey stands at one end of a 24-m long, 125-kg plank. Where could a fulcrum be placed below the plank such that the system would be balanced?

It would balance on its center of mass. Setup a coordinate system for this problem. We place our origin on the free end of the plank with the $+x$-axis oriented toward the monkey. With this choice, the center of the plank lies at $(x_1, y_1) = (12 \text{ m}, 0)$ and the monkey is at $(x_2, y_2) = (24 \text{ m}, 0)$. Use the formula for the center of mass with $N = 2$ objects:

$$x_{cm} = \frac{m_1 x_1 + m_2 x_2}{m_1 + m_2} = \frac{(125)(12) + (25)(24)}{125 + 25} = \frac{2100}{150} = 14 \text{ m} \quad \text{and} \quad y_{cm} = 0$$

Example: As illustrated below, a circular hole with a radius of 3.0 m is cut out of an otherwise uniform circular solid disc with a radius of 6.0 m. Where is the center of mass of this object?

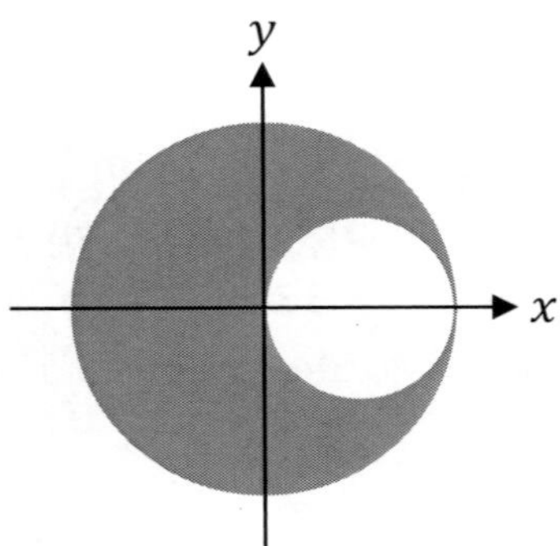

One way to solve this problem is to visualize the complete circle as the sum of the missing piece plus the shape with the hole cut out of it:

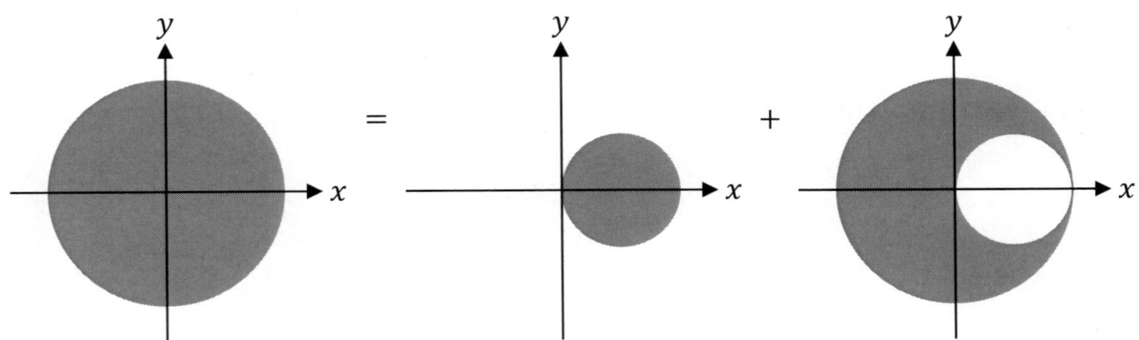

Let's write an equation for the center of mass of the large circle. (We'll only do this for x, since all three shapes obviously have $y_{cm} = 0$.)

$$x_{cm} = \frac{m_1 x_1 + m_2 x_2}{m_1 + m_2}$$

Here, $x_{cm} = 0$, since the large circle is centered about the origin. That's the not the center of mass we're solving for. What we're solving for is x_2 (the center of mass of the shape with the hole cut out of it). The symbol x_1 represents the center of mass of the small circle: $x_1 = 3.0$ m since the center of the small circle lies 3.0 m from the origin. Let's plug in these values for x_{cm} and x_1:

$$0 = \frac{3m_1 + m_2 x_2}{m_1 + m_2}$$

Since these object have uniform density, the masses will be proportional to the areas. Recall that the area of a circle is pi times radius squared (Chapter 5). The masses of the small (m_1) and large (m_L) circles can be expressed as:

$$m_1 = \sigma\pi R_S^2 = \sigma\pi(3)^2 = 9\sigma\pi$$
$$m_L = \sigma\pi R_L^2 = \sigma\pi(6)^2 = 36\sigma\pi$$

Notation: We're using a subscript L for the large circle, but subscripts 1 for the small circle and 2 for the original shape. The Greek symbol lowercase sigma (σ) is a constant of proportionality (mass is proportional to area).

The mass m_2 of the piece with the hole cut out of it can be found by subtracting the mass of the small circle (m_1) from the mass of the large circle (m_L):

$$m_2 = m_L - m_1 = 36\sigma\pi - 9\sigma\pi = 27\sigma\pi$$

Substitute these expressions for mass into the previous equation for the center of mass of the large circle:

$$0 = \frac{3(9\sigma\pi) + (27\sigma\pi)x_2}{9\sigma\pi + 27\sigma\pi} = \frac{3(9\sigma\pi) + (27\sigma\pi)x_2}{36\sigma\pi}$$

Multiply both sides by the denominator $36\sigma\pi$: It cancels on the right side and vanishes on the left side (since it's multiplying zero). The concept behind this algebra is the following: If you want a fraction to equal zero, just set the numerator equal to zero (since zero out of anything equals zero).

$$0 = 3(9\sigma\pi) + (27\sigma\pi)x_2$$
$$0 = 27\sigma\pi + 27\sigma\pi x_2$$

Divide both sides by $\sigma\pi$ and these constants will cancel out, too:

$$0 = 27 + 27x_2$$
$$-27 = 27x_2$$
$$x_2 = -\frac{27}{27} = -1.0 \text{ m}$$

164. A 150-g banana lies 250 cm from a 350-g bunch of bananas. Where is the center of mass of the system?

Answer: 175 cm from the 150-g banana (75 cm from the bunch of bananas)

165. Three bananas have the following masses and coordinates. Find the location of the center of mass of the system.

- 200 g at (7.0 m, 1.0 m)
- 500 g at (6.0 m, 3.0 m)
- 800 g at (2.0 m, –4.0 m)

Want help? Check the hints section at the back of the book.

Answer: (4.0 m, –1.0 m)

166. The T-shaped object shown below consists of a 20.0-cm long handle and a 4.0-cm wide end. Each piece has uniform density (but the densities of the two pieces differ). The mass of the handle is 6.0 kg, while the mass of the end is 18.0 kg. Where could a fulcrum be placed below the object such that the object would be balanced?

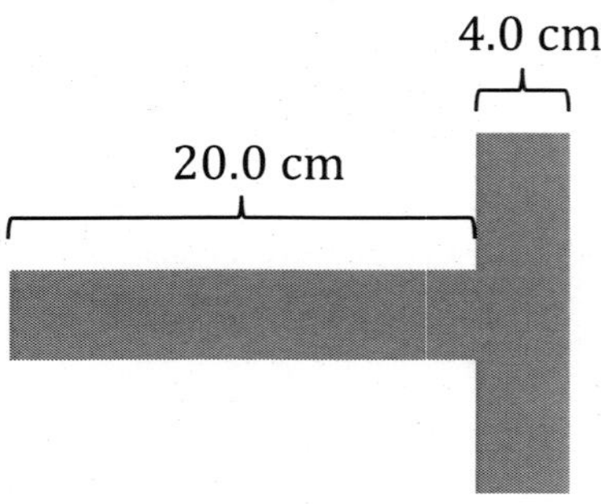

Want help? Check the hints section at the back of the book.

Answer: 1.0 cm left of where the handle meets the end

167. Find the center of mass of the system illustrated below in gray, where each gray square has the same uniform density and an edge length of 20 m.

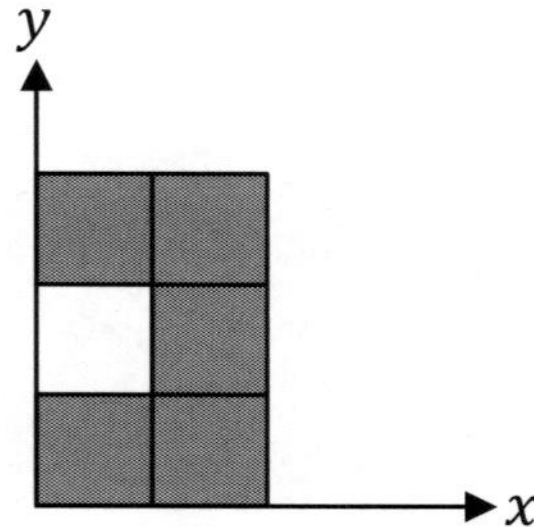

Want help? Check the hints section at the back of the book.

Answer: (22 m, 30 m)

168. As illustrated below, a circular hole with a radius of 2.0 m is cut out of an otherwise uniform circular solid disc with a radius of 8.0 m. Where is the center of mass of this object?

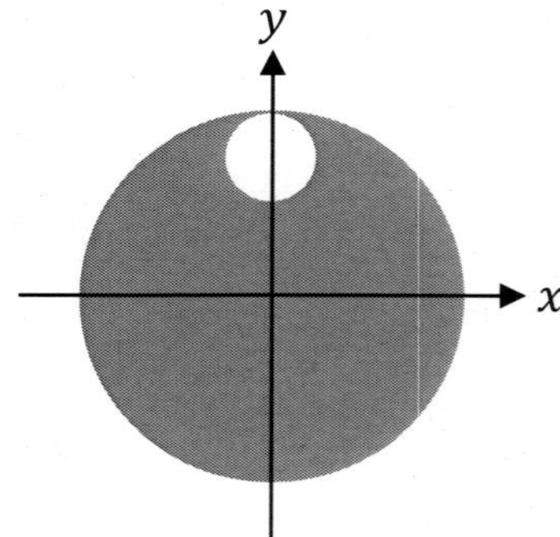

Want help? Check the hints section at the back of the book.

Answer: $(0, -\frac{2}{5}\text{ m})$

Strategy for Performing the Center of Mass Integral

To find the center of mass of a continuous object like a rod or triangle, follow these steps:

1. Draw the object. Draw and label a representative dm somewhere within the object. **Don't** draw dm at the origin or on an axis (unless the object is a rod and every point of the rod lies on an axis, then you have no choice). Visualize integrating over every possible dm in the object. Draw a horizontal line connecting dm to the y-axis and label this distance as x. Similarly, draw a vertical line connecting dm to the x-axis and label this distance as y. See the diagrams in the examples that follow.
2. Begin with the following center of mass integrals:
$$x_{cm} = \frac{1}{m}\int x\,dm \quad , \quad y_{cm} = \frac{1}{m}\int y\,dm \quad , \quad z_{cm} = \frac{1}{m}\int z\,dm$$
3. Make one of the following substitutions for dm, depending on the geometry:
    - $dm = \lambda ds$ for an arc length (like a rod or circular arc).
    - $dm = \sigma dA$ for a surface area (like a triangle, disc, or thin spherical shell).
    - $dm = \rho dV$ for a 3D solid (like a solid cube or a solid hemisphere).
4. Choose the appropriate coordinate system and make a substitution for ds, dA, or dV from Step 3 using the strategy from Chapter 26:
    - For a straight line parallel to the x-axis, $ds = dx$.
    - For a straight line parallel to the y-axis, $ds = dy$.
    - For a circular arc of radius R, $ds = Rd\theta$.
    - For a solid polygon like a rectangle or triangle, $dA = dxdy$.
    - For a solid semicircle (**not** a circular arc) or pie slice, $dA = rdrd\theta$.
    - For a very thin spherical shell of radius R, $dA = R^2 \sin\theta \, d\theta d\varphi$.
    - For a solid polyhedron like a cube, $dV = dxdydz$.
    - For a solid cylinder or cone, $dV = r_c dr_c d\theta dz$.
    - For a portion of a solid sphere like a hemisphere, $dV = r^2 \sin\theta \, drd\theta d\varphi$.
5. Is the density uniform or non-uniform?
    - If the density is uniform, you can pull λ, σ, or ρ out of the integral.
    - If the density is non-uniform, leave λ, σ, or ρ in the integral.
6. Which coordinate system did you choose in Step 4?
    - Cartesian coordinates: Leave x, y, and z as they are.
    - 2D polar coordinates: Replace x and y with the following expressions:
    $$x = r\cos\theta \quad , \quad y = r\sin\theta$$
    - Cylindrical coordinates: Replace x and y with the following expressions:
    $$x = r_c\cos\theta \quad , \quad y = r_c\sin\theta$$
    - Spherical coordinates: Replace x, y, and z with the following expressions:
    $$x = r\sin\theta\cos\varphi \quad , \quad y = r\sin\theta\sin\varphi \quad , \quad z = r\cos\theta$$

7. An integral over ds is a single integral, an integral over dA is a double integral, and an integral over dV is a triple integral. Set the limits of each integration variable that map out the region of integration, as illustrated in Chapter 26. Perform the integral using techniques from this chapter and also from Chapters 2, 9, and 26.
8. Perform the following integral to determine the total mass (m) of the object:

$$m = \int dm$$

Make the same substitutions as you made in Steps 3-6. Use the same limits of integration as you used in Step 7.
9. When you finish with Step 8, substitute your expression for m into your original center of mass integrals. Simplify the resulting expression.
10. The **center of mass** of the system lies at (x_{cm}, y_{cm}, z_{cm}).

Symbols and SI Units

Symbol	Name	SI Units
dm	differential mass element	kg
m	total mass of the object	kg
x, y, z	Cartesian coordinates of dm	m, m, m
r, θ	2D polar coordinates of dm	m, rad
r_c, θ, z	cylindrical coordinates of dm	m, rad, m
r, θ, φ	spherical coordinates of dm	m, rad, rad
λ	linear mass density (for an arc)	kg/m
σ	mass density for a surface area	kg/m^2
ρ	mass density for a 3D solid	kg/m^3
ds	differential arc length	m
dA	differential area element	m^2
dV	differential volume element	m^3
x_{cm}	x-coordinate of the center of mass	m
y_{cm}	y-coordinate of the center of mass	m

Note: The symbols λ, σ, and ρ are the lowercase Greek letters lambda, sigma, and rho.

Example: A non-uniform rod has one end at the origin, the other end at $(3.0\text{ m}, 0)$, and non-uniform density $\lambda = \beta x$, where β is a constant. Where is the center of mass of the rod?

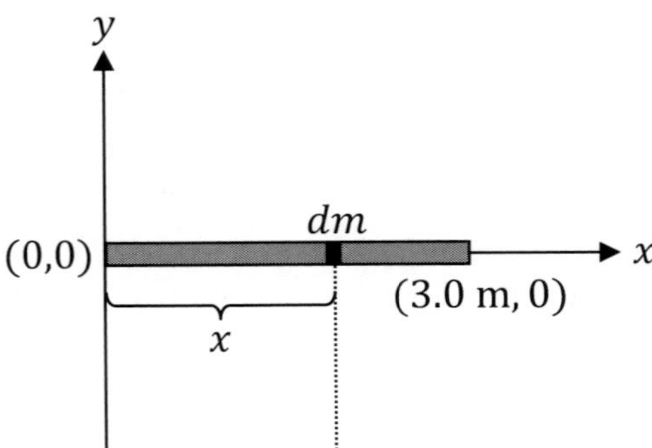

Begin with a labeled diagram. Draw a representative dm. Label x for that dm. When we perform the integration, we effectively integrate over every dm that makes up the rod. Begin with the center of mass integral:

$$x_{cm} = \frac{1}{m}\int x\, dm$$

For a thin rod, we write $dm = \lambda ds$.

$$x_{cm} = \frac{1}{m}\int x\lambda\, ds$$

For a rod lying along the x-axis, we work with Cartesian coordinates and write the differential arc length as $ds = dx$. The limits of integration correspond to the length of the rod: $0 \leq x \leq 3.0$ m.

$$x_{cm} = \frac{1}{m}\int_{x=0}^{3} x\lambda\, dx$$

Since the rod has non-uniform density, we **<u>can't</u>** pull λ out of the integral. Instead, we substitute the equation for λ given in the problem: $\lambda = \beta x$.

$$x_{cm} = \frac{1}{m}\int_{x=0}^{3} x(\beta x)\, dx$$

Since β is a constant, we can pull it out of the integral.

$$x_{cm} = \frac{\beta}{m}\int_{x=0}^{3} x^2\, dx = \frac{\beta}{m}\left[\frac{x^3}{3}\right]_{x=0}^{3} = \frac{\beta}{m}\left(\frac{3^3}{3} - \frac{0^3}{3}\right) = \frac{9\beta}{m}$$

Now integrate to find the total mass of the rod, using the same substitutions as before:

$$m = \int dm = \int \lambda\, ds = \int_{x=0}^{3} (\beta x)\, dx = \beta\int_{x=0}^{3} (x)\, dx = \beta\left[\frac{x^2}{2}\right]_{x=0}^{3} = \beta\left(\frac{3^2}{2} - \frac{0^2}{2}\right) = \frac{9\beta}{2}$$

Substitute this expression for m into the equation for x_{cm}.

$$x_{cm} = \frac{9\beta}{m} = 9\beta \div m = 9\beta \div \left(\frac{9\beta}{2}\right) = 9\beta\left(\frac{2}{9\beta}\right) = 2.0\text{ m}$$

This result, $x_{cm} = 2.0$ m, should make sense conceptually. If the rod were uniform, the center of mass would be at its geometric center, 1.5 m from the origin. However, the rod is non-uniform. Since the non-uniform density is proportional to x (according to $\lambda = \beta x$), there is more mass at the right end of the rod (where x has a larger value). Therefore, the center of mass should be to the right of the geometric center. Indeed, the center of mass ($x_{cm} = 2.0$ m) is to the right of the geometric center, as expected.

Example: Find the location of the center of mass of the uniform solid triangle illustrated below.

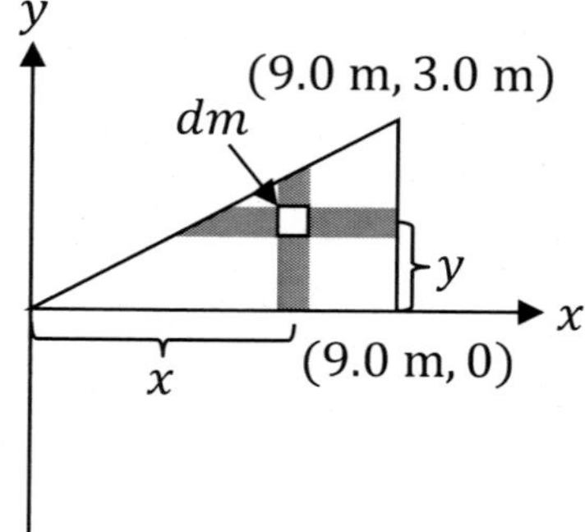

Begin with a labeled diagram. Draw a representative dm. Label x and y for that dm. When we perform the integration, we effectively integrate over every dm that makes up the solid triangle. Begin with the center of mass integrals:

$$x_{cm} = \frac{1}{m}\int x\,dm \quad , \quad y_{cm} = \frac{1}{m}\int y\,dm$$

For a solid triangle (as opposed to a wire bent into a triangle), we write $dm = \sigma dA$.

$$x_{cm} = \frac{1}{m}\int x\sigma\,dA \quad , \quad y_{cm} = \frac{1}{m}\int y\sigma\,dA$$

For a triangle, we work with Cartesian coordinates and write the differential area element as $dA = dxdy$. As with the example involving a triangle from Chapter 26, if we let x vary from 0 to 9.0 m, we need to find the equation of the hypotenuse for the upper limit of y. That line has a slope equal to $\frac{y_2-y_1}{x_2-x_1} = \frac{3-0}{9-0} = \frac{1}{3}$ and a y-intercept of 0. Since the general equation for a line is $y = mx + b$, the equation for this line is $y = \frac{1}{3}x + 0 = \frac{x}{3}$. For a given value of x, y will vary from 0 to $y = \frac{x}{3}$. See the vertical gray band in the diagram above. Since the problem states that the triangle has **uniform** density, we can pull σ out of the integrals.

$$x_{cm} = \frac{\sigma}{m}\int_{x=0}^{9}\int_{y=0}^{\frac{x}{3}} x\,dy\,dx \quad , \quad y_{cm} = \frac{\sigma}{m}\int_{x=0}^{9}\int_{y=0}^{\frac{x}{3}} y\,dy\,dx$$

We must perform the y-integration first because y has x in its upper limit. In the double integral for x_{cm}, we can pull x out of the y-integral, but not the x-integral.

$$x_{cm} = \frac{\sigma}{m} \int_{x=0}^{9} x \int_{y=0}^{\frac{x}{3}} dy\, dx = \frac{\sigma}{m} \int_{x=0}^{9} x \left(\int_{y=0}^{\frac{x}{3}} dy \right) dx = \frac{\sigma}{m} \int_{x=0}^{9} x \left([y]_{y=0}^{y=\frac{x}{3}} \right) dx$$

$$x_{cm} = \frac{\sigma}{m} \int_{x=0}^{9} x \left(\frac{x}{3} - 0\right) dx = \frac{\sigma}{m} \int_{x=0}^{9} \frac{x^2}{3} dx = \frac{\sigma}{m} \left[\frac{x^3}{(3)(3)}\right]_{x=0}^{x=9} = \frac{\sigma}{m}\left(\frac{9^3}{9} - \frac{0^3}{9}\right) = 9^2 \frac{\sigma}{m} = \frac{81\sigma}{m}$$

In the similar double integral for y_{cm}, we must again integrate over y first. This double integral has y in the integrand, and y can't come out of the y-integration. Compare these two similar double integrals closely.

$$y_{cm} = \frac{\sigma}{m} \int_{x=0}^{9} \int_{y=0}^{\frac{x}{3}} y\, dy\, dx = \frac{\sigma}{m} \int_{x=0}^{9} \left(\int_{y=0}^{\frac{x}{3}} y\, dy \right) dx = \frac{\sigma}{m} \int_{x=0}^{9} \left(\left[\frac{y^2}{2}\right]_{y=0}^{y=\frac{x}{3}} \right) dx$$

$$y_{cm} = \frac{\sigma}{m} \int_{x=0}^{9} \left[\frac{\left(\frac{x}{3}\right)^2}{2} - \frac{0^2}{2} \right] dx = \frac{\sigma}{m} \int_{x=0}^{9} \frac{x^2}{(9)(2)} dx = \frac{\sigma}{18m} \left[\frac{x^3}{3}\right]_{x=0}^{x=9} = \frac{\sigma}{m}\left(\frac{9^3}{54} - \frac{0^3}{54}\right) = \frac{27}{2}\frac{\sigma}{m}$$

Now integrate to find the total mass of the triangle, using the same substitutions as before:

$$m = \int dm = \int \sigma\, dA = \sigma \int_{x=0}^{9} \int_{y=0}^{\frac{x}{3}} dy\, dx = \sigma \int_{x=0}^{9} \left(\int_{y=0}^{\frac{x}{3}} dy \right) dx = \sigma \int_{x=0}^{9} \left([y]_{y=0}^{y=\frac{x}{3}} \right) dx$$

$$m = \sigma \int_{x=0}^{9} \left(\frac{x}{3} - 0\right) dx = \sigma \int_{x=0}^{9} \frac{x}{3} dx = \sigma \left[\frac{x^2}{(3)(2)}\right]_{x=0}^{x=9} = \sigma\left(\frac{9^2}{6} - \frac{0^2}{6}\right) = \frac{81\sigma}{6} = \frac{27\sigma}{2}$$

Substitute this expression for m into the equations for x_{cm} and y_{cm}.

$$x_{cm} = \frac{81\sigma}{m} = 81\sigma \div m = 81\sigma \div \left(\frac{27\sigma}{2}\right) = 81\sigma\left(\frac{2}{27\sigma}\right) = (3)(2) = 6.0 \text{ m}$$

$$y_{cm} = \frac{27}{2}\frac{\sigma}{m} = \frac{27\sigma}{2} \div m = \frac{27\sigma}{2} \div \left(\frac{27\sigma}{2}\right) = \frac{27\sigma}{2}\left(\frac{2}{27\sigma}\right) = 1.0 \text{ m}$$

The center of mass of the triangle lies at (6.0 m, 1.0 m). If you mark this location on the diagram on the previous page, it should agree with your expectations.

Example: A solid uniform semicircle (i.e. one-half of a solid circle) with a 3.0-cm radius has its straight side lying in the xy plane and extends in the positive y-direction. The center of the "full" circle would be at the origin. Determine the location of its center of mass.

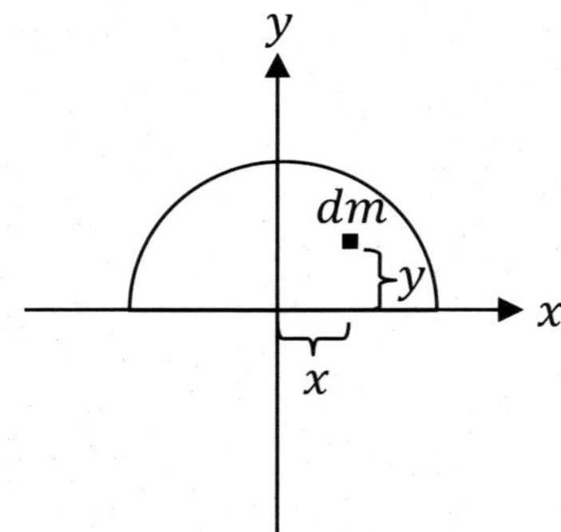

Begin with a labeled diagram. Draw a representative dm. Label x and y for that dm. When we perform the integration, we effectively integrate over every dm that makes up the solid semicircle. It should be clear from the symmetry of the semicircle that $x_{cm} = 0$. Begin with the center of mass integral:

$$y_{cm} = \frac{1}{m}\int y\, dm$$

For a solid semicircle (as opposed to a semicircular arc), we write $dm = \sigma dA$.

$$y_{cm} = \frac{1}{m}\int y\sigma\, dA$$

For a solid semicircle, use $dA = rdrd\theta$. This requires integrating over both r and θ. Unlike in the previous example, r and θ each have constant limits:

- $0 \leq r \leq 3.0$ cm
- $0 \leq \theta \leq \pi$ (since it's only half of a circle)

For any value of θ, the variable r ranges between 0 and 3.0 cm. Contrast this with the previous example where the upper limit of y depended on the value of x. Substitute $dA = rdrd\theta$ into the center of mass integral. We can pull σ out of the integral because the problem declares the density of the semicircle to be **<u>uniform</u>**.

$$y_{cm} = \frac{1}{m}\int y\sigma\, dA = \frac{\sigma}{m}\int y\, dA = \frac{\sigma}{m}\int_{r=0}^{3}\int_{\theta=0}^{\pi} y\,(rdrd\theta) = \frac{\sigma}{m}\int_{r=0}^{3}\int_{\theta=0}^{\pi} yr\, dr\, d\theta$$

We **<u>can't</u>** pull y out of the integration because y is a variable: The value of y is different for every dm that we integrate over (see the picture above). Instead, we use the substitution $y = r\sin\theta$ for 2D polar coordinates from Chapter 26. Since all of the limits are constant, we can do these integrals in any order.

$$y_{cm} = \frac{\sigma}{m}\int_{r=0}^{3}\int_{\theta=0}^{\pi} (r\sin\theta)\, r\, dr\, d\theta = \frac{\sigma}{m}\int_{r=0}^{3}\left(\int_{\theta=0}^{\pi} r^2\sin\theta\, d\theta\right) dr$$

When integrating over θ, treat the independent variable r as a constant. This means that you can pull r out of the θ integral (but be careful not to pull r out of the r integral).

$$y_{cm} = \frac{\sigma}{m} \int_{r=0}^{3} r^2 \left(\int_{\theta=0}^{\pi} \sin\theta \, d\theta \right) dr$$

The integral over the sine function is given in Chapter 9.

$$y_{cm} = \frac{\sigma}{m} \int_{r=0}^{3} r^2 [-\cos\theta]_{\theta=0}^{\pi} \, dr = \frac{\sigma}{m} \int_{r=0}^{3} r^2 [-\cos\pi - (-\cos 0)] \, dr$$

$$y_{cm} = \frac{\sigma}{m} \int_{r=0}^{3} r^2 [-(-1) + 1] \, dr = \frac{\sigma}{m} \int_{r=0}^{3} r^2 (1 + 1) \, dr = \frac{\sigma}{m} \int_{r=0}^{3} 2r^2 \, dr = \frac{2\sigma}{m} \int_{r=0}^{3} r^2 \, dr$$

$$y_{cm} = \frac{2\sigma}{m} \left[\frac{r^3}{3} \right]_{r=0}^{3} = \frac{2\sigma}{m} \left(\frac{3^3}{3} - \frac{0^3}{3} \right) = \frac{2\sigma}{m} \left(\frac{27}{3} \right) = \frac{2\sigma}{m} (9) = \frac{18\sigma}{m}$$

Now integrate to find the total mass of the semicircle, using the same substitutions as before:

$$m = \int dm = \int \sigma \, dA = \sigma \int_{r=0}^{3} \int_{\theta=0}^{\pi} r \, dr \, d\theta = \sigma \int_{r=0}^{3} r \left(\int_{\theta=0}^{\pi} d\theta \right) dr = \sigma \int_{r=0}^{3} r[\theta]_{\theta=0}^{\pi} \, dr$$

$$m = \sigma \int_{r=0}^{3} r(\pi - 0) \, dr = \sigma \int_{r=0}^{3} \pi r \, dr = \sigma\pi \int_{r=0}^{3} r \, dr = \sigma\pi \left[\frac{r^2}{2} \right]_{r=0}^{3} = \sigma\pi \left(\frac{3^2}{2} - \frac{0^2}{2} \right)$$

$$m = \sigma\pi \left(\frac{9}{2} \right) = \frac{9\pi\sigma}{2}$$

Substitute this expression for m into the equation for y_{cm}.

$$y_{cm} = \frac{18\sigma}{m} = 18\sigma \div m = 18\sigma \div \left(\frac{9\pi\sigma}{2} \right) = 18\sigma \left(\frac{2}{9\pi\sigma} \right) = \frac{4}{\pi} \text{ cm}$$

We can check whether or not this result ($y_{cm} = \frac{4}{\pi}$ cm) seems reasonable. Look at the diagram on the previous page. There is clearly more mass near the bottom than at the top, so y_{cm} should be less than half the radius. Half the radius is $\frac{R}{2} = \frac{3}{2} = 1.5$ cm. Compare that with $\frac{4}{\pi}$ cm, which is 1.27 cm when expressed as a decimal. Indeed, 1.27 cm is less than 1.5 cm, as expected.

169. A non-uniform rod has one end at the origin, the other end at $(0, 5.0 \text{ m})$, and non-uniform density $\lambda = \beta\sqrt{y}$, where β is a constant. Where is the center of mass of the rod?

Want help? Check the hints section at the back of the book.

Answer: $(0, 3.0 \text{ m})$

170. Find the location of the center of mass of the uniform solid triangle illustrated below.

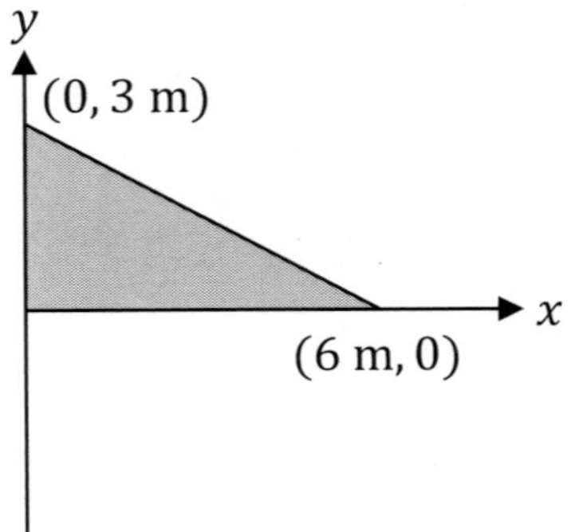

Want help? Check the hints section at the back of the book.

Answer: (2.0 m, 1.0 m)

171. A thin uniform ring with a 6.0-cm radius lies in the xy plane centered about the origin. A monkey cuts the thin ring in half along the x-axis and discards the negative y-portion, such that only a thin semicircular arc remains. Determine the location of its center of mass. **Note**: Unlike the similar example, this is an arc length (one-half the circumference of a circle), and **<u>not</u>** a solid semicircle.

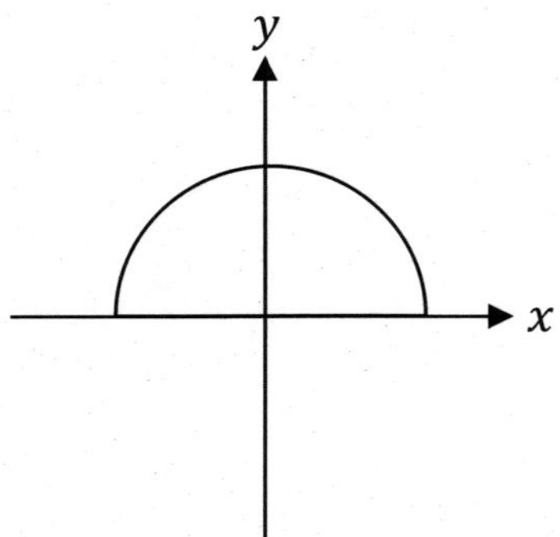

Want help? Check the hints section at the back of the book.

Answer: $\left(0, \frac{12}{\pi}\text{ cm}\right)$

172. Monkey Isle is the region bounded by the curve $y = x^2$ and the line $y = 4$. Monkey Isle has uniform density $\sigma = 2$. SI units have been suppressed. A monkey pirate buried a treasure in the center of mass of Monkey Isle. Find the (x, y) coordinates of the buried treasure.

Want help? Check the hints section at the back of the book.

Answer: $\left(0, \frac{12}{5} \text{ m}\right)$

173. The shaded semicircular ring illustrated below has an inner radius of 7.0 cm, an outer radius of 14.0 cm, and a non-uniform mass density $\sigma(r) = \beta r$, where β is a constant. Find the location of its center of mass.

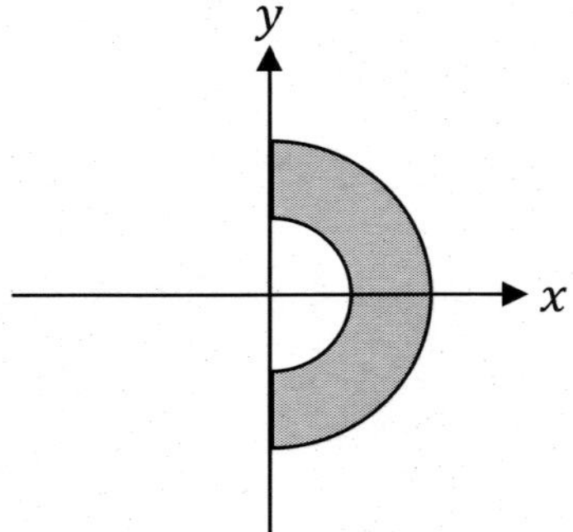

Want help? Check the hints section at the back of the book.

Answer: $\left(\frac{45}{2\pi}\text{ cm}, 0\right)$

28 UNIFORM ANGULAR ACCELERATION

Relevant Terminology

Angular velocity – the instantaneous rate at which angle is swept out as measured from the center of a circle.
Tangential velocity – the component of velocity tangential to a curved path.
Arc length – the distance traveled along a curved path.
Angular acceleration – the instantaneous rate at which angular velocity is changing.
Tangential acceleration – the instantaneous rate at which speed is changing.
Centripetal acceleration – the instantaneous rate at which the direction is changing.
Acceleration – the instantaneous rate at which velocity is changing.
Uniform angular acceleration – motion for which the angular acceleration is constant, meaning that the angular velocity changes at a constant rate.

Important Distinctions

Angular and tangential quantities have different units. Pay attention to the units to distinguish between similar quantities:

- **Angular displacement** ($\Delta\theta$) is in rad while arc length (Δs) is in m.
- **Angular velocity** (ω) is in rad/s while tangential velocity (v_T) is in m/s.
- **Angular acceleration** (α) is in rad/s^2 while tangential acceleration (a_T) is in m/s^2.

Equations of Uniform Angular Acceleration

The equations of **uniform angular acceleration** are:

$$\Delta\theta = \omega_0 t + \frac{1}{2}\alpha t^2 \quad , \quad \omega = \omega_0 + \alpha t \quad , \quad \omega^2 = \omega_0^2 + 2\alpha\Delta\theta$$

In terms of the **tangential** quantities:

$$\Delta s = v_{T0} t + \frac{1}{2}a_T t^2 \quad , \quad v_T = v_{T0} + a_T t \quad , \quad v_T^2 = v_{T0}^2 + 2a_T\Delta s$$

Angular and **tangential** quantities are related by the radius (R) of the circle:

$$\Delta s = R\Delta\theta \quad , \quad v_T = R\omega \quad , \quad a_T = R\alpha$$

Tangential acceleration describes how speed changes, whereas **centripetal** acceleration (Chapter 16) describes how the direction of velocity changes. Recall the equation for centripetal acceleration:

$$a_c = \frac{v^2}{R}$$

The magnitude of the **total** acceleration is found from its components:

$$a = \sqrt{a_T^2 + a_c^2}$$

Calculus-based Equations for Angular Acceleration

In terms of calculus, angular velocity (ω) and angular acceleration (α) are given by:

$$\omega = \frac{d\theta}{dt} \quad , \quad \alpha = \frac{d\omega}{dt}$$

Symbols and SI Units

Symbol	Name	SI Units
$\Delta\theta$	angular displacement	rad
ω_0	initial angular velocity	rad/s
ω	final angular velocity	rad/s
α	angular acceleration	rad/s^2
t	time	s
Δs	arc length	m
v_{T0}	initial tangential velocity	m/s
v_T	final tangential velocity	m/s
a_T	tangential acceleration	m/s^2
a_c	centripetal acceleration	m/s^2
a	acceleration	m/s^2

Note: The symbol for angular acceleration (α) is the lowercase Greek letter alpha.

Notes Regarding Units

The units can help you determine which quantities are given in a problem. For example, only the symbol α can be expressed in rad/s^2, and a quantity expressed in m/s is limited to the symbols v_{T0} and v_T.

In the three equations involving radius ($\Delta s = R\Delta\theta$, $v_T = R\omega$, and $a_T = R\alpha$), the angular quantity **must** be expressed in radians ($\Delta\theta$ in rad, ω in rad/s, and α in rad/s^2). However, in the equations of uniform angular acceleration, the units are more flexible: There you may use radians or revolutions, provided that you're consistent. Recall that 1 rev $= 2\pi$ rad.

Uniform Angular Acceleration Strategy

An object experiences uniform angular acceleration if its angular acceleration remains constant throughout its motion. To solve a problem with uniform angular acceleration, follow these steps:

1. Identify the unknown symbol and known symbols.
2. Choose the right equations to solve for the unknown. Think about which symbol you're solving for and which symbols you know to help you choose the right equations.
    - The equations of **uniform angular acceleration** are:
    $$\Delta\theta = \omega_0 t + \frac{1}{2}\alpha t^2$$
    $$\omega = \omega_0 + \alpha t$$
    $$\omega^2 = \omega_0^2 + 2\alpha\Delta\theta$$
    The above equations are handy when you know multiple angular quantities ($\Delta\theta$, ω_0, ω, and α).
    - The same equations can be written in terms of **tangential** quantities:
    $$\Delta s = v_{T0} t + \frac{1}{2}a_T t^2$$
    $$v_T = v_{T0} + a_T t$$
    $$v_T^2 = v_{T0}^2 + 2a_T\Delta s$$
    The above equations are handy when you know multiple tangential quantities (Δs, v_{T0}, v_T, and a_T).
    - When you need to mix and match **angular** and **tangential** quantities, the following equations are helpful:
    $$\Delta s = R\Delta\theta$$
    $$v_T = R\omega$$
    $$a_T = R\alpha$$
    These three equations only work when radians are used. If the angular quantity involves revolutions, convert revolutions (rev) to radians (rad) first: $1 \text{ rev} = 2\pi$ rad.
    - If a question asks you to solve for the **number of revolutions**, solve for $\Delta\theta$ and, if necessary, convert from radians to revolutions.
    - To find "the" acceleration (as opposed to angular acceleration, tangential acceleration, or centripetal acceleration), use the following formula:
    $$a = \sqrt{a_T^2 + a_c^2}$$
    Recall the formula for **centripetal** acceleration:
    $$a_c = \frac{v^2}{R}$$

Example: A monkey runs in a circle with uniform angular acceleration. Beginning from rest, the monkey completes 4.0 revolutions in a total time of 8.0 s. What is the monkey's angular acceleration?

The unknown we are looking for is α. List the knowns. Note that 4.0 revolutions refers to $\Delta\theta$, since angle can be measured in revolutions (or degrees or radians). It's not necessary to convert to radians if we only use the three purely angular equations (but if we use any equation with radius, we will need to convert to radians first).

$$\alpha = ? \quad , \quad \Delta\theta = 4.0 \text{ rev} \quad , \quad \omega_0 = 0 \quad , \quad t = 8.0 \text{ s}$$

(We also know that $v_{T0} = 0$, since $v_{T0} = R\omega_0$.) Based on this, it would be simplest to use the first equation of uniform angular acceleration:

$$\Delta\theta = \omega_0 t + \frac{1}{2}\alpha t^2$$

Plug the knowns into this equation. To avoid clutter, suppress the units until the end.

$$4 = 0(8) + \frac{1}{2}\alpha(8)^2$$

$$4 = 32\alpha$$

$$\alpha = \frac{1}{8} \text{ rev/s}^2$$

The answer came out in rev/s^2 (instead of rad/s^2) because we put $\Delta\theta$ in revolutions (rather than radians). When working with purely angular equations, you can get away with this, as long as you're consistent (use all revolutions or all radians; don't mix and match). However, with some equations (like those involving an R), you must use radians. When in doubt, use radians, not revolutions.

Example: One monkey gets inside of a 50-cm diameter vertical tire. Another monkey gives the tire a push and lets go. The tire completes 6.0 revolutions before it topples over. Its uniform angular deceleration is $-\frac{1}{3}$ rev/s^2. What is the initial speed of the tire?

The unknown we are looking for is v_{T0}. List the knowns. Note that $R = \frac{D}{2} = 25 \text{ cm} = \frac{1}{4}$ m.

$$v_{T0} = ? \quad , \quad R = \frac{1}{4} \text{ m} \quad , \quad \Delta\theta = 6.0 \text{ rev} \quad , \quad \omega = 0 \quad , \quad \alpha = -\frac{1}{3} \text{ rev/s}^2$$

First solve for ω_0:

$$\omega^2 = \omega_0^2 + 2\alpha\Delta\theta$$

$$0 = \omega_0^2 + 2\left(-\frac{1}{3}\right)6 = \omega_0^2 - 4$$

$$\omega_0 = \sqrt{4} \text{ rev/s} = 2.0 \text{ rev/s} = 4\pi \text{ rad/s}$$

Use the radius to find the initial tangential velocity (where ω_0 **must** be in rad/s):

$$v_{T0} = R\omega_0 = \frac{1}{4}(4\pi) = \pi \text{ m/s}$$

174. A monkey drives a bananamobile in a circle with an initial angular speed of 8.0 rev/s and a uniform angular acceleration of 16.0 rev/s^2, completing 6.0 revolutions.

Based on the questions below, list the three symbols that you know along with their values and SI units.

________________ ________________ ________________

(A) How much time does this take?

(B) What is the final angular speed?

Want help? Check the hints section at the back of the book.

Answers: $\frac{1}{2}$ s, 16 rev/s

175. A monkey drives a bananamobile in a circle with an initial angular speed of $\frac{1}{5}$ rev/s and a uniform angular acceleration of $\frac{1}{20}$ rev/s^2 for one minute.

Based on the questions below, list the three symbols that you know along with their values and SI units.

____________________ ____________________ ____________________

(A) How many revolutions does the bananamobile complete?

(B) If the radius of the circle is 15 m, what is the final speed of the bananamobile?

Want help? Check the hints section at the back of the book.

Answer: 102 rev, 96π m/s

29 THE VECTOR PRODUCT

Relevant Terminology

Matrix – an array of numbers arranged in (horizontal) rows and (vertical) columns.
Determinant – an operation applied to a matrix resulting in a single number (see below).
Cofactor – the number in a matrix where a particular row and column intersect. One cofactor of a 3×3 matrix is demonstrated visually in the next section.

Determinants

To find the **determinant** of a 2×2 matrix, multiply the numbers on the main diagonal (top left times bottom right) and subtract the product of the numbers on the cross diagonal (top right times bottom left) as follows:†

$$\begin{vmatrix} a & b \\ c & d \end{vmatrix} = ad - bc$$

One way to find the **determinant** of a 3×3 matrix is to use the method of **cofactors**. The elements of the top row of the matrix serve as cofactors. Blocking out the row and column of a cofactor yields its **submatrix**. For example, we have illustrated the submatrix of the cofactor a in the 3×3 matrix below:

$$\begin{pmatrix} a & b & c \\ d & e & f \\ g & h & i \end{pmatrix}$$

$$= ad - cb$$

The **determinant** of a 3×3 matrix can be found by treating each element of the top row as a cofactor, multiplying each cofactor by the determinant of its corresponding 2×2 submatrix, and adding these factors together with alternating signs.

$$\begin{vmatrix} a & b & c \\ d & e & f \\ g & h & i \end{vmatrix} = a\begin{vmatrix} e & f \\ h & i \end{vmatrix} - b\begin{vmatrix} d & f \\ g & i \end{vmatrix} + c\begin{vmatrix} d & e \\ g & h \end{vmatrix} = aei - afh - bdi + bfg + cdh - ceg$$

An alternative method for finding the determinant of a 3×3 matrix involves adding two columns to the right of the matrix by repeating the two leftmost columns, multiplying down the three diagonals to the right and then down the three diagonals to the left, and adding the terms together with negative signs for the leftward multiplications as follows:

$$\begin{vmatrix} a & b & c & a & b \\ d & e & f & d & e \\ g & h & i & g & h \end{vmatrix} = aei + bfg + cdh - ceg - afh - bdi$$

† It's a good habit to work top **down** across both diagonals, writing $ad - bc$, as opposed to multiplying up the cross diagonal and writing $ad - cb$. With numbers, it doesn't matter, since $bc = cb$. However, in higher-level math courses, you encounter differential operators in determinants, and then it *does* matter.

Vector Product Equations

Recall from Chapter 10 that any vector can be expressed in terms of its components (A_x, A_y, and A_z) and Cartesian unit vectors ($\hat{\mathbf{x}}$, $\hat{\mathbf{y}}$, and $\hat{\mathbf{z}}$) in the following form:

$$\vec{\mathbf{A}} = A_x\hat{\mathbf{x}} + A_y\hat{\mathbf{y}} + A_z\hat{\mathbf{z}}$$

The **vector** product (or **cross** product) $\vec{\mathbf{A}} \times \vec{\mathbf{B}}$ is defined according to a 3×3 determinant formed by placing the Cartesian unit vectors along the top row, the Cartesian components of the first vector along the middle row, and the Cartesian components of the second vector along the bottom row:

$$\vec{\mathbf{A}} \times \vec{\mathbf{B}} = \begin{vmatrix} \hat{\mathbf{x}} & \hat{\mathbf{y}} & \hat{\mathbf{z}} \\ A_x & A_y & A_z \\ B_x & B_y & B_z \end{vmatrix} = \hat{\mathbf{x}} \begin{vmatrix} A_y & A_z \\ B_y & B_z \end{vmatrix} - \hat{\mathbf{y}} \begin{vmatrix} A_x & A_z \\ B_x & B_z \end{vmatrix} + \hat{\mathbf{z}} \begin{vmatrix} A_x & A_y \\ B_x & B_y \end{vmatrix}$$

$$\vec{\mathbf{A}} \times \vec{\mathbf{B}} = (A_yB_z - A_zB_y)\hat{\mathbf{x}} - (A_xB_z - A_zB_x)\hat{\mathbf{y}} + (A_xB_y - A_yB_x)\hat{\mathbf{z}}$$

$$\vec{\mathbf{A}} \times \vec{\mathbf{B}} = A_yB_z\hat{\mathbf{x}} - A_zB_y\hat{\mathbf{x}} + A_zB_x\hat{\mathbf{y}} - A_xB_z\hat{\mathbf{y}} + A_xB_y\hat{\mathbf{z}} - A_yB_x\hat{\mathbf{z}}$$

The magnitude of the vector product, $\|\vec{\mathbf{A}} \times \vec{\mathbf{B}}\|$, turns out to have a form that is similar to the scalar product (Chapter 20), except involving a sine instead of a cosine:

$$\|\vec{\mathbf{A}} \times \vec{\mathbf{B}}\| = AB \sin\theta$$

The angle θ is the angle between the two vectors $\vec{\mathbf{A}}$ and $\vec{\mathbf{B}}$.

Essential Concepts

The vector product $\vec{\mathbf{A}} \times \vec{\mathbf{B}}$ between two vectors $\vec{\mathbf{A}}$ and $\vec{\mathbf{B}}$ represents the area of the parallelogram formed by joining $\vec{\mathbf{A}}$ and $\vec{\mathbf{B}}$ tip-to-tail, as illustrated below. Whereas the scalar product $\vec{\mathbf{A}} \cdot \vec{\mathbf{B}}$ results in a **scalar** (meaning that $\vec{\mathbf{A}} \cdot \vec{\mathbf{B}}$ has <u>**no**</u> direction, only a magnitude), the vector product $\vec{\mathbf{A}} \times \vec{\mathbf{B}}$ results in a **vector** (meaning that $\vec{\mathbf{A}} \times \vec{\mathbf{B}}$ includes direction). The direction of the vector product is perpendicular to the plane containing $\vec{\mathbf{A}}$ and $\vec{\mathbf{B}}$ (the plane of the parallelogram illustrated below).

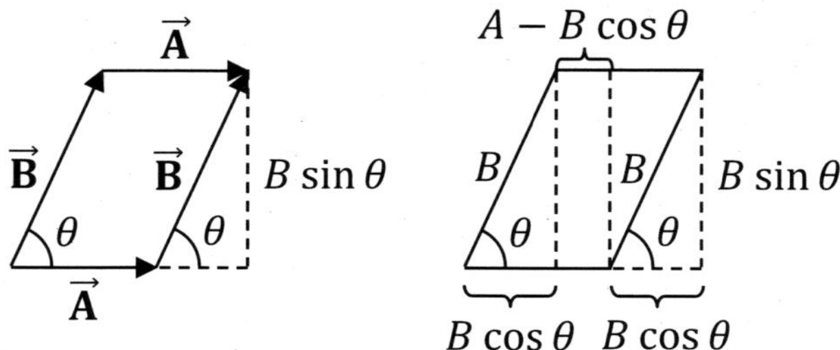

By dividing the parallelogram illustrated above into two right triangles and the thin rectangle between them, it can be seen that the area of the parallelogram is:

$$A_p = 2A_t + A_r = 2\frac{1}{2}(B\cos\theta)(B\sin\theta) + (A - B\cos\theta)(B\sin\theta) = AB\sin\theta$$

This area, $AB\sin\theta$, equals the magnitude of the vector product: $\|\vec{\mathbf{A}} \times \vec{\mathbf{B}}\| = AB\sin\theta$.

Notation

The cross ($\times$) is used to designate the vector (or cross) product, as in $\vec{\mathbf{A}} \times \vec{\mathbf{B}}$, which results in a vector, whereas the dot ($\cdot$) is used to designate the scalar (or dot) product, as in $\vec{\mathbf{A}} \cdot \vec{\mathbf{B}}$. When multiplying ordinary numbers, we use the cross and dot interchangeably. For example, $3 \times 2 = 3 \cdot 2 = (3)(2) = 6$. However, you may **not** swap the cross and dot in the context of vector multiplication: $\vec{\mathbf{A}} \times \vec{\mathbf{B}}$ is much different from $\vec{\mathbf{A}} \cdot \vec{\mathbf{B}}$.

Anti-commutativity of the Vector Product

The vector product is anti-commutative because $\vec{\mathbf{B}} \times \vec{\mathbf{A}} = -\vec{\mathbf{A}} \times \vec{\mathbf{B}}$. It's also helpful to remember that the vector product of any vector with itself equals zero: $\vec{\mathbf{A}} \times \vec{\mathbf{A}} = 0$.

Vector Product between Unit Vectors

The vector product $\hat{\mathbf{x}} \times \hat{\mathbf{y}}$ can be performed by identifying $\vec{\mathbf{A}} = \hat{\mathbf{x}}$ and $\vec{\mathbf{B}} = \hat{\mathbf{y}}$. Compare the general expression $\vec{\mathbf{A}} = A_x\hat{\mathbf{x}} + A_y\hat{\mathbf{y}} + A_z\hat{\mathbf{z}}$ to the special case $\vec{\mathbf{A}} = \hat{\mathbf{x}}$ to see that in this case $A_x = 1$ and $A_y = A_z = 0$. Similarly, for $\vec{\mathbf{B}} = \hat{\mathbf{y}}$, $B_x = B_z = 0$ and $B_y = 1$. We can evaluate $\hat{\mathbf{x}} \times \hat{\mathbf{y}}$ in determinant form using these components:

$$\hat{\mathbf{x}} \times \hat{\mathbf{y}} = \begin{vmatrix} \hat{\mathbf{x}} & \hat{\mathbf{y}} & \hat{\mathbf{z}} \\ A_x & A_y & A_z \\ B_x & B_y & B_z \end{vmatrix} = \begin{vmatrix} \hat{\mathbf{x}} & \hat{\mathbf{y}} & \hat{\mathbf{z}} \\ 1 & 0 & 0 \\ 0 & 1 & 0 \end{vmatrix} = \hat{\mathbf{z}}$$

You can similarly work out the vector product between other pairs of Cartesian unit vectors:

$$\hat{\mathbf{x}} \times \hat{\mathbf{x}} = 0 \quad , \quad \hat{\mathbf{y}} \times \hat{\mathbf{y}} = 0 \quad , \quad \hat{\mathbf{z}} \times \hat{\mathbf{z}} = 0$$
$$\hat{\mathbf{x}} \times \hat{\mathbf{y}} = \hat{\mathbf{z}} \quad , \quad \hat{\mathbf{y}} \times \hat{\mathbf{z}} = \hat{\mathbf{x}} \quad , \quad \hat{\mathbf{z}} \times \hat{\mathbf{x}} = \hat{\mathbf{y}}$$
$$\hat{\mathbf{y}} \times \hat{\mathbf{x}} = -\hat{\mathbf{z}} \quad , \quad \hat{\mathbf{z}} \times \hat{\mathbf{y}} = -\hat{\mathbf{x}} \quad , \quad \hat{\mathbf{x}} \times \hat{\mathbf{z}} = -\hat{\mathbf{y}}$$

Memorization Tip

Here is one way to remember which of the cross products among unit vectors are positive and which are negative. The vector product is said to be cyclic. That is, the unit vectors follow a circular order, according to the following sequence, $\hat{\mathbf{x}}, \hat{\mathbf{y}}, \hat{\mathbf{z}}, \hat{\mathbf{x}}, \hat{\mathbf{y}}$ ... What this means is that $\hat{\mathbf{x}} \times \hat{\mathbf{y}}$ equals $+\hat{\mathbf{z}}$, $\hat{\mathbf{y}} \times \hat{\mathbf{z}}$ equals $+\hat{\mathbf{x}}$, and $\hat{\mathbf{z}} \times \hat{\mathbf{x}}$ equals $+\hat{\mathbf{y}}$ because they follow this cyclic order (where $\hat{\mathbf{x}}$ comes after $\hat{\mathbf{z}}$). If the unit vectors proceed in the reverse order, the result is negative, as in $\hat{\mathbf{z}} \times \hat{\mathbf{y}} = -\hat{\mathbf{x}}$.

Strategy for Finding the Vector Product

To compute the vector product $\vec{\mathbf{C}} = \vec{\mathbf{A}} \times \vec{\mathbf{B}}$, follow these steps:

- If you know the components of the given vectors, find the following determinant.
$$\vec{\mathbf{C}} = \vec{\mathbf{A}} \times \vec{\mathbf{B}} = \begin{vmatrix} \hat{\mathbf{x}} & \hat{\mathbf{y}} & \hat{\mathbf{z}} \\ A_x & A_y & A_z \\ B_x & B_y & B_z \end{vmatrix} = \hat{\mathbf{x}} \begin{vmatrix} A_y & A_z \\ B_y & B_z \end{vmatrix} - \hat{\mathbf{y}} \begin{vmatrix} A_x & A_z \\ B_x & B_z \end{vmatrix} + \hat{\mathbf{z}} \begin{vmatrix} A_x & A_y \\ B_x & B_y \end{vmatrix}$$
Once you find the vector product, if you would like to determine the magnitude of the vector product, you can use the three-dimensional generalization of the Pythagorean theorem:
$$C = \|\vec{\mathbf{A}} \times \vec{\mathbf{B}}\| = \sqrt{C_x^2 + C_y^2 + C_z^2}$$
(C_x is the coefficient of $\hat{\mathbf{x}}$, C_y is the coefficient of $\hat{\mathbf{y}}$, and C_z is the coefficient of $\hat{\mathbf{z}}$ in the expression for $\vec{\mathbf{C}}$ resulting from the determinant.)
- If you know the magnitudes of the given vectors along with the angle between the two vectors, you can find the magnitude of the vector product using trig:
$$C = \|\vec{\mathbf{A}} \times \vec{\mathbf{B}}\| = AB \sin\theta$$

Example: Find the vector product between $\vec{\mathbf{A}} = 2\,\hat{\mathbf{x}} - \hat{\mathbf{y}} + 3\,\hat{\mathbf{z}}$ and $\vec{\mathbf{B}} = \hat{\mathbf{x}} + 4\,\hat{\mathbf{y}} - 2\,\hat{\mathbf{z}}$.

Compare $\vec{\mathbf{A}} = 2\,\hat{\mathbf{x}} - \hat{\mathbf{y}} + 3\,\hat{\mathbf{z}}$ to $\vec{\mathbf{A}} = A_x\hat{\mathbf{x}} + A_y\hat{\mathbf{y}} + A_z\hat{\mathbf{z}}$ to see that $A_x = 2$, $A_y = -1$, and $A_z = 3$. Similarly, $B_x = 1$, $B_y = 4$, and $B_z = -2$. Plug these values into the determinant form of the vector product:

$$\vec{\mathbf{A}} \times \vec{\mathbf{B}} = \begin{vmatrix} \hat{\mathbf{x}} & \hat{\mathbf{y}} & \hat{\mathbf{z}} \\ A_x & A_y & A_z \\ B_x & B_y & B_z \end{vmatrix} = \begin{vmatrix} \hat{\mathbf{x}} & \hat{\mathbf{y}} & \hat{\mathbf{z}} \\ 2 & -1 & 3 \\ 1 & 4 & -2 \end{vmatrix} = \hat{\mathbf{x}} \begin{vmatrix} -1 & 3 \\ 4 & -2 \end{vmatrix} - \hat{\mathbf{y}} \begin{vmatrix} 2 & 3 \\ 1 & -2 \end{vmatrix} + \hat{\mathbf{z}} \begin{vmatrix} 2 & -1 \\ 1 & 4 \end{vmatrix}$$

$$\vec{\mathbf{A}} \times \vec{\mathbf{B}} = \hat{\mathbf{x}}[(-1)(-2) - (3)(4)] - \hat{\mathbf{y}}[(2)(-2) - (3)(1)] + \hat{\mathbf{z}}[(2)(4) - (-1)(1)]$$

$$\vec{\mathbf{A}} \times \vec{\mathbf{B}} = \hat{\mathbf{x}}(2 - 12) - \hat{\mathbf{y}}(-4 - 3) + \hat{\mathbf{z}}(8 + 1)$$

$$\vec{\mathbf{A}} \times \vec{\mathbf{B}} = -10\hat{\mathbf{x}} - (-7)\hat{\mathbf{y}} + 9\hat{\mathbf{z}}$$

$$\vec{\mathbf{A}} \times \vec{\mathbf{B}} = -10\hat{\mathbf{x}} + 7\hat{\mathbf{y}} + 9\hat{\mathbf{z}}$$

Example: $\vec{\mathbf{A}}$ has a magnitude of 4 and $\vec{\mathbf{B}}$ has a magnitude of 8. The angle between $\vec{\mathbf{A}}$ and $\vec{\mathbf{B}}$ is 60°. Find the magnitude of $\vec{\mathbf{C}}$, where $\vec{\mathbf{C}} = \vec{\mathbf{A}} \times \vec{\mathbf{B}}$.

Use the trigonometric form of the vector product.

$$C = \|\vec{\mathbf{A}} \times \vec{\mathbf{B}}\| = AB \sin\theta$$

$$C = (4)(8)\sin 60° = 32\left(\frac{\sqrt{3}}{2}\right) = 16\sqrt{3}$$

176. Find the vector product between $\vec{\mathbf{A}} = 5\,\hat{\mathbf{x}} + 2\,\hat{\mathbf{y}} + 4\,\hat{\mathbf{z}}$ and $\vec{\mathbf{B}} = 3\,\hat{\mathbf{x}} + 4\,\hat{\mathbf{y}} + 6\,\hat{\mathbf{z}}$.

Answer: $-4\,\hat{\mathbf{x}} - 18\,\hat{\mathbf{y}} + 14\,\hat{\mathbf{z}}$

177. Find the vector product between $\vec{\mathbf{A}} = 3\,\hat{\mathbf{x}} - 2\,\hat{\mathbf{z}}$ and $\vec{\mathbf{B}} = 8\,\hat{\mathbf{x}} - \hat{\mathbf{y}} - 4\,\hat{\mathbf{z}}$.

Want help? Check the hints section at the back of the book.

Answer: $-2\,\hat{\mathbf{x}} - 4\,\hat{\mathbf{y}} - 3\,\hat{\mathbf{z}}$

178. Given $\vec{\mathbf{A}} = -\hat{\mathbf{y}} + \hat{\mathbf{z}}$ and $\vec{\mathbf{B}} = \hat{\mathbf{x}} - \hat{\mathbf{z}}$, find $\vec{\mathbf{C}}$ where $\vec{\mathbf{C}} = \vec{\mathbf{A}} \times \vec{\mathbf{B}}$.

Answer: $\vec{\mathbf{C}} = \hat{\mathbf{x}} + \hat{\mathbf{y}} + \hat{\mathbf{z}}$

179. $\vec{\mathbf{A}}$ has a magnitude of 3 and $\vec{\mathbf{B}}$ has a magnitude of 4. The angle between $\vec{\mathbf{A}}$ and $\vec{\mathbf{B}}$ is $150°$. Find the magnitude of $\vec{\mathbf{C}}$, where $\vec{\mathbf{C}} = \vec{\mathbf{A}} \times \vec{\mathbf{B}}$.

Want help? Check the hints section at the back of the book.

Answer: 6

30 TORQUE

Relevant Terminology

Lever arm – the perpendicular distance connecting the line of force to the axis of rotation (see the illustration below).
Rigid body – an object which preserves its shape when it rotates.
Torque – the vector product between $\vec{\mathbf{r}}$ (see the illustration below) and force. A net torque causes a rigid body to change its angular momentum.
Fulcrum – a point of support upon which a rigid body may rotate.
Hinge – a joint upon which a rigid body may pivot (allowing the rigid body to rotate).

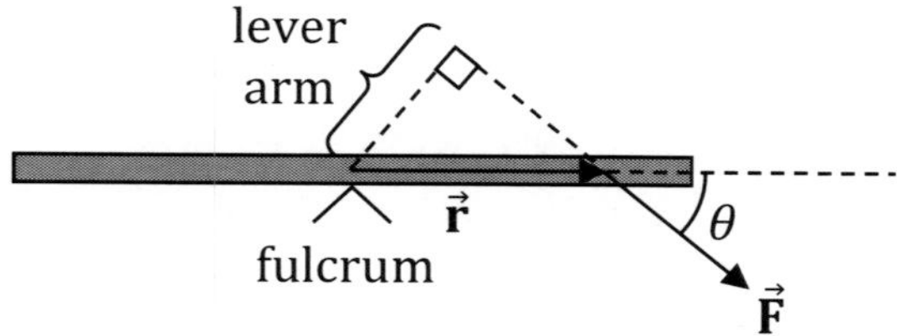

Essential Concepts

Torque ($\vec{\boldsymbol{\tau}}$) is the **vector product** (Chapter 29) between the vector $\vec{\mathbf{r}}$ and force ($\vec{\mathbf{F}}$).

$$\vec{\boldsymbol{\tau}} = \vec{\mathbf{r}} \times \vec{\mathbf{F}}$$

In the torque equation, $\vec{\mathbf{r}}$ extends from the axis of rotation to the point where $\vec{\mathbf{F}}$ is applied (as shown below). The magnitude of the torque is given by the appropriate equation from Chapter 29, where θ is the angle between $\vec{\mathbf{r}}$ and $\vec{\mathbf{F}}$.

$$\tau = r\,F\sin\theta$$

Torque depends upon the lever arm.

- Geometrically, **lever arm** is found like the <u>**top**</u> illustration above. Imagine a long line extending along the force. Connect the axis of rotation to the line of force with a perpendicular line. This perpendicularly connecting line is termed the "lever arm."
- Practically, **lever arm** can be found by multiplying r (the magnitude of $\vec{\mathbf{r}}$ in the diagram above) by the sine of θ (the angle between $\vec{\mathbf{r}}$ and $\vec{\mathbf{F}}$).

$$\text{lever arm} = r\sin\theta$$

A net force causes an object to change momentum (resulting in acceleration). A net torque causes a rigid body to change angular momentum (resulting in angular acceleration).

Strategy to Find Torque

How you find torque depends on whether you're working with vector expressions or if you're just working with the magnitudes of the vectors:

- If you know the components of $\vec{\mathbf{r}}$ and $\vec{\mathbf{F}}$ (in the form $\vec{\mathbf{r}} = x\hat{\mathbf{x}} + y\hat{\mathbf{y}} + z\hat{\mathbf{z}}$ and $\vec{\mathbf{F}} = F_x\hat{\mathbf{x}} + F_y\hat{\mathbf{y}} + F_z\hat{\mathbf{z}}$), find the following determinant (from Chapter 29).

$$\vec{\boldsymbol{\tau}} = \vec{\mathbf{r}} \times \vec{\mathbf{F}} = \begin{vmatrix} \hat{\mathbf{x}} & \hat{\mathbf{y}} & \hat{\mathbf{z}} \\ x & y & z \\ F_x & F_y & F_z \end{vmatrix} = \hat{\mathbf{x}} \begin{vmatrix} y & z \\ F_y & F_z \end{vmatrix} - \hat{\mathbf{y}} \begin{vmatrix} x & z \\ F_x & F_z \end{vmatrix} + \hat{\mathbf{z}} \begin{vmatrix} x & y \\ F_x & F_y \end{vmatrix}$$

  Once you find the vector product, if you would like to determine the magnitude of the torque, you can use the three-dimensional generalization of the Pythagorean theorem:

$$\tau = \left\|\vec{\mathbf{r}} \times \vec{\mathbf{F}}\right\| = \sqrt{\tau_x^2 + \tau_y^2 + \tau_z^2}$$

  (τ_x is the coefficient of $\hat{\mathbf{x}}$, τ_y is the coefficient of $\hat{\mathbf{y}}$, and τ_z is the coefficient of $\hat{\mathbf{z}}$ in the expression for $\vec{\boldsymbol{\tau}}$ resulting from the determinant.)

- If you know the magnitudes of $\vec{\mathbf{r}}$ and $\vec{\mathbf{F}}$ along with the angle between $\vec{\mathbf{r}}$ and $\vec{\mathbf{F}}$, you can find the magnitude of the vector product using trig. Draw both $\vec{\mathbf{r}}$ and $\vec{\mathbf{F}}$, where:
  - $\vec{\mathbf{r}}$ extends from the axis of rotation to the point where $\vec{\mathbf{F}}$ is applied.
  - $\vec{\mathbf{F}}$ is the force that is exerting the specified torque.

  Use the following equation to find the magnitude of the torque.

$$\tau = \left\|\vec{\mathbf{r}} \times \vec{\mathbf{F}}\right\| = r\,F\sin\theta$$

  In the above equation, θ is the angle between $\vec{\mathbf{r}}$ and $\vec{\mathbf{F}}$.

Symbols and Units

Symbol	Quantity	Units
r	distance from the axis of rotation to the point where the force is applied	m
F	force	N
θ	the angle between $\vec{\mathbf{r}}$ and $\vec{\mathbf{F}}$	° or rad
τ	torque	Nm

Note: The symbol for torque (τ) is the lowercase Greek letter tau.

Notes Regarding Units

The SI units of torque are Nm (a Newton times a meter). It's best not to reverse the order, since if you write mN (instead of Nm) it may be confused with milliNewtons.

Although a Nm equals a Joule, we **don't** express torque in Joules since torque is **not** a measure of work or energy. (The distinction has to do with radians. If a torque acts on a rigid body that rotates, the torque times the angle equals rotational work. Rotational work is measured in Joules, and the radians from the angle make the difference. Unlike rotational work, torque is measured in Nm, **not** Joules.)

The SI units of torque can be broken down into $\text{kg} \cdot \text{m}^2/\text{s}^2$, since a Newton equals a $\text{kg} \cdot \text{m}/\text{s}^2$.

Example: In the diagram below, a 20-kg box is atop a plank. The box is 4.0 m from the fulcrum. What torque does the weight of the box exert on the plank?

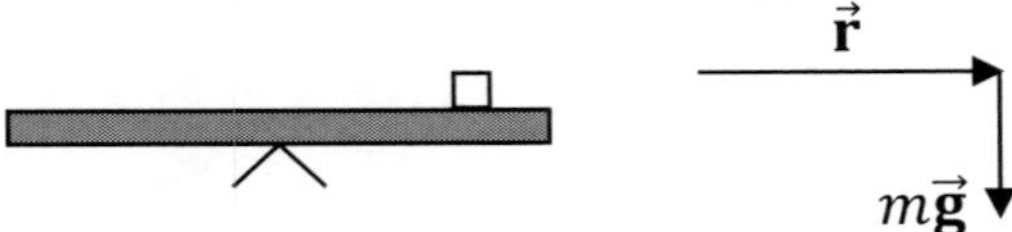

Draw a picture showing $\vec{\mathbf{r}}$ and $\vec{\mathbf{F}}$. The specified force is the weight ($m\vec{\mathbf{g}}$) of the box, while $\vec{\mathbf{r}}$ extends from the fulcrum to the white box. In this example, $r = 4.0$ m, $F = mg = (20)(9.81) \approx (20)(10) = 200$ N, and $\theta = 90°$ ($\vec{\mathbf{r}}$ and $m\vec{\mathbf{g}}$ are perpendicular):

$$\tau = r\,F \sin\theta = (4)(200)\sin 90° = (4)(200)(1) = 800 \text{ Nm}$$

Example: In the diagram below, a monkey hammers a nail into a horizontal pole 3.0 m to the right of the fulcrum. The monkey ties a string around the nail and pulls with a force of 80 N in the direction shown. What torque does the monkey exert on the pole?

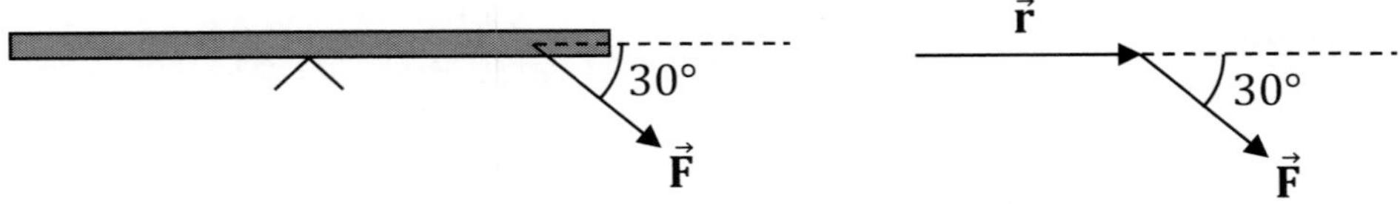

Draw a picture showing $\vec{\mathbf{r}}$ and $\vec{\mathbf{F}}$. The specified force is the monkey's pull ($\vec{\mathbf{F}}$). In this example, $r = 3.0$ m, $F = 80$ N, and $\theta = 30°$:

$$\tau = r\,F \sin\theta = (3)(80)\sin 30° = (3)(80)\left(\frac{1}{2}\right) = 120 \text{ Nm}$$

Example: As shown below, one end of a 4.0-m long, 9.0-kg rod is connected to a floor by a hinge, which allows the rod to rotate. (The rod won't stay in the position shown: It will rotate until it reaches the ground, where the rod will be horizontal.) Find the instantaneous torque exerted on the rod due to the weight of the rod in the position shown below.

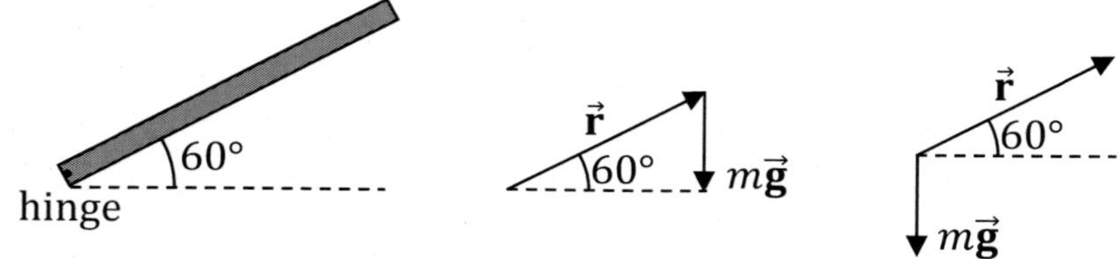

Draw a picture showing $\vec{\mathbf{r}}$ and $\vec{\mathbf{F}}$. The specified force is the weight ($m\vec{\mathbf{g}}$) of the rod, while $\vec{\mathbf{r}}$ extends from the fulcrum to $m\vec{\mathbf{g}}$. In this example, $r = \frac{L}{2} = \frac{4}{2} = 2.0$ m (the distance from the hinge to the **center** of the rod, since the center of the rod is the point where gravity acts "on average"), $F = mg = (9)(9.81) \approx (9)(10) = 90$ N, and $\theta = 90° + 60° = 150°$. It's easier to get θ correct when you draw $\vec{\mathbf{r}}$ and $\vec{\mathbf{F}}$ tail to tail (see the right diagram above):

$$\tau = r\,F \sin\theta = (2)(90)\sin 150° = (2)(90)\left(\frac{1}{2}\right) = 90 \text{ Nm}$$

Example: The force $\vec{\mathbf{F}} = 4\,\hat{\mathbf{x}} - 3\,\hat{\mathbf{y}} + 2\,\hat{\mathbf{z}}$ exerts a torque for which $\vec{\mathbf{r}} = 5\,\hat{\mathbf{x}} + 2\,\hat{\mathbf{y}} + \hat{\mathbf{z}}$. Find the torque vector.

Compare $\vec{\mathbf{r}} = 5\,\hat{\mathbf{x}} + 2\,\hat{\mathbf{y}} + \hat{\mathbf{z}}$ to $\vec{\mathbf{r}} = x\hat{\mathbf{x}} + y\hat{\mathbf{y}} + z\hat{\mathbf{z}}$ to see that $x = 5$, $y = 2$, and $z = 1$. Similarly, compare $\vec{\mathbf{F}} = 4\,\hat{\mathbf{x}} - 3\,\hat{\mathbf{y}} + 2\,\hat{\mathbf{z}}$ to $\vec{\mathbf{F}} = F_x\hat{\mathbf{x}} + F_y\hat{\mathbf{y}} + F_z\hat{\mathbf{z}}$ to see that $F_x = 4$, $F_y = -3$, and $F_z = 2$. Plug these values into the determinant form of the vector product:

$$\vec{\boldsymbol{\tau}} = \vec{\mathbf{r}} \times \vec{\mathbf{F}} = \begin{vmatrix} \hat{\mathbf{x}} & \hat{\mathbf{y}} & \hat{\mathbf{z}} \\ x & y & z \\ F_x & F_y & F_z \end{vmatrix} = \begin{vmatrix} \hat{\mathbf{x}} & \hat{\mathbf{y}} & \hat{\mathbf{z}} \\ 5 & 2 & 1 \\ 4 & -3 & 2 \end{vmatrix}$$

$$\vec{\boldsymbol{\tau}} = \hat{\mathbf{x}}\begin{vmatrix} 2 & 1 \\ -3 & 2 \end{vmatrix} - \hat{\mathbf{y}}\begin{vmatrix} 5 & 1 \\ 4 & 2 \end{vmatrix} + \hat{\mathbf{z}}\begin{vmatrix} 5 & 2 \\ 4 & -3 \end{vmatrix}$$

$$\vec{\boldsymbol{\tau}} = \hat{\mathbf{x}}[(2)(2) - (1)(-3)] - \hat{\mathbf{y}}[(5)(2) - (1)(4)] + \hat{\mathbf{z}}[(5)(-3) - (2)(4)]$$

$$\vec{\boldsymbol{\tau}} = \hat{\mathbf{x}}(4 + 3) - \hat{\mathbf{y}}(10 - 4) + \hat{\mathbf{z}}(-15 - 8)$$

$$\vec{\boldsymbol{\tau}} = 7\hat{\mathbf{x}} - 6\hat{\mathbf{y}} - 23\hat{\mathbf{z}}$$

180. As illustrated below, a 30-kg white box is at the very right end of a 60-kg plank. The plank is 20.0 m long and the fulcrum is 4.0-m from the left end.

(A) Find the torque exerted on the system due to the weight of the white box.

(B) Find the torque exerted on the system due to the weight of the plank.

Answers: 4800 Nm, 3600 Nm

181. As illustrated below, one end of a 6.0-m long, 7.0-kg rod is connected to a wall by a hinge, which allows the rod to rotate. (The rod won't stay in the position shown: It will rotate until it reaches the wall, where the rod will be vertical.) Find the instantaneous torque exerted on the rod due to the weight of the rod in the position shown below.

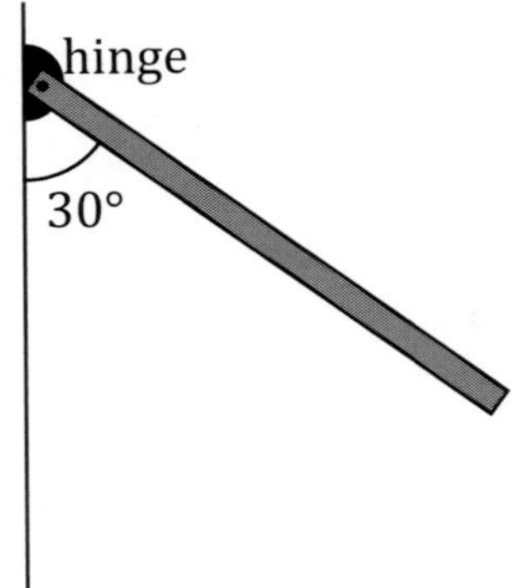

Want help? Check the hints section at the back of the book.

Answer: 105 Nm

182. As illustrated below, a 60-kg monkey hangs from one end of a 12.0-m long rod, while an 80-kg monkey stands 4.0 m from the opposite end. A fulcrum rests beneath the center of the rod.

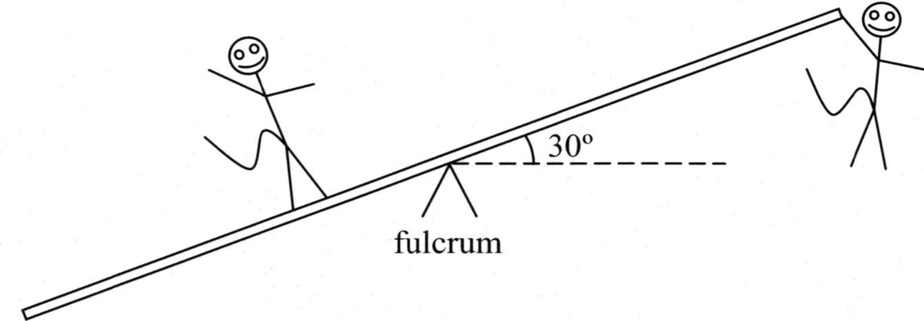

(A) What torque is exerted by the hanging monkey?

(B) What torque is exerted by the standing monkey?

Want help? Check the hints section at the back of the book.

Answers: $1800\sqrt{3}$ Nm, $-800\sqrt{3}$ Nm

183. A door is 200 cm tall and 50 cm wide. One vertical side of the door is connected to a doorway via hinges. The doorknobs are located near the other end of the door, 45 cm from the hinges. The door is presently ajar (partway open). Consider the following ways that a monkey might attempt to close the door.

(A) A monkey grabs the two doorknobs (one in each hand) and pulls with a force of 90 N directly away from the hinges. What torque does the monkey exert on the door, if any?

(B) A monkey pushes with a force of 60 N on one doorknob, pushing perpendicular to the plane of the door. What torque does the monkey exert on the door, if any?

(C) A monkey pushes on the geometric center of the door with a force of 80 N, pushing in a direction 30° from the normal (that is, the line perpendicular to the plane of the door). What torque does the monkey exert on the door, if any?

Want help? Check the hints section at the back of the book.

Answers: 0, 27 Nm, $10\sqrt{3}$ Nm

184. The force $\vec{\mathbf{F}} = 3\,\hat{\mathbf{x}} + 2\,\hat{\mathbf{y}} + 4\,\hat{\mathbf{z}}$ exerts a torque for which $\vec{\mathbf{r}} = 6\,\hat{\mathbf{x}} + \hat{\mathbf{y}} - 5\hat{\mathbf{z}}$. Find the torque vector.

Want help? Check the hints section at the back of the book.

Answer: $\vec{\boldsymbol{\tau}} = 14\,\hat{\mathbf{x}} - 39\,\hat{\mathbf{y}} + 9\,\hat{\mathbf{z}}$

31 STATIC EQUILIBRIUM

Essential Concepts

Static equilibrium means that the system is stationary (at rest). Since the system isn't moving, all components of the acceleration equal zero ($a_x = 0$ and $a_y = 0$) and the angular acceleration equals zero (α). We solve static equilibrium problems by setting acceleration equal to zero in Newton's second law (recall Chapter 14) and also setting the sum of the torques equal to zero (torque was the subject of Chapter 30). Conceptually, the reasons for this are:

- Setting the net force equal to zero ($\sum F_x = 0$ and $\sum F_y = 0$) ensures that the center of mass of the system doesn't accelerate.
- Setting the net torque equal to zero ($\sum \tau$) ensures that the system doesn't acquire angular acceleration.

If there is a **hinge** in the problem, the hinge may experience a force. From an engineering perspective, it's important to calculate things like the tension in a cord or the force exerted on a hinge in order to ensure that the materials (the cord or hingepin, for example) can withstand those forces. We call F_H and θ_H the magnitude and direction of the **hingepin force**, which can be related to horizontal (H) and vertical (V) components. It's easier to work with H and V when solving a problem, and to find F_H and θ_H later using the Pythagorean theorem and an inverse tangent (see below).

Static Equilibrium Equations

For a system in **static equilibrium**, set the components of acceleration equal to zero in Newton's second law and also set the net torque equal to zero:

$$\sum F_x = 0 \quad , \quad \sum F_y = 0 \quad , \quad \sum \tau = 0$$

Recall that **torque** (τ) is defined by the equation below. See Chapter 30 for a description of how to find r, F, and θ.

$$\tau = r\,F \sin\theta$$

If you know the horizontal (H) and vertical (V) components of the **hingepin force**, you can find the magnitude of the hingepin force (F_H) from the Pythagorean theorem and the direction of the hingepin force (θ_H) with an inverse tangent:

$$F_H = \sqrt{H^2 + V^2} \quad , \quad \theta_H = \tan^{-1}\left(\frac{V}{H}\right)$$

If you know F_H and θ_H, you can find H and V using trig.

$$H = F_H \cos\theta_H \quad , \quad V = F_H \sin\theta_H$$

The two pairs of equations above are vector equations from Chapter 10.

Symbols and Units

Symbol	Quantity	Units
r	distance from the axis of rotation to the point where the force is applied	m
F	force	N
θ	the angle between $\vec{\mathbf{r}}$ and $\vec{\mathbf{F}}$	° or rad
τ	torque	Nm
F_H	magnitude of the hingepin force	N
θ_H	direction of the hingepin force	° or rad
H	horizontal component of the hingepin force	N
V	vertical component of the hingepin force	N

Strategy for Static Equilibrium Problems

To solve a problem with a system in static equilibrium, follow these steps:

1. Draw an **extended free-body diagram** (FBD). Unlike the FBD's that we drew in Chapter 14, an extended free-body diagram shows the rigid body's shape (for example, a rod) and shows where each force acts on the object. This is helpful for working out the torques. Draw and label forces like we did in Chapter 14, but draw each force where it acts on the object. Label the direction for positive torques.
2. If there is a **hingepin**, draw the horizontal (H) and vertical (V) components of the hingepin force. This is done in one of the examples that follow.
3. Identify the **axis of rotation**. This point is needed to determine the torques. Since the object isn't rotating, you can actually choose the axis anywhere you want.
4. Set the sums of the **forces** and **torques** equal to zero:
$$\sum F_x = 0 \quad , \quad \sum F_y = 0 \quad , \quad \sum \tau = 0$$
5. Rewrite the left-hand side of each force sum ($\sum F_x$ and $\sum F_y$) in terms of the x- and y-components of the forces acting on each object, like we did in Chapter 14.
6. Rewrite the left-hand side of the torque sum ($\sum \tau$) in terms of the torques, using the equation $\tau = r\, F \sin\theta$. See Chapter 30. Note that torques causing clockwise rotations will have a different **sign** than those causing counterclockwise rotations.
7. Carry out the algebra to solve for the unknown(s).

Example: As illustrated below, a 20-kg white box rests on a plank, 6.0 m to the right of the fulcrum. The fulcrum rests beneath the center of the plank. Where should a 30-kg black box be placed in order for the system to be in static equilibrium?

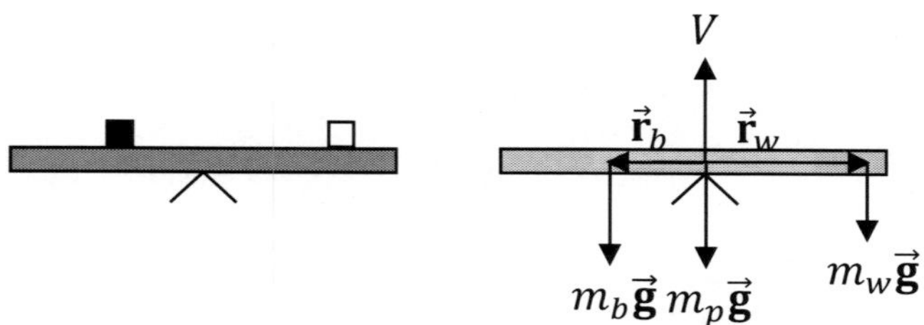

Begin by drawing an extended free-body diagram for the plank. See the diagram at the top right. In the extended free-body diagram, we draw the forces (the weight of each block) where they act on the plank. The fulcrum exerts an upward support force, V, which prevents the system from falling. Sum the forces and torques for the system. Note that $r_V = 0$ and $r_p = 0$ since V and $m_p g$ (the weight of the plank) act on the axis of rotation (on average, in the case of the plank's weight), such that V and $m_p g$ do not exert torques on the system. For the torques exerted by the two weights, $\theta = 90°$ since $\vec{\mathbf{r}}$ is horizontal and weight is vertical. We choose clockwise to be the positive rotation direction, such that the torque exerted by $m_b g$ is negative. The torques exerted by the three forces are:

$$\tau_V = r_V F_V \sin\theta = (0)V \sin\theta = 0$$
$$\tau_p = r_p F_p \sin\theta = (0)m_p g \sin\theta = 0$$
$$\tau_b = -r_b F_b \sin\theta = -r_b m_b g \sin 90°$$
$$\tau_w = r_w F_w \sin\theta = r_w m_w g \sin 90°$$

Note that no forces have x-components in this example (neither $\vec{\mathbf{r}}_b$ nor $\vec{\mathbf{r}}_w$ is a force).

$$\sum F_y = 0 \quad , \quad \sum \tau = 0$$
$$V - m_p g - m_b g - m_w g = 0 \quad , \quad r_w m_w g \sin 90° - r_b m_b g \sin 90° = 0$$

We can answer the question by solving for r_b:

$$r_w m_w g = r_b m_b g$$
$$r_b = \frac{r_w m_w}{m_b} = \frac{(6)(20)}{30} = 4.0 \text{ m}$$

The black block should be placed 4.0 m to the left of the fulcrum.

Example: As illustrated below, a mallet is balanced on a fulcrum in static equilibrium. The mallet has a 30.0-cm long handle and a 6.0-cm wide end. Each piece has uniform density (but the densities of the two pieces differ). The total mass of the mallet is 12.0 kg. The fulcrum is 3.0 cm from the right edge of the handle. A monkey saws the mallet in two pieces at the fulcrum. Determine the mass of each piece.

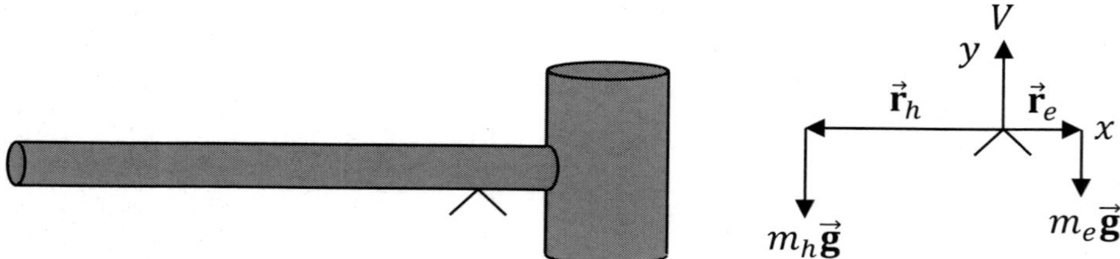

Begin by drawing an extended free-body diagram for the mallet. See the diagram at the top right. It's very similar to the previous example, since there is a weight on each side. We use m_h and r_h for the handle and m_e and r_e for the end (but as we'll see later, m_h and m_e are **<u>not</u>** the two masses that the problem asks for). Sum the forces and torques for the system. For the torques exerted by the two weights, $\theta = 90°$ since $\vec{\mathbf{r}}$ is horizontal and weight is vertical. We choose clockwise to be the positive rotation direction, such that the torque exerted by $m_h g$ is negative. The torques exerted by the three forces are:

$$\tau_V = r_V F_V \sin\theta = (0)V \sin\theta = 0$$
$$\tau_h = -r_h F_h \sin\theta = -r_h m_h g \sin 90°$$
$$\tau_e = r_e F_e \sin\theta = r_e m_e g \sin 90°$$

Note that no forces have x-components in this example (neither $\vec{\mathbf{r}}_h$ nor $\vec{\mathbf{r}}_e$ is a force).

$$\sum F_y = 0 \quad , \quad \sum \tau = 0$$
$$V - m_h g - m_e g = 0 \quad , \quad r_h m_h g \sin 90° - r_e m_e g \sin 90° = 0$$

The second equation will help us solve the problem:

$$r_h m_h g = r_e m_e g$$
$$r_h m_h = r_e m_e$$

The quantity r_h is measured from the fulcrum to the center of the handle: It's half the length of the handle minus 3.0 cm. The quantity while r_e is measured from the fulcrum to the of the mallet's end: It's half the width of the mallet plus 3.0 cm.

$$r_h = \frac{L}{2} - 3 = \frac{30}{2} - 3 = 15 - 3 = 12 \text{ cm}$$
$$r_e = \frac{W}{2} + 3 = \frac{6}{2} + 3 = 3 + 3 = 6.0 \text{ cm}$$

Plug these values into the prior equation:

$$12 m_h = 6 m_e$$
$$m_h = \frac{6 m_e}{12} = \frac{m_e}{2}$$

Now we use the fact that the two masses, m_h and m_e, add up to the total mass of the mallet, given as 12.0 kg in the problem:

$$m_h + m_e = m = 12$$

Substitute the expression $m_h = \frac{m_e}{2}$ that we found previously into the above equation:

$$\frac{m_e}{2} + m_e = 12$$

$$\frac{m_e}{2} + \frac{2m_e}{2} = \frac{3m_e}{2} = 12$$

$$m_e = \frac{2}{3} 12 = 8.0 \text{ kg}$$

Since the total mass is 12.0 kg, the handle's mass must be:

$$m_h = 12 - m_e = 12 - 8 = 4.0 \text{ kg}$$

However, m_h and m_e are **not** the masses of the two pieces. They are the masses of the handle and the mallet's end, but part of the handle is to the right of the fulcrum. When the monkey saws the mallet into two pieces at the fulcrum, he creates a left piece of mass m_L which will have less mass than the handle and a right piece of mass m_R which will have more mass than the mallet's end (because a piece of the handle is attached to the mallet's end to form the right piece). The fulcrum is 3.0 cm from the right end of the handle, and the handle is 30.0 cm long. This means that the left piece has 90% of the mass of the handle (divide 27.0 cm by 30.0 cm to come up with 90%), while the right piece has 10% of the mass of the handle (3.0 cm divided by 30.0 cm makes 10%) plus the mass of the mallet's end. Recall from math that 90% equals 0.9 when expressed as a decimal, and similarly 10% is 0.1. Therefore, $m_L = 0.9\, m_h$ represents that 90% of the handle lies to the left of the fulcrum, and $m_R = 0.1\, m_h + m_e$ represents that 10% of the handle plus the mallet's end lie to the right of the fulcrum.

$$m_L = 0.9\, m_h = (0.9)(4) = 3.6 \text{ kg}$$

$$m_R = 0.1\, m_h + m_e = (0.1)(4) + 8 = 0.4 + 8 = 8.4 \text{ kg}$$

As a check, we can confirm that $m_L + m_R = 3.6 + 8.4 = 12.0$ kg add up to the total mass of the mallet. The left end has a mass of 3.6 kg, while the right end has a mass of 8.4 kg. These are the final answers.

Are you surprised that the two ends **don't** have the same mass? The notion that the masses "should" be the same is a common **misconception** among physics students. The weights **aren't** equal. Instead, the torques are equal. The system is balanced when the two torques cancel out. Each torque involves an expression of the form rmg (with different r's and m's). Since the left end has a larger r and the right end has a smaller r, the masses must be **different** in order for the torques to be equal: The left piece must have a smaller mass to compensate for its larger r, while the right piece has a larger mass because it is closer (on average) to the fulcrum.

Example: As shown below, a 20-kg box of bananas is suspended from a rope in static equilibrium. The vertical rope is connected to two other ropes which are supported from the ceiling. Determine the tension in each cord.

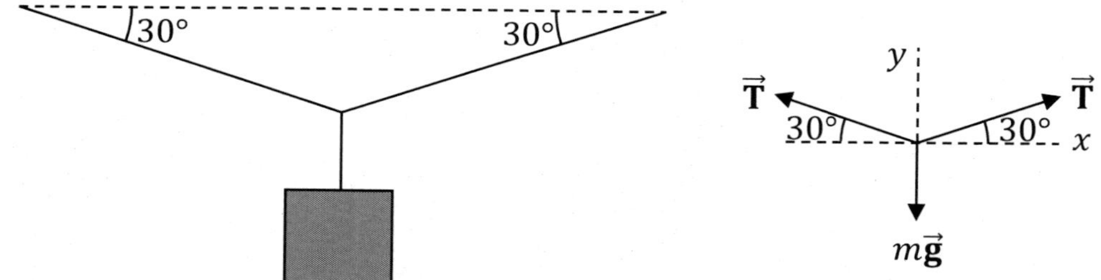

The trick to this problem is to draw a FBD for the knot (where the three cords meet). Since the system is in static equilibrium, we don't need to know the mass of the knot (the right-hand side of Newton's second law equals zero since the acceleration is zero).

Since the system is in static equilibrium, it's not rotating, so we are free to choose the axis of rotation anywhere we like. Let's choose the axis of rotation to pass through the knot, perpendicular to the page. With this choice, all of the torques are zero since each force pulls directly away from the knot (it's like pushing on the hinges to try to open a door). The sum of the torques simply states that zero equals zero in this problem.

Resolve each force into components following the technique from Chapter 14. In this problem, the two tensions are equal since the FBD is symmetric (both angles equal 30°).

$$\sum F_x = 0 \quad , \quad \sum F_y = 0$$

$$T\cos 30° - T\cos 30° = 0 \quad , \quad T\sin 30° + T\sin 30° - mg = 0$$

In this example, the left equation is just an identity ($T = T$). Use the right-hand side to solve for tension.

$$T\sin 30° + T\sin 30° = mg$$

$$2\,T\sin 30° = mg$$

$$2\,T\frac{1}{2} = mg$$

$$T = mg = (20)(9.81) \approx (20)(10) = 200 \text{ N}$$

There are three answers to this problem:

- The top left tension equals 200 N.
- The top right tension equals 200 N by symmetry.
- The bottom tension also equals 200 N. It equals the weight of the box of bananas:

  $$mg = (20)(9.81) \approx (20)(10) = 200 \text{ N}$$

  If you draw a FBD for the box of bananas, there will be an upward tension ($\mathbf{T}_b$) and a downward weight ($m\vec{\mathbf{g}}$). These must be equal for the box to be in static equilibrium.

Example: As illustrated below, a uniform boom (the object that looks like a rod or pole) is supported by a horizontal tie rope that connects to a wall. The system is in static equilibrium. The boom has a mass of 50 kg and is 10-m long. The lower end of the boom is connected to a hingepin. The tie rope can sustain a maximum tension of $750\sqrt{3}$ N. (A) What maximum load can the boom support without snapping the tie rope? (B) Find the magnitude and direction of the maximum force exerted on the hingepin.

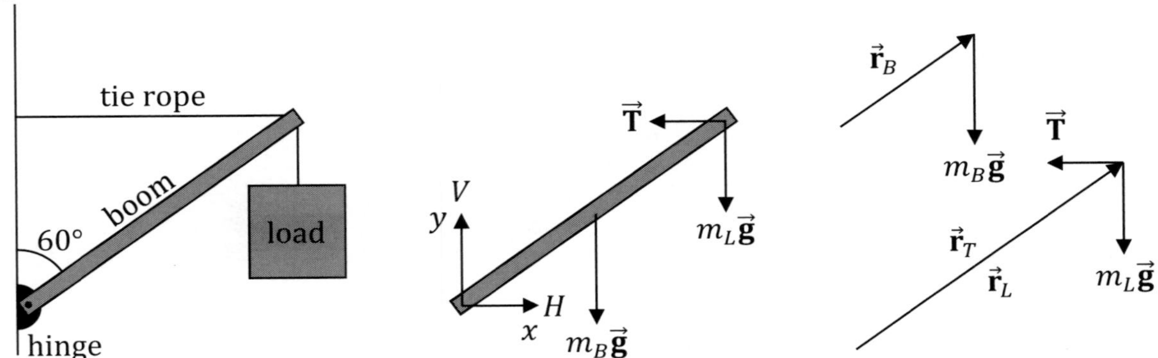

(A) Begin by drawing an extended free-body diagram for the boom. See the center diagram above. In the extended free-body diagram, we draw the forces where they act on the boom. Tension pulls to the left, along the tie rope. The weights pull downward. The weight of the boom effectively acts at the center of the boom (on average). In the diagram, H and V are the horizontal and vertical components of the hingepin force. The values of r and θ needed in the torque equations are tabulated below (join each $\vec{\mathbf{r}}$ to its corresponding force tail to tail to see the angles correctly).

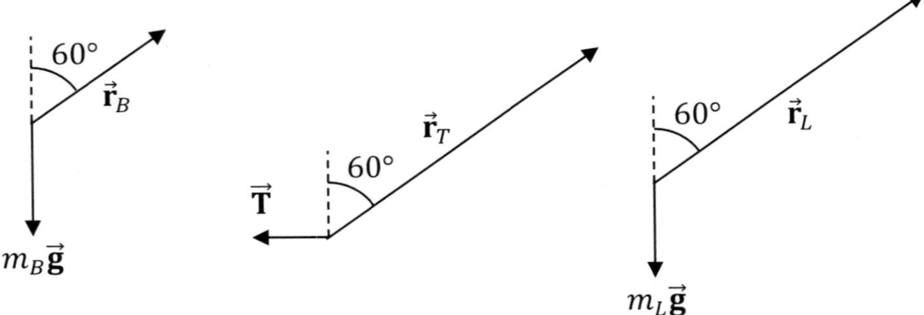

$$r_V = 0 \quad , \quad r_H = 0 \quad , \quad r_B = \frac{L}{2} = \frac{10}{2} = 5.0 \text{ m} \quad , \quad r_T = L = 10.0 \text{ m} \quad , \quad r_L = L = 10.0 \text{ m}$$

$$\theta_B = 180° - 60° = 120° \quad , \quad \theta_T = 90° + 60° = 150° \quad , \quad \theta_L = 180° - 60° = 120°$$

Sum the forces and torques for the system. We choose clockwise to be the positive rotation direction, such that the torque exerted by the tension is negative.

$$\sum F_x = 0 \quad , \quad \sum F_y = 0 \quad , \quad \sum \tau = 0$$

$$H - T = 0 \quad , \quad V - m_B g - m_L g = 0 \quad , \quad r_B m_B g \sin 120° + r_L m_L g \sin 120° - r_T T \sin 150° = 0$$

We can answer the first question by solving for m_L in the torque sum:

$$r_B m_B g \sin 120° + r_L m_L g \sin 120° - r_T T \sin 150° = 0$$

$$r_L m_L g \sin 120° = r_T T \sin 150° - r_B m_B g \sin 120°$$

Use the maximum tension ($750\sqrt{3}$ N) to find the maximum load that the boom can support:

$$(10)m_L(9.81)\left(\frac{\sqrt{3}}{2}\right) = (10)(750\sqrt{3})\left(\frac{1}{2}\right) - (5)(50)(9.81)\left(\frac{\sqrt{3}}{2}\right)$$

$$(10)m_L(10)\left(\frac{\sqrt{3}}{2}\right) \approx (10)(750\sqrt{3})\left(\frac{1}{2}\right) - (5)(50)(10)\left(\frac{\sqrt{3}}{2}\right)$$

$$50\sqrt{3}\, m_L \approx 3750\sqrt{3} - 1250\sqrt{3} = 2500\sqrt{3}$$

$$m_L \approx \frac{2500\sqrt{3}}{50\sqrt{3}} = \frac{2500}{50} = 50 \text{ kg}$$

The boom can support a maximum load of 50 kg.

(B) Solve for the horizontal and vertical components of the hingepin force from the expressions that we obtained previously from summing the forces:

$$H - T = 0 \quad , \quad V - m_B g - m_L g = 0$$

Use the maximum tension and maximum load to find the maximum components of the hingepin force.

$$H = T = 750\sqrt{3} \text{ N}$$

$$V = m_B g + m_L g = (50)(9.81) + (50)(9.81)$$

$$V \approx (50)(10) + (50)(10) = 500 + 500 = 1000 \text{ N}$$

Use the Pythagorean theorem to find the magnitude of the maximum hingepin force:

$$F_H = \sqrt{H^2 + V^2} = \sqrt{50^2 + 50^2} = 50\sqrt{1^2 + 1^2} = 50\sqrt{2} \text{ N}$$

Use an inverse tangent to determine the direction of the maximum hingepin force:

$$\theta_H = \tan^{-1}\left(\frac{V}{H}\right) = \tan^{-1}\left(\frac{50}{50}\right) = \tan^{-1}(1) = 45°$$

185. In the diagram below, a 60-kg black box rests on a plank, 5.0 m to the left of the fulcrum. The fulcrum rests beneath the center of the plank. Where should a 75-kg white box be placed in order for the system to be in static equilibrium?

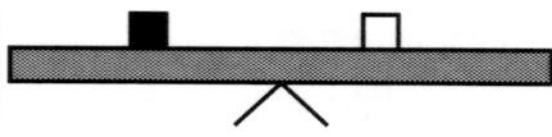

186. In the diagram below, a 20-kg white box rests on a plank, 2.0 m to the right of the fulcrum, while a black box rests 5.0 m to the left of the fulcrum. The fulcrum rests beneath the center of the plank. The system is in static equilibrium. Find the mass of the black box.

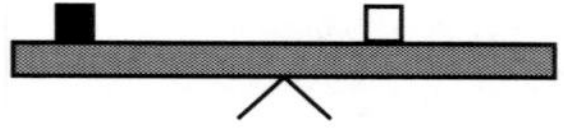

Want help? Check the hints section at the back of the book.

Answers: 4.0 m to the right of the fulcrum, 8.0 kg

187. In the diagram below, a 100-kg black box rests on a plank, 4.0 m to the left of the fulcrum, and a 50-kg white box rests 16.0 m to the left of the fulcrum. The fulcrum rests beneath the center of the plank. Where should a 60-kg gray box be placed in order for the system to be in static equilibrium?

188. In the diagram below, a 30-kg white box rests on a plank, 2.5 m to the left of the fulcrum, and a 40-kg black box rests 7.5 m to the right of the fulcrum. The fulcrum rests beneath the center of the plank. Where should a 25-kg gray box be placed in order for the system to be in static equilibrium?

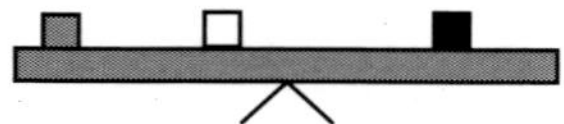

Want help? Check the hints section at the back of the book.

Answers: 20.0 m to the right of the fulcrum, 9.0 m to the left of the fulcrum

189. As illustrated below, a 40-kg monkey is suspended from a rope in static equilibrium. The monkey's rope is connected to two other ropes which are supported from the ceiling. Determine the tension in each cord.

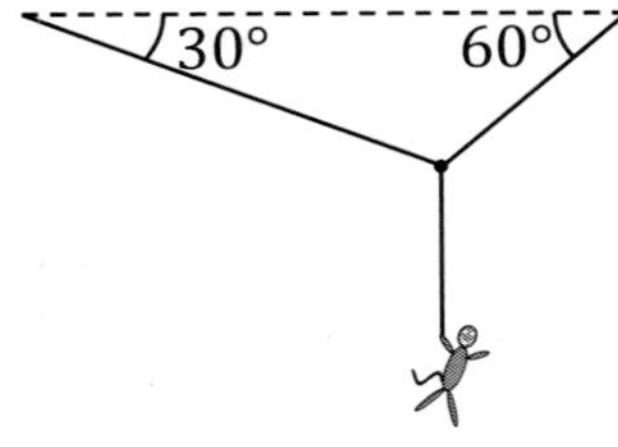

Want help? Check the hints section at the back of the book.

Answers: 200 N (left), $200\sqrt{3}$ N (right), 400 N (bottom)

190. As illustrated below, a uniform boom is supported by a tie rope that connects to a wall. The system is in static equilibrium. The boom has a mass of 50 kg and is 10-m long. The lower end of the boom is connected to a hingepin. The tie rope is perpendicular to the boom and can sustain a maximum tension of 600 N.

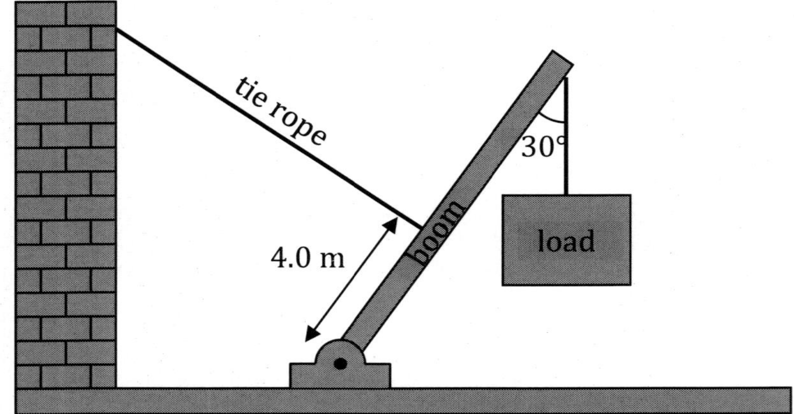

(A) What maximum load can the boom support without snapping the tie rope?

(B) Find the maximum horizontal and vertical components of the force exerted on the hingepin.

Want help? Check the hints section at the back of the book.

Answers: 23 kg, $300\sqrt{3}$ N, 430 N

32 MOMENT OF INERTIA

Relevant Terminology

Momentum – mass times velocity.
Rigid body – an object that doesn't change shape when it rotates.
Inertia – the natural tendency of any object to maintain constant momentum.
Moment of inertia – the natural tendency of a rigid body to maintain constant angular momentum.

Essential Concepts

The moment of inertia of a rigid body is a measure of its tendency to resist changes to its angular momentum. An object's moment of inertia depends on how its mass is distributed about the axis of rotation:

- **More mass further from the axis of rotation** results in a **larger moment of inertia**, as illustrated in the following conceptual example.
- The greater an object's moment of inertia, the **harder it is to change the object's angular momentum**.

Conceptual Example

Consider three different ways of rotating a rod illustrated below.

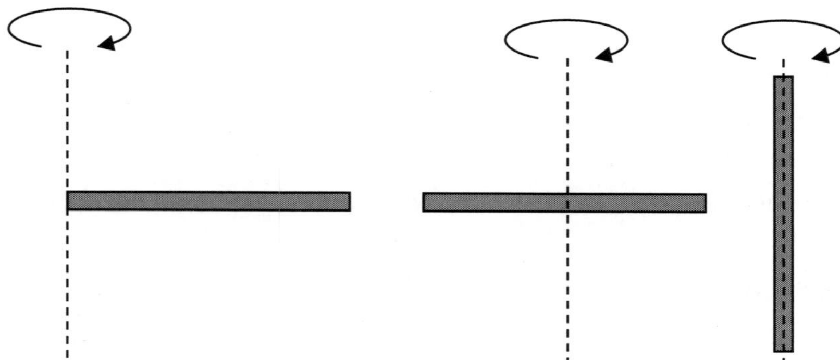

In which case is it easier to change the angular speed of the rod? Grab a rod or yardstick and test it out.

- **left** figure: It is hardest to change the angular speed in this case. See how long it takes you to complete one revolution. More mass is far from the axis of rotation in this case. The moment of inertia is **greatest** for this axis of rotation.
- **middle** figure: This case is in between the other two. You can spin it faster than you can in the left picture, but not as fast as in the right picture.

- **right** figure: It is easiest to change the angular speed in this case. If you grab the rod between your hands and give your hands a quick flick, you can easily make it spin rapidly, completing multiple revolutions per second. All of the mass is very close to the axis of rotation in this case. The moment of inertia is **smallest** for this axis of rotation.

Parallel-Axis Theorem

Sometimes, an object isn't rotating about an axis listed in a table of moments of inertia. When that's the case, you may be able to apply the parallel-axis theorem. If you know the moment of inertia of an object about an axis passing through its center of mass, you can use the parallel-axis theorem to find the moment of inertia (I) about an axis parallel to the axis that passes through the center of mass.

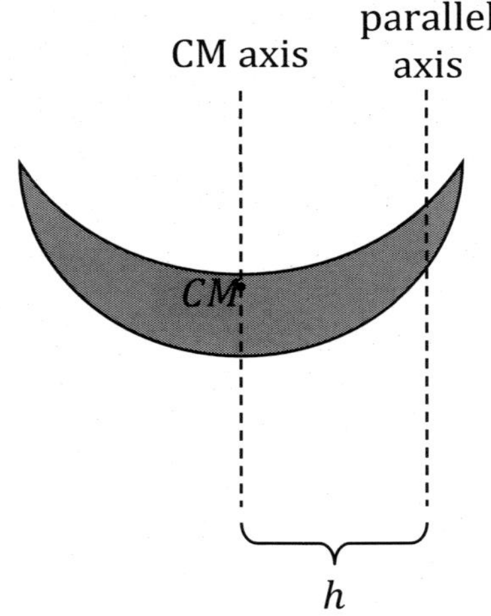

The equation for the **parallel-axis theorem** is:

$$I_p = I_{CM} + mh^2$$

Here is what these symbols mean:

- I_{CM} is the moment of inertia of the object about an axis passing through its **center of mass**. You might find this in a table of moments of inertia (see pages 333-336).
- I_p is the moment of inertia of the object about an axis that is **parallel** to the axis used for I_{CM}.
- m is the mass of the object.
- h is the distance between the two parallel axes, as illustrated above.

Moment of Inertia Equations

The equation for moment of inertia (I) depends on:

- the shape of the object
- the axis about which the object is rotating

For a **pointlike** object traveling in a circle, use the following equation for moment of inertia, where R is the radius of the circle that the object is traveling in:

$$I = mR^2$$

For a system of objects that share a common axis of rotation, find the moment of inertia of each object and then add the moments of inertia together.

$$I = I_1 + I_2 + \cdots + I_N$$

The following pages of tables show a variety of common geometric objects about common axes of rotation. **Note**: If the axis you need is parallel to an axis that passes through the center of mass of the object, you can apply the **parallel-axis theorem** (discussed on the previous page).

$$I_p = I_{CM} + mh^2$$

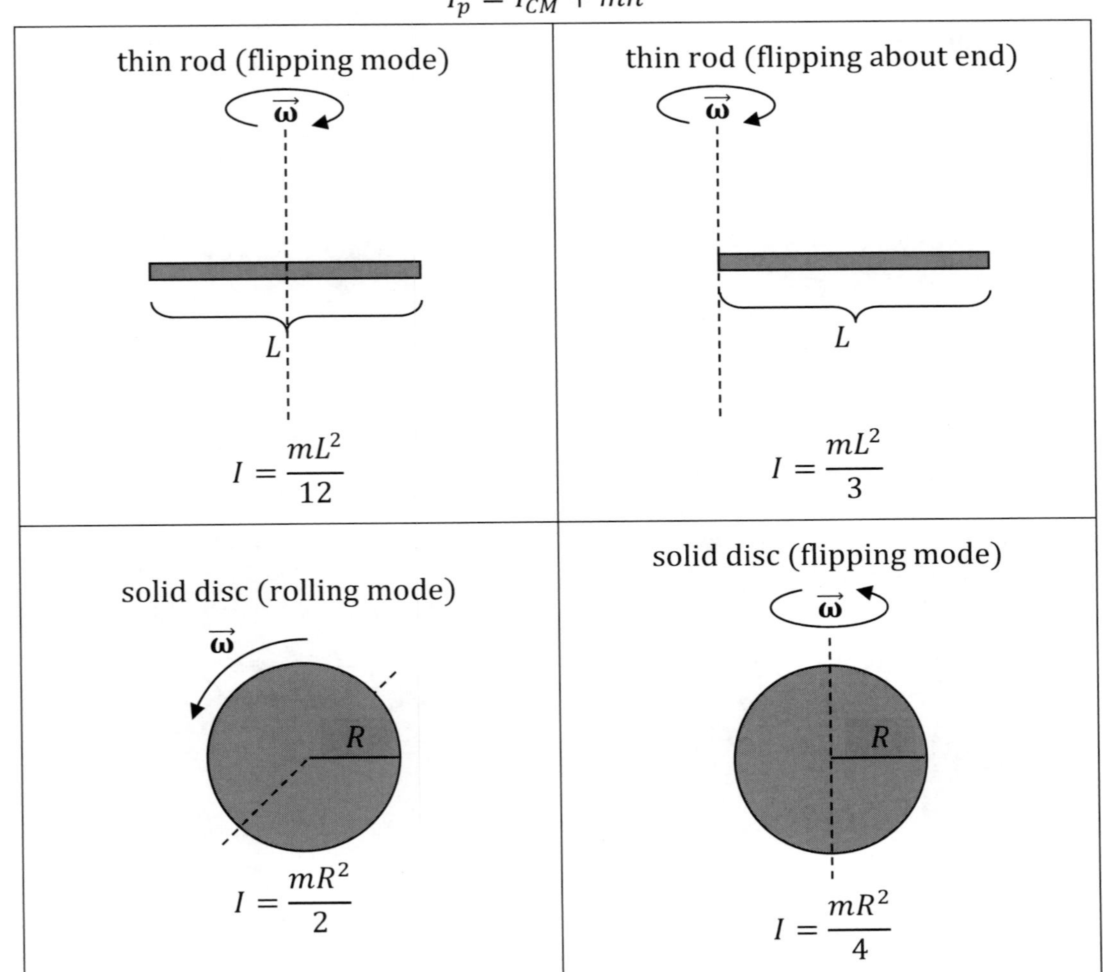

thin ring (rolling mode) 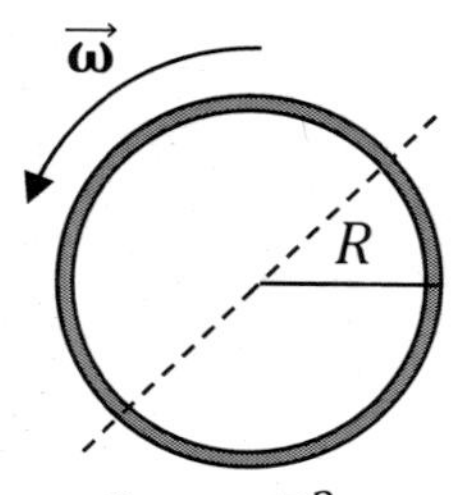 $I = mR^2$	thin ring (flipping mode) 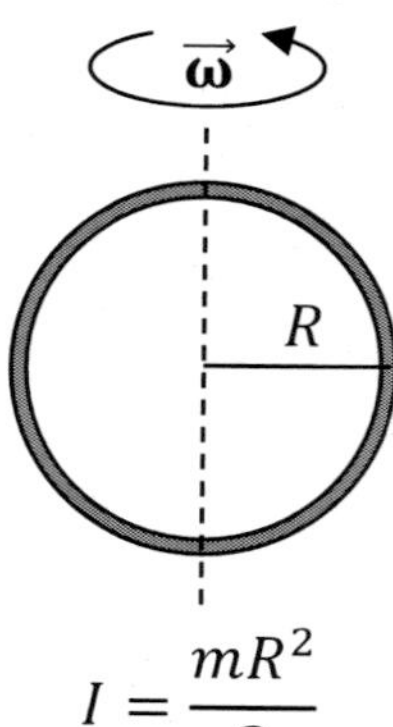 $I = \frac{mR^2}{2}$
thick ring (rolling mode) 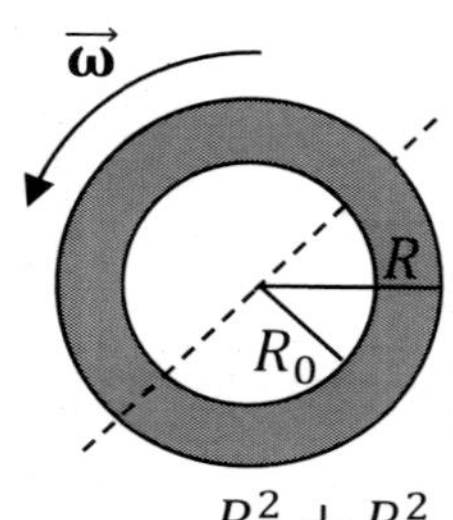 $I = m\frac{R_0^2 + R^2}{2}$	thick ring (flipping mode) 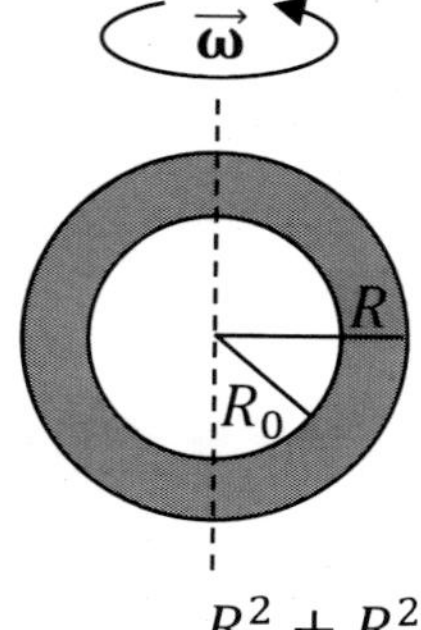 $I = m\frac{R_0^2 + R^2}{4}$
solid cylinder (rolling mode) 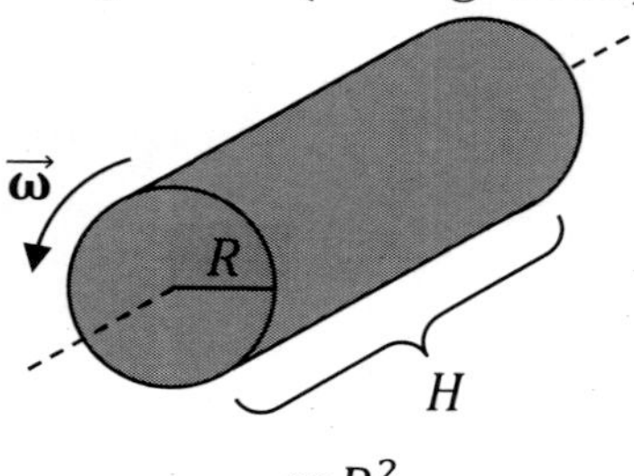 $I = \frac{mR^2}{2}$	thin hollow cylinder (rolling mode) 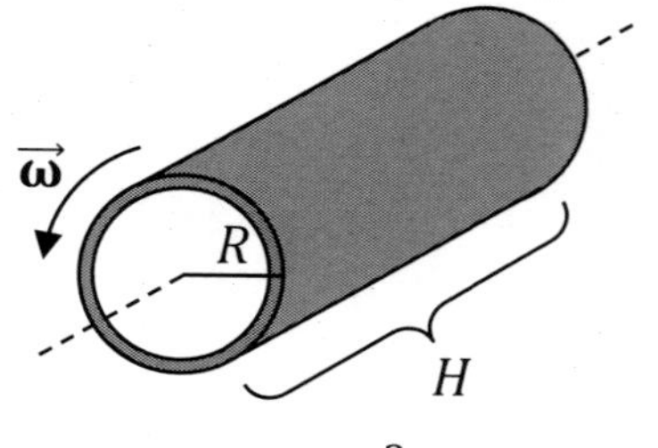 $I = mR^2$

thick hollow cylinder (rolling mode) 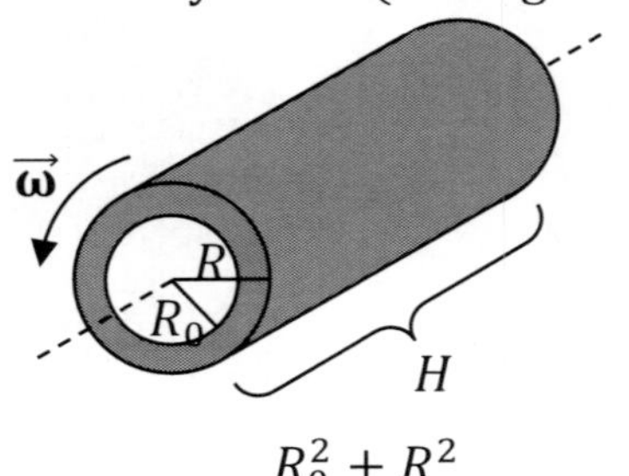 $I = m\frac{R_0^2 + R^2}{2}$	solid sphere (rolling mode) 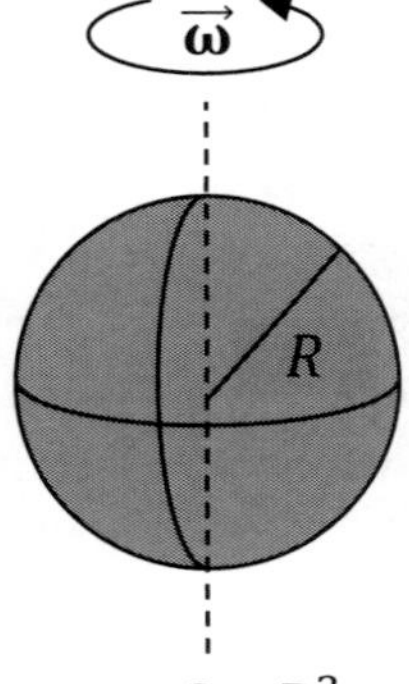 $I = \frac{2mR^2}{5}$
thin hollow sphere (rolling mode) 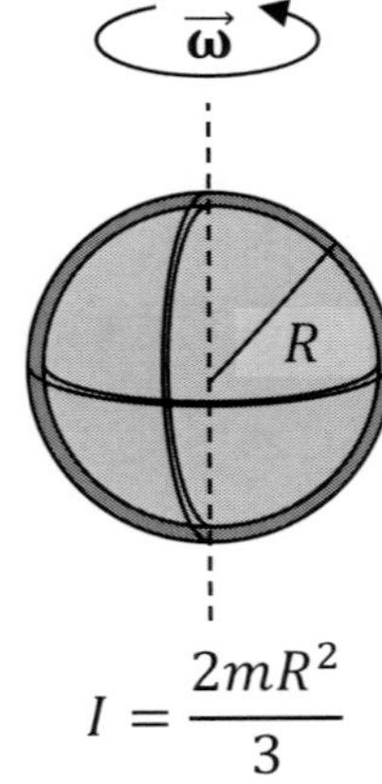 $I = \frac{2mR^2}{3}$	thick hollow sphere (rolling mode) 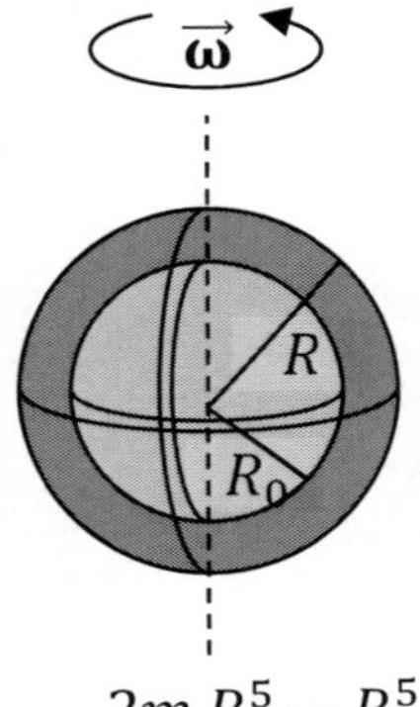 $I = \frac{2m}{5}\frac{R^5 - R_0^5}{R^3 - R_0^3}$
solid cube about an axis through its center and perpendicular to a face 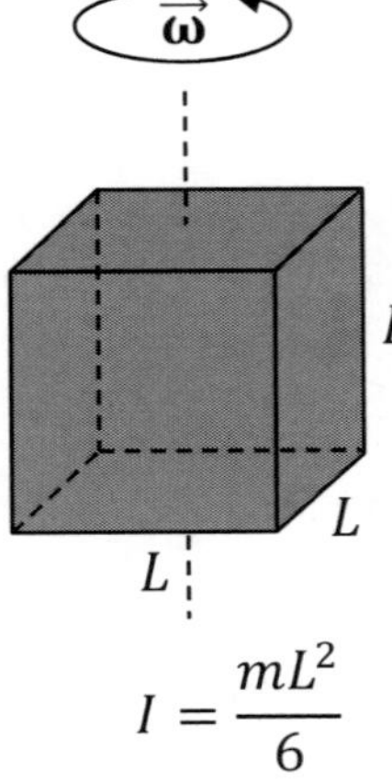 $I = \frac{mL^2}{6}$	solid rectangle about an axis through its center and perpendicular to the rectangle 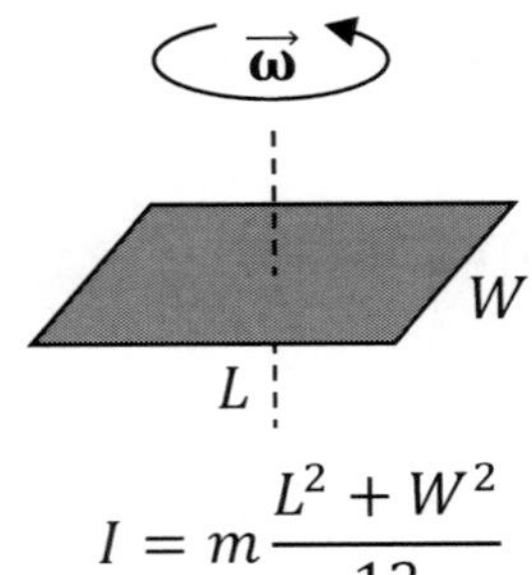 $I = m\frac{L^2 + W^2}{12}$

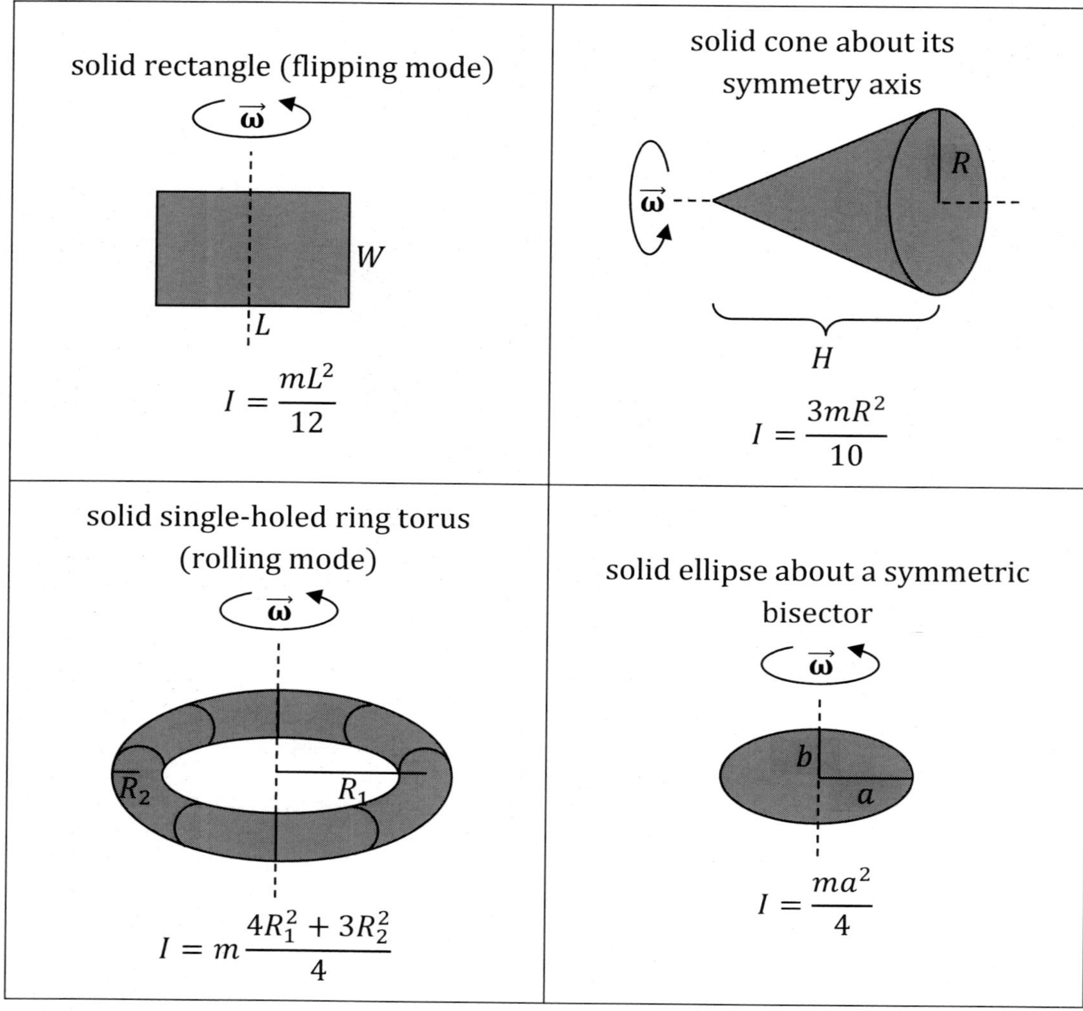

For a shape or axis of rotation that you don't find in the table, you can derive an equation for the moment of inertia by applying calculus:

$$I = \int r_\perp^2 \, dm$$

Here, $r_\perp^2$ is the shortest distance from each dm to the axis of rotation. See pages 343-354.

Notes Regarding Units

The SI units of moment of inertia (I) follow from the equations for moment of inertia. All of the equations involve mass and a distance squared, such as $I = mR^2$, $I = \frac{2}{5}mR^2$, or $I = \frac{mL^2}{12}$. Since the SI unit of mass is the kilogram (kg) and the SI unit of distance is the meter (m), it follows that moment of inertia's SI units are kg·m^2 (since distance is squared in the formulas).

Symbols and SI Units

Symbol	Name	SI Units
m	mass	kg
R	radius	m
R_0	inner radius	m
L	length	m
W	width	m
I	moment of inertia	kg·m^2
I_{CM}	moment of inertia about an axis passing through the center of mass	kg·m^2
I_p	moment of inertia about an axis parallel to the axis used for I_{CM}	kg·m^2
h	distance between the two parallel axes used for I_{CM} and I_p	m

Strategy for finding the Moment of Inertia of a System of Objects

How you find moment of inertia depends on the situation:

1. For a single **pointlike** object traveling in a circle (or for an object that is small compared to the radius of the circle it is traveling in), use the following formula:
$$I = mR^2$$
2. For a common geometric shape like a disc, rod, or sphere, see if you can find the formula in a **table** (see pages 333-336). Otherwise, see pages 343-354.
3. If the axis you need is parallel to an axis that passes through the center of mass of the object, you can apply the **parallel-axis theorem** (see page 339):
$$I_p = I_{CM} + mh^2$$
4. If there are two or more objects that share a common axis of rotation, find the moment of inertia of each object and then add the moments of inertia together:
$$I = I_1 + I_2 + \cdots + I_N$$
If all of the objects are pointlike and share a common axis of rotation, the formula for the moment of the inertia of the system becomes:
$$I = m_1R_1^2 + m_2R_2^2 + \cdots + m_NR_N^2$$

Example: As shown below, four 5.0-kg masses are joined together in the shape of a square with approximately massless rods. The 5.0-kg masses are small in size compared to the length of the rods. Determine the moment of inertia of the system about the y-axis.

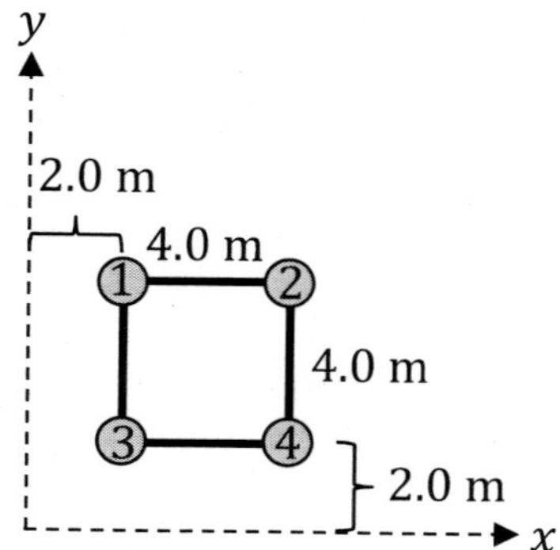

Since the rods are very light (approximately massless compared to the 5.0-kg masses), we can neglect them. Since the masses are small compared to the length of the rods, we can treat them as pointlike objects. Use the formula for the moment of inertia of four pointlike objects sharing a common axis of rotation (the y-axis):

$$I = m_1 R_1^2 + m_2 R_2^2 + m_3 R_3^2 + m_4 R_4^2$$

All of the masses are the same (5 kg). Each R is the distance of the corresponding mass from the axis of rotation (the y-axis). The left two masses are 2.0 m from the y-axis, while the right two masses are 2.0 m + 4.0 m = 6.0 m from the y-axis:

$$I = (5)(2)^2 + (5)(6)^2 + (5)(2)^2 + (5)(6)^2 = 20 + 180 + 20 + 180 = 400 \text{ kg·m}^2$$

Example: The yo-yo illustrated below consists of a hollow cylinder with a mass of 50 g and a radius of 3.0 cm between two solid discs. Each solid disc has a mass of 25 g and a radius of 8.0 cm. Determine the moment of inertia of the yo-yo about its natural axis.

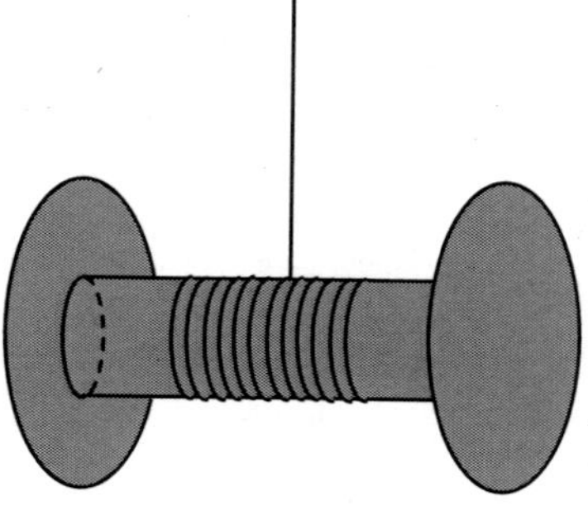

All three objects share the same axis of rotation (the axis of the cylinder). Look up the moments of inertia for a hollow cylinder and solid disc in rolling mode in the table on pages 333-336: $I_c = m_c R_c^2$ and $I_d = \frac{1}{2} m_d R_d^2$. Add these moments of inertia together.

$$I = m_c R_c^2 + \frac{1}{2} m_d R_d^2 + \frac{1}{2} m_d R_d^2 = (50)(3)^2 + \frac{1}{2}(25)(8)^2 + \frac{1}{2}(25)(8)^2 = 2050 \text{ g·cm}^2$$

Note: If you leave mass in grams (g) and radius in centimeters (cm), moment of inertia comes out in g·cm^2 instead of kg·m^2. If you want SI units, convert g to kg and cm to m first.

Example: The uniform rod shown below rotates about the axis p. The rod has a mass of $\frac{1}{2}$ kg and a length of 2.0 m. Determine its moment of inertia about the axis p.

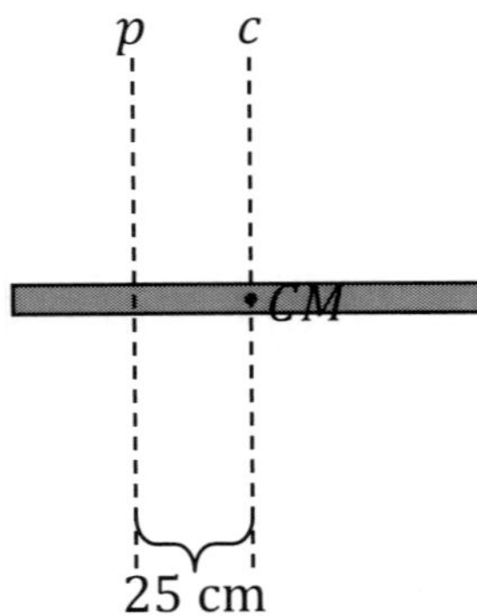

The strategy for this problem is to first look up the formula for the moment of inertia about the axis through the center of the rod and then apply the parallel-axis theorem. According to the table, the moment of inertia about the axis c is:

$$I_{CM} = \frac{mL^2}{12} = \frac{\left(\frac{1}{2}\right)(2)^2}{12} = \frac{1}{6} \text{ kg·m}^2$$

Now we apply the parallel-axis theorem with $h = 25 \text{ cm} = 0.25 \text{ m} = \frac{1}{4}$ m:

$$I_p = I_{CM} + mh^2 = \frac{1}{6} + \left(\frac{1}{2}\right)\left(\frac{1}{4}\right)^2 = \frac{1}{6} + \frac{1}{32} = \frac{16}{96} + \frac{3}{96} = \frac{19}{96} \text{ kg·m}^2$$

Example: The diagram below shows a uniform rod that is welded onto a solid sphere. The rod has a mass of 9.0 kg and length of 6.0 m, while the sphere has a mass of 5.0 kg and a radius of 2.0 m. Determine the moment of inertia about the axis shown.

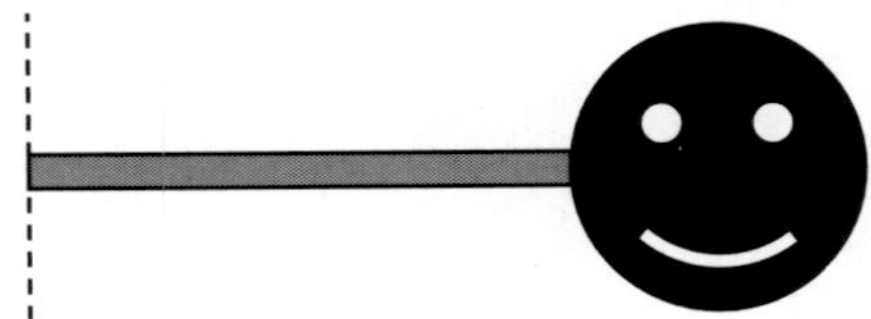

The rod is rotating about its end. This moment of inertia is listed in the table:

$$I_r = \frac{1}{3}m_r L^2 = \frac{1}{3}(9)(6)^2 = 108 \text{ kg·m}^2$$

We first find the solid sphere's moment of inertia about an axis through its center:

$$I_{CM} = \frac{2}{5}m_s R^2 = \frac{2}{5}(5)(2)^2 = 8.0 \text{ kg·m}^2$$

Now we apply the parallel-axis theorem where $h = L + R = 6.0 + 2.0 = 8.0$ m (that's the distance between the desired axis and the center of the sphere):

$$I_p = I_{CM} + m_s h^2 = 8 + (5)(8)^2 = 328 \text{ kg·m}^2$$

Add the moments of inertia of the rod and sphere (both about axis p) together:

$$I = I_r + I_p = 108 + 328 = 436 \text{ kg·m}^2$$

191. As illustrated below, three 6.0-kg masses are joined together with approximately massless rods in the shape of an equilateral triangle with an edge length of $L = 4.0$ m. The 6.0-kg masses are small in size compared to the length of the rods. Determine the moment of inertia of the system about (A) the x-axis, (B) the y-axis, and (C) the z-axis (which is perpendicular to the page and passes through the origin).

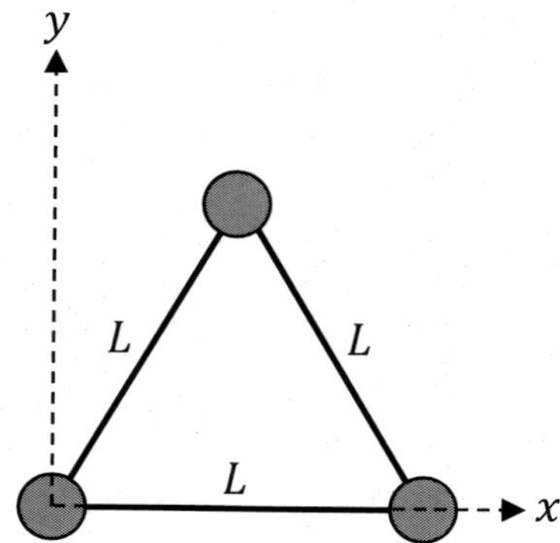

Want help? Check the hints section at the back of the book.

Answer: 72 kg·m^2, 120 kg·m^2, 192 kg·m^2

192. Three bananas are joined together with approximately massless rods. The bananas are small in size compared to the length of the rods. The masses and the (x, y) coordinates of the bananas are listed below. Determine the moment of inertia of the system about (A) the x-axis, (B) the y-axis, and (C) the z-axis (which is perpendicular to the page and passes through the origin).

- A 400-g banana has coordinates (3.0 m, 0).
- A 300-g banana has coordinates (0, 2.0 m).
- A 500-g banana has coordinates (−4.0 m, 1.0 m).

Want help? Check the hints section at the back of the book.

Answer: $\frac{17}{10}$ kg·m^2, $\frac{58}{5}$ kg·m^2, $\frac{133}{10}$ kg·m^2

193. The diagram below shows a uniform rod that is welded onto two hollow spheres. The rod has a mass of 6.0 kg and length of 8.0 m, the left sphere has a mass of 9.0 kg and a radius of 2.0 m, and the right sphere has a mass of 3.0 kg and a radius of 1.0 m. Determine the moment of inertia about the axis shown, which bisects the rod.

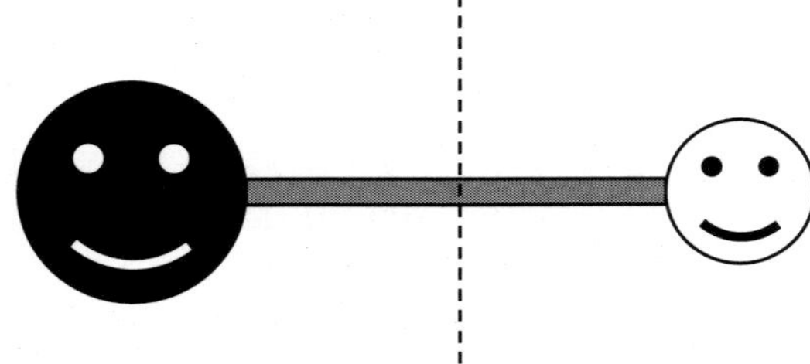

Want help? Check the hints section at the back of the book.

Answer: 457 $kg \cdot m^2$

Strategy for Performing the Moment of Inertia Integral

To find the moment of inertia of a continuous object like a rod or disc, follow these steps:

1. Draw the object. Draw and label a representative dm somewhere within the object. **Don't** draw dm at the origin or on an axis (unless the object is a rod and every point of the rod lies on an axis, then you have no choice).
2. Draw and label the specified axis of rotation.
3. Draw the shortest possible line connecting dm to the specified axis of rotation. (This line will be **perpendicular** to the axis of rotation). Label the length of this line as $r_\perp$. See the diagrams in the examples that follow.
4. Begin with the formula for the moment of inertia integral: $I = \int r_\perp^2 \, dm$
5. Make one of the following substitutions for dm, depending on the geometry:
   - $dm = \lambda ds$ for an arc length (like a rod or circular arc).
   - $dm = \sigma dA$ for a surface area (like a triangle, disc, or thin spherical shell).
   - $dm = \rho dV$ for a 3D solid (like a solid cube or a solid hemisphere).
6. Choose the appropriate coordinate system and make a substitution for ds, dA, or dV from Step 3 using the strategy from Chapter 26:
   - For a straight line parallel to the x-axis, $ds = dx$.
   - For a straight line parallel to the y-axis, $ds = dy$.
   - For a circular arc of radius R, $ds = R d\theta$.
   - For a solid polygon like a rectangle or triangle, $dA = dxdy$.
   - For a solid semicircle (**not** a circular arc) or pie slice, $dA = r dr d\theta$.
   - For a very thin spherical shell of radius R, $dA = R^2 \sin\theta \, d\theta d\varphi$.
   - For a solid polyhedron like a cube, $dV = dxdydz$.
   - For a solid cylinder or cone, $dV = r_c dr_c d\theta dz$.
   - For a portion of a solid sphere like a hemisphere, $dV = r^2 \sin\theta \, dr d\theta d\varphi$.
7. Is the density uniform or non-uniform?
   - If the density is uniform, you can pull λ, σ, or ρ out of the integral.
   - If the density is non-uniform, leave λ, σ, or ρ in the integral.
8. Look at the $r_\perp$ that you drew in Step 3. Express $r_\perp$ in terms of coordinates listed in Step 6. Study the examples that follow to learn how to write this equation.
9. Which coordinate system did you choose in Step 6?
   - Cartesian coordinates: Leave x, y, and z as they are.
   - 2D polar coordinates: Replace x and y with the following expressions:
     $$x = r\cos\theta \quad , \quad y = r\sin\theta$$
   - Cylindrical coordinates: Replace x and y with the following expressions:
     $$x = r_c\cos\theta \quad , \quad y = r_c\sin\theta$$
   - Spherical coordinates: Replace x, y, and z with the following expressions:
     $$x = r\sin\theta\cos\varphi \quad , \quad y = r\sin\theta\sin\varphi \quad , \quad z = r\cos\theta$$

10. An integral over ds is a single integral, an integral over dA is a double integral, and an integral over dV is a triple integral. Set the limits of each integration variable that map out the region of integration, as illustrated in Chapter 26. Perform the integral using techniques from this chapter and also from Chapters 2, 9, and 26.
11. Perform the following integral to determine the total mass (m) of the object: $m = \int dm$. Make the same substitutions as you made in Steps 5-9. Use the same limits of integration as you used in Step 10.
12. When you finish with Step 11, substitute your expression into the equation from your original moment of inertia integral. Simplify the resulting expression. Your final answer should involve the mass of the object and a distance squared.

Symbols and SI Units

Symbol	Name	SI Units
dm	differential mass element	kg
m	total mass of the object	kg
$r_\perp$	shortest distance from dm to the axis of rotation	m
x, y, z	Cartesian coordinates of dm	m, m, m
r, θ	2D polar coordinates of dm	m, rad
r_c, θ, z	cylindrical coordinates of dm	m, rad, m
r, θ, φ	spherical coordinates of dm	m, rad, rad
λ	linear mass density (for an arc)	kg/m
σ	mass density for a surface area	kg/m^2
ρ	mass density for a 3D solid	kg/m^3
ds	differential arc length	m
dA	differential area element	m^2
dV	differential volume element	m^3
I	moment of inertia	kg·m^2

Note: The symbols λ, σ, and ρ are the lowercase Greek letters lambda, sigma, and rho.

Example: The rod illustrated below has non-uniform linear mass density $\lambda(x) = \beta x$, where β is a positive constant. Derive an equation for the moment of inertia of the rod about the line $x = \frac{L}{3}$ in terms of the total mass of the rod, m, and the length of the rod, L.

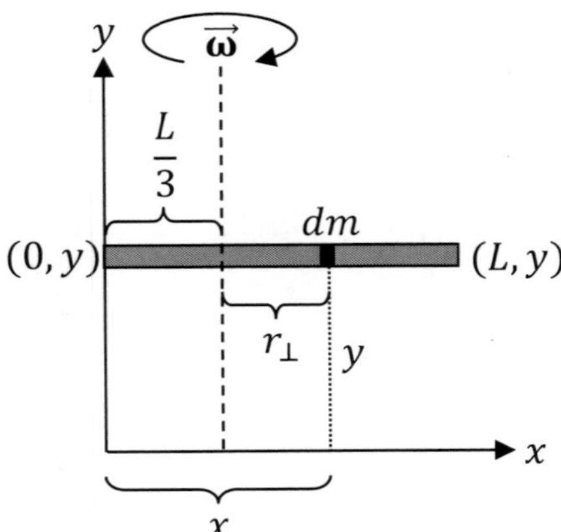

Draw a representative dm. Label $r_\perp$ for that dm (where $r_\perp$ is the shortest distance from dm to the axis of rotation). When we perform the integration, we effectively integrate over every dm that makes up the rod. Begin with the moment of inertia integral.

$$I = \int r_\perp^2 \, dm$$

For a thin rod, we write $dm = \lambda ds$.

$$I = \int r_\perp^2 \lambda \, ds$$

Since the rod has non-uniform density, we **can't** pull λ out of the integral. Instead, we substitute the equation for λ given in the problem: $\lambda = \beta x$.

$$I = \int r_\perp^2 \beta x \, ds$$

We can pull β out of the integral since β is a constant. For a rod lying along the x-axis, we work with Cartesian coordinates and write the differential arc length as $ds = dx$. The limits of integration correspond to the length of the rod: $0 \le x \le L$.

$$I = \beta \int_{x=0}^{L} r_\perp^2 x \, dx$$

Looking at the diagram above, it should be clear that $r_\perp = x - \frac{L}{3}$ for the representative dm. Note that if we had drawn dm to the left of the axis of rotation, we would have found that $r_\perp = \frac{L}{3} - x$. This distinction is not important since $r_\perp$ is **squared**: $\left(x - \frac{L}{3}\right)^2 = \left(\frac{L}{3} - x\right)^2$.

$$I = \beta \int_{x=0}^{L} \left(x - \frac{L}{3}\right)^2 x \, dx$$

One way to perform this integral is to foil out $\left(x - \frac{L}{3}\right)^2$.

$$I = \beta \int_{x=0}^{L} \left(x^2 - \frac{2xL}{3} + \frac{L^2}{9}\right) x\, dx$$

Now multiply by the x.

$$I = \beta \int_{x=0}^{L} \left(x^3 - \frac{2x^2L}{3} + \frac{xL^2}{9}\right) dx = \beta \left[\frac{x^4}{4} - \frac{2x^3L}{9} + \frac{x^2L^2}{18}\right]_{x=0}^{L}$$

$$I = \beta \left(\frac{L^4}{4} - \frac{2L^4}{9} + \frac{L^4}{18}\right) = \beta \left(\frac{9L^4}{36} - \frac{8L^4}{36} + \frac{2L^4}{36}\right) = \frac{3\beta L^4}{36} = \frac{\beta L^4}{12}$$

Now integrate to find the total mass of the rod, using the same substitutions as before:

$$m = \int dm = \int \lambda\, ds = \int_{x=0}^{L} (\beta x) dx = \beta \int_{x=0}^{L} (x) dx = \beta \left[\frac{x^2}{2}\right]_{x=0}^{L} = \beta \left(\frac{L^2}{2} - \frac{0^2}{2}\right) = \frac{\beta L^2}{2}$$

Solve for β in terms of m in the above equation.

$$\beta = \frac{2m}{L^2}$$

Substitute this expression for β into the equation for moment of inertia.

$$I = \frac{\beta L^4}{12} = \left(\frac{2M}{L^2}\right) \frac{L^4}{12} = \frac{ML^2}{6}$$

Example: The right triangle illustrated below has uniform density. Derive an equation for its moment of inertia about the y-axis in terms of b, h, and/or m.

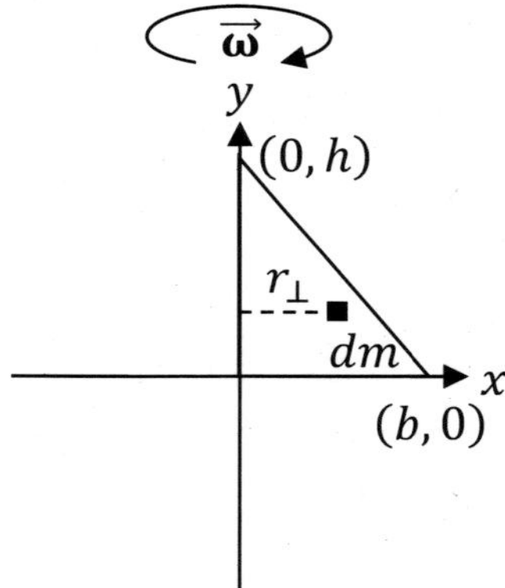

Draw a representative dm. Label $r_\perp$ for that dm (where $r_\perp$ is the shortest distance from dm to the axis of rotation). Begin with the moment of inertia integral.

$$I = \int r_\perp^2\, dm$$

For a solid triangle (as opposed to a wire bent into a triangle), we write $dm = \sigma dA$.

$$I = \int r_\perp^2 \sigma\, dA$$

Since the triangle has **uniform** density, we can pull σ out of the integral.

$$I = \sigma \int r_\perp^2\, dA$$

For a triangle, we work with Cartesian coordinates and write the differential area element as $dA = dxdy$. As with the example involving a triangle from Chapter 26, if we let x vary from 0 to b, we need to find the equation of the hypotenuse for the upper limit of y. That line has a slope equal to $\frac{y_2-y_1}{x_2-x_1} = \frac{0-h}{b-0} = -\frac{h}{b}$ and a y-intercept of h. The equation for this line is therefore $y = -\frac{h}{b}x + h$. For a given value of x, y will vary from 0 to $y = -\frac{hx}{b} + h$.

$$I = \sigma \int_{x=0}^{b} \int_{y=0}^{-\frac{hx}{b}+h} r_\perp^2 \, dxdy$$

As illustrated on the previous page, every dm lies a distance $r_\perp = x$ from the axis of rotation (the y-axis).

$$I = \sigma \int_{x=0}^{b} \int_{y=0}^{-\frac{hx}{b}+h} x^2 \, dxdy$$

We must perform the y-integration first because y has x in its upper limit. In the double integral, we can pull x out of the y-integral (but not out of the x-integral).

$$I = \sigma \int_{x=0}^{b} x^2 \left(\int_{y=0}^{-\frac{hx}{b}+h} dy \right) dx = \sigma \int_{x=0}^{b} x^2 \left(-\frac{hx}{b} + h\right) dx = \sigma \int_{x=0}^{b} \left(-\frac{hx^3}{b} + hx^2\right) dx$$

$$I = \sigma \left[-\frac{hx^4}{4b} + \frac{hx^3}{3}\right]_{x=0}^{b} = \sigma\left(-\frac{hb^4}{4b} + \frac{hb^3}{3}\right) = \sigma hb^3\left(-\frac{1}{4} + \frac{1}{3}\right) = \frac{\sigma hb^3}{12}$$

Now integrate to find the total mass of the triangle, using the same substitutions as before:

$$m = \int dm = \int \sigma \, dA = \sigma \int_{x=0}^{b} \int_{y=0}^{-\frac{hx}{b}+h} dx\, dy = \sigma \int_{x=0}^{b} \left(\int_{y=0}^{-\frac{hx}{b}+h} dy \right) dx = \sigma \int_{x=0}^{b} \left(-\frac{hx}{b} + h\right) dx$$

$$m = \sigma \left[-\frac{hx^2}{2b} + hx\right]_{x=0}^{b} = \sigma\left(-\frac{hb^2}{2b} + hb\right) = \sigma\left(-\frac{hb}{2} + hb\right) = \frac{\sigma hb}{2}$$

Solve for σ in terms of m in the above equation.

$$\sigma = \frac{2m}{hb}$$

Substitute this expression for σ into the equation for moment of inertia.

$$I = \frac{\sigma hb^3}{12} = \left(\frac{2m}{hb}\right)\frac{hb^3}{12} = \frac{mb^2}{6}$$

Example: Derive an equation for the moment of inertia of a solid uniform sphere about an axis that passes through its center in terms of its radius, R, and total mass, m.

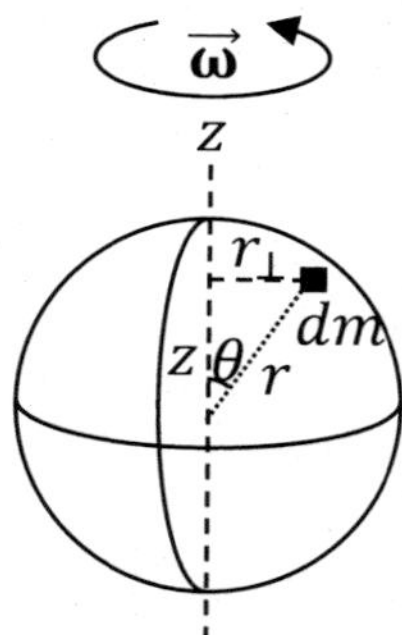

Draw a representative dm. Label $r_\perp$ for that dm (where $r_\perp$ is the shortest distance from dm to the axis of rotation). Begin with the moment of inertia integral.

$$I = \int r_\perp^2 \, dm$$

For a solid sphere (as opposed to a very thin hollow sphere), we write $dm = \rho dV$.

$$I = \int r_\perp^2 \rho \, dV$$

Since the solid sphere has **uniform** density, we can pull ρ out of the integral.

$$I = \rho \int r_\perp^2 \, dV$$

For a solid sphere, use $dV = r^2 \sin\theta \, dr d\theta d\varphi$ (Chapter 26). This requires integrating over r, θ, and φ. Unlike in the previous example, r, θ, and φ each have constant limits:

- $0 \le r \le R$
- $0 \le \theta \le \pi$
- $0 \le \varphi \le 2\pi$

Integrating over $0 \le r \le R$ sweeps out a line from the origin. Integrating over $0 \le \varphi \le 2\pi$ sweeps the line (from the r-integration) into a cone. It's then only necessary to integrate over $0 \le \theta \le \pi$ (and **not** up to 2π) to sweep the cone into a full sphere. Substitute $dV = r^2 \sin\theta \, dr d\theta d\varphi$ into the moment of inertia integral.

$$I = \rho \int_{r=0}^{R} \int_{\varphi=0}^{2\pi} \int_{\theta=0}^{\pi} r_\perp^2 r^2 \sin\theta \, dr d\theta d\varphi$$

As illustrated above, every dm lies a distance $r_\perp = r\sin\theta$ from the axis of rotation (the z-axis). Find the right triangle above: r is the hypotenuse and $r_\perp$ is opposite to θ.

$$I = \rho \int_{r=0}^{R} \int_{\varphi=0}^{2\pi} \int_{\theta=0}^{\pi} (r^2 \sin^2\theta) r^2 \sin\theta \, dr d\theta d\varphi = \rho \int_{r=0}^{R} \int_{\varphi=0}^{2\pi} \int_{\theta=0}^{\pi} r^4 \sin^3\theta \, dr d\theta d\varphi$$

We can do these integrals in any order since all of the limits are constants. We can pull r out of the angular integrals (but not the r-integral), and we can pull $\sin\theta$ out of the r and φ integrals (but not the θ-integral).

$$I = \rho \int_{r=0}^{R} r^4 dr \int_{\theta=0}^{\pi} \sin^3\theta \, d\theta \int_{\varphi=0}^{2\pi} d\varphi = \rho \left[\frac{r^5}{5}\right]_{r=0}^{R} [\varphi]_{\varphi=0}^{2\pi} \int_{\theta=0}^{\pi} \sin^3\theta \, d\theta = \frac{2\pi\rho R^5}{5} \int_{\theta=0}^{\pi} \sin^3\theta \, d\theta$$

The 'trick' to performing the θ-integration is to use the following substitution:

$$\sin^3\theta = \sin\theta \sin^2\theta = \sin\theta\,(1 - \cos^2\theta)$$

(Here, we used the fact that $\sin^2\theta + \cos^2\theta = 1$ to write $\sin^2\theta = 1 - \cos^2\theta$.) After we plug this expression into the previous integral, we will separate the integral into two terms.

$$I = \frac{2\pi\rho R^5}{5} \int_{\theta=0}^{\pi} \sin^3\theta \, d\theta = \frac{2\pi\rho R^5}{5} \left(\int_{\theta=0}^{\pi} \sin\theta \, d\theta - \int_{\theta=0}^{\pi} \sin\theta \cos^2\theta \, d\theta \right)$$

Recall from Chapter 26 that an integral can often be made simpler with a substitution. With the second integral, we define $u = \cos\theta$ such that $du = -\sin\theta \, d\theta$. When $\theta = 0$, the new variable u equals $\cos 0 = 1$, and when $\theta = \pi$, the variable u equals $\cos\pi = -1$.

$$I = \frac{2\pi\rho R^5}{5} \left([-\cos\theta]_{\theta=0}^{\pi} + \int_{u=1}^{-1} u^2 du \right) = \frac{2\pi\rho R^5}{5} \left(-\cos\pi + \cos 0 + \left[\frac{u^3}{3}\right]_{u=1}^{-1} \right)$$

$$I = \frac{2\pi\rho R^5}{5} \left[-(-1) + 1 + \frac{(-1)^3}{3} - \frac{(1)^3}{3} \right] = \frac{2\pi\rho R^5}{5} \left(1 + 1 - \frac{1}{3} - \frac{1}{3} \right) = \frac{2\pi\rho R^5}{5} \frac{4}{3} = \frac{8\pi\rho R^5}{15}$$

Now integrate to find the total mass of the sphere, using the same substitutions as before:

$$m = \int dm = \rho \int_{r=0}^{R} \int_{\varphi=0}^{2\pi} \int_{\theta=0}^{\pi} r^2 \sin\theta \, dr d\varphi d\theta = \rho \int_{r=0}^{R} r^2 dr \int_{\theta=0}^{\pi} \sin\theta \, d\theta \int_{\varphi=0}^{2\pi} d\varphi$$

$$m = \rho \left[\frac{r^3}{3}\right]_{r=0}^{R} [-\cos\theta]_{\theta=0}^{\pi} [\varphi]_{\varphi=0}^{2\pi} = \frac{2\pi\rho R^3}{3} [-\cos\pi - (-\cos 0)] = \frac{2\pi\rho R^3}{3} (1 + 1)$$

$$m = \frac{4\pi\rho R^3}{3}$$

Solve for ρ in terms of m in the above equation.

$$\rho = \frac{3m}{4\pi R^3}$$

Substitute this expression for ρ into the equation for moment of inertia.

$$I = \frac{8\pi\rho R^5}{15} = \left(\frac{3m}{4\pi R^3}\right) \frac{8\pi R^5}{15} = \frac{2mR^2}{5}$$

We could have found this answer, $I = \frac{2mR^2}{5}$, from the table on pages 333-336. However, you can't always find the formula for the moment of inertia that you need from a table, so it's a valuable skill to be able perform the moment of inertia integral.

194. A rod with endpoints at the origin and the point $(0, L)$ has non-uniform linear mass density $\lambda(y) = \beta\sqrt{y}$, where β is a positive constant. Derive an equation for the moment of inertia of the rod about the x-axis in terms of the total mass of the rod, m, and the length of the rod, L.

Want help? Check the hints section at the back of the book.

Answer: $I = \frac{3mL^2}{7}$

195. The right triangle illustrated below has uniform density. Derive an equation for its moment of inertia about the x-axis in terms of b, h, and/or m.

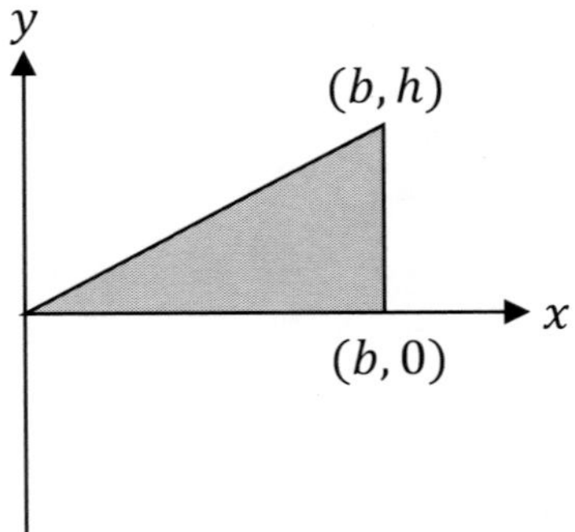

Want help? Check the hints section at the back of the book.

Answer: $I = \frac{mh^2}{6}$

196. Derive an equation for the moment of inertia of a solid uniform disc lying in the xy plane and centered about the origin about the y-axis in terms of its radius, R, and total mass, m.

Want help? Check the hints section at the back of the book.

Answer: $I = \frac{mR^2}{4}$

197. Derive an equation for the moment of inertia of a solid sphere with non-uniform density $\rho = \beta r$, where β is a positive constant, about an axis that passes through its center in terms of its radius, R, and total mass, m.

Want help? Check the hints section at the back of the book.

Answer: $I = \frac{4mR^2}{9}$

198. Derive an equation for the moment of inertia of a very thin uniform hollow sphere about an axis that passes through its center in terms of its radius, R, and total mass, m.

Want help? Check the hints section at the back of the book.

Answer: $I = \frac{2}{3}mR^2$

33 A PULLEY ROTATING WITHOUT SLIPPING

Essential Concepts

When we first encountered pulley problems back in Chapter 14, we ignored the rotational inertia of the pulley. We solved the problems in Chapter 14 as if the cord slipped over the pulley without friction, which isn't too realistic. In this chapter, we will explore the more realistic case where **the cord rotates with the pulley without slipping**. The way to solve a problem where the cord rotates with the pulley without slipping is to sum the torques in addition to applying Newton's second law. The sum of the torques equals moment of inertia times angular acceleration (see below).

Relevant Equations

A net force causes acceleration (this is Newton's second law, discussed in Chapter 14). Similarly, a net torque causes angular acceleration. The net torque ($\sum \tau$) acting on the system equals moment of inertia (I) times angular acceleration (α).

$$\sum F_x = ma_x \quad , \quad \sum F_y = ma_y \quad , \quad \sum \tau = I\alpha$$

Recall the equation for torque (τ) from Chapter 30, where $\vec{\mathbf{r}}$ extends from the axis of rotation to the point where $\vec{\mathbf{F}}$ is applied and θ is the angle between $\vec{\mathbf{r}}$ and $\vec{\mathbf{F}}$.

$$\tau = r\,F \sin\theta$$

Symbols and Units

Symbol	Name	Units
m	mass	kg
a	acceleration	m/s^2
F	force	N
I	moment of inertia	kg·m^2
α	angular acceleration	rad/s^2
τ	torque	N·m
r	distance from the axis of rotation to $\vec{\mathbf{F}}$	m
θ	the angle between $\vec{\mathbf{r}}$ and $\vec{\mathbf{F}}$	° or rad

Strategy for a Pulley that Rotates without Slipping

To solve a problem where a cord rotates with a pulley without slipping, follow these steps:

1. Draw an **extended free-body diagram** (FBD) for the pulley (as we did in Chapter 31), and also draw a FBD for any other objects in the problem (like we did in Chapter 14). Draw and label forces like we did in Chapter 14 and Chapter 31. On any ordinary FBD's, draw and label $+x$ in the direction of the object's acceleration and make $+y$ perpendicular to $+x$. For the pulley's extended FBD, draw and label the direction for positive rotation that is consistent with your choice of $+x$ in your other FBD's (see the example that follows).
2. Sum the **torques** for the pulley and also sum the components of the **forces** acting on any other objects in the problem:

$$\sum F_{1x} = m_1 a_x \quad , \quad \sum F_{1y} = m_1 a_y \quad , \quad \sum F_{2x} = m_2 a_x \quad , \quad \sum F_{2y} = m_2 a_y \quad , \quad \sum \tau = I\alpha$$

3. Rewrite the left-hand side of each force sum ($\sum F_x$ and $\sum F_y$) in terms of the x- and y-**components of the forces** acting on each object, like we did in Chapter 14.
4. Rewrite the left-hand side of the torque sum ($\sum \tau = 0$) in terms of the **torques**, using the equation $\tau = r\,F \sin\theta$, like we did in Chapter 31. Note that torques causing clockwise rotations will have a different **sign** than those causing counterclockwise rotations. Whether clockwise or counterclockwise is positive should be consistent with your choice of $+x$ in the other FBD's (see the example that follows).
5. Substitute the correct expression for **moment of inertia** from Chapter 32 for I in the torque sum.
6. Replace $R_p\alpha$ with a_x in the torque sum (R_p is the radius of the pulley) using the equation $a_x = R_p\alpha$. Since the cord is tangential to the pulley, this is really the equation $a_T = R_p\alpha$ that we learned in chapter 28, where $a_T = a_x$ since the cord is moving along the x-axis with the same acceleration as the objects attached to it (assuming that the cord doesn't stretch).
7. Include any substitutions that we learned in Chapter 14 which may be relevant to the problem. For example, for a problem with friction, $f = \mu N$.
8. Carry out the algebra to solve for the unknowns.

Tip for Finding the Torques Exerted by Tension

When applying the torque equation ($\tau = r\,F \sin\theta$) to determine the torques exerted by the tension forces, the angle θ will **always** equal 90° regardless of the orientation of the cord since tension always pulls on the pulley along a tangent (see below).

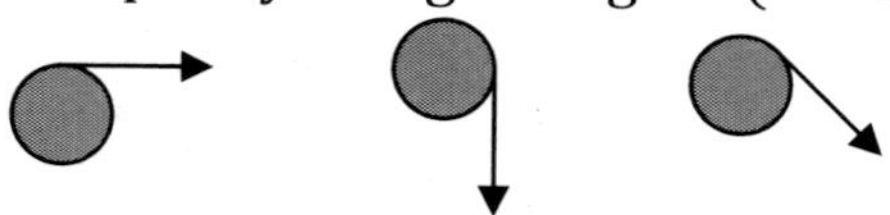

Example: Atwood's machine, shown below, is constructed by suspending a 60-g banana (on the right) and a 30-g banana (on the left) from the two ends of a cord that passes over a pulley. The cord rotates with the pulley without slipping. The pulley is a solid disc with a mass of 20 g. Determine the acceleration of the system and the tension in each cord.

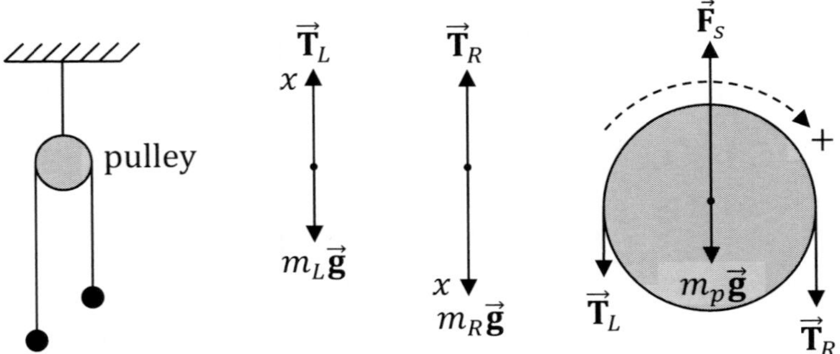

Draw and label a FBD (free-body diagram) for each banana. Each banana has weight pulling downward and tension pulling upward. The tensions are different due to friction between the pulley and cord. Newton's third law still applies:

- The left weight exerts a tension $\vec{\mathbf{T}}_L$ on the pulley, and the pulley exerts an equal force in the opposite direction back on the left weight.
- The right weight exerts a tension $\vec{\mathbf{T}}_R$ on the pulley, and the pulley exerts an equal force in the opposite direction back on the right weight.

These equal and opposite forces are shown in the extended FBD for the pulley, which also includes the weight of the pulley and a support force. Since the right banana is heavier, we choose $+x$ down for the right banana and up for the left banana. For consistency, the pulley must have clockwise rotation as its positive direction of motion. Apply Newton's second law to each banana:

$$\sum F_{Lx} = m_L a_x \quad , \quad \sum F_{Rx} = m_R a_x$$

$$T_L - m_L g = m_L a_x \quad , \quad m_R g - T_R = m_R a_x$$

Sum the torques for the pulley. The pulley's weight and the support force don't exert torques (they don't affect the pulley's rotation). The right tension creates a positive torque while the left tension creates a negative torque (counterclockwise). For the tensions, $\theta = 90°$ in the torque equation ($\tau = r\,F\sin\theta$) since tension pulls tangentially to the pulley. Since the pulley is a solid disc, its moment of inertia is $I = \frac{1}{2}m_p R_p^2$ (Chapter 32).

$$\sum \tau = I\alpha$$

$$R_p T_R \sin 90° - R_p T_L \sin 90° = \frac{1}{2}m_p R_p^2 \alpha$$

$$R_p T_R - R_p T_L = \frac{1}{2}m_p R_p^2 \alpha$$

Divide both sides by R_p:

$$T_R - T_L = \frac{1}{2} m_p R_p \alpha$$

Use the substitution $a_x = R_p \alpha$ to eliminate α:

$$T_R - T_L = \frac{1}{2} m_p a_x$$

We are now in a position to combine the equation from summing the torques with the two equations from summing the components of the forces. Below are the three equations that we have obtained thus far:

$$T_L - m_L g = m_L a_x$$
$$m_R g - T_R = m_R a_x$$
$$T_R - T_L = \frac{1}{2} m_p a_x$$

Add all three equations together to cancel tension:

$$T_L - m_L g + m_R g - T_R + T_R - T_L = m_L a_x + m_R a_x + \frac{1}{2} m_p a_x$$
$$-m_L g + m_R g = m_L a_x + m_R a_x + \frac{1}{2} m_p a_x$$
$$(m_R - m_L) g = \left(m_L + m_R + \frac{m_p}{2}\right) a_x$$

$$a_x = \frac{m_R - m_L}{m_L + m_R + \frac{m_p}{2}} g = \frac{60 - 30}{60 + 30 + \frac{20}{2}}(9.81) \approx \frac{60 - 30}{60 + 30 + 10}(10) = \frac{30}{100}(10) = 3.0 \text{ m/s}^2$$

Plug the acceleration into the previous equations to solve for the tension forces:

$$T_L - m_L g = m_L a_x \quad , \quad m_R g - T_R = m_R a_x$$
$$T_L = m_L a_x + m_L g \quad , \quad m_R g - m_R a_x = T_R$$

Convert the masses from grams to kilograms (since the SI unit of kg is needed to help make a Newton for force):

$$60 \text{ g} = 60 \text{ g} \times \frac{1 \text{ kg}}{1000 \text{ g}} = 0.060 \text{ kg}$$
$$30 \text{ g} = 30 \text{ g} \times \frac{1 \text{ kg}}{1000 \text{ g}} = 0.030 \text{ kg}$$

Plug the masses and accelerations into the tension equations:

$$T_L = m_L a_x + m_L g = m_L(a_x + g) = (0.03)(3 + 9.81) \approx (0.03)(3 + 10) = 0.39 \text{ N}$$
$$T_R = m_R g - m_R a_x = m_R(g - a_x) = (0.06)(9.81 - 3) \approx (0.06)(10 - 3) = 0.42 \text{ N}$$

We can check our answer for consistency by plugging these into the pulley equation:

$$T_R - T_L = \frac{1}{2} m_p a_x$$

The left-hand side equals $T_R - T_L = 0.42 \text{ N} - 0.39 \text{ N} = 0.03 \text{ N}$. The right-hand side equals $\frac{1}{2} m_p a_x = \frac{1}{2}(0.02)(3) = 0.03 \text{ N}$ (after converting the pulley's mass from 20 g to 0.02 kg). Since both sides equal 0.03 N, everything checks out.

199. As illustrated below, a 50-kg monkey is connected to a 40-kg box of bananas by a cord that passes over a pulley. The 20-kg pulley is a solid disc. The cord rotates with the pulley without slipping. The coefficient of friction between the box and ground is $\frac{1}{2}$.

(A) Draw and label a FBD for the monkey and box of bananas, and draw an extended FBD for the pulley. Also label x- and y-axes, and the positive sense of rotation, on your FBD's.

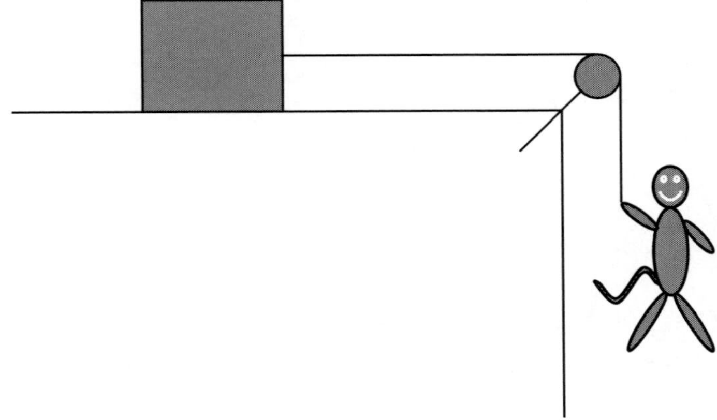

(B) Sum the x- and y-components of the forces acting on the monkey and box of bananas. Rewrite each sum in symbols in terms of the forces labeled in your FBD's.

(C) Sum the torques acting on the pulley in symbols in terms of the forces labeled in your extended FBD.

(D) Solve for the acceleration of the system.

Want help? Check the hints section at the back of the book.

Answers: 3.0 m/s^2

200. As illustrated below, two monkeys are connected by a cord. The 20-kg pulley is a solid disc. The cord rotates with the pulley without slipping. The surface is frictionless.

(A) Draw and label a FBD for each monkey and draw an extended FBD for the pulley. Also label x- and y-axes, and the positive sense of rotation, on your FBD's.

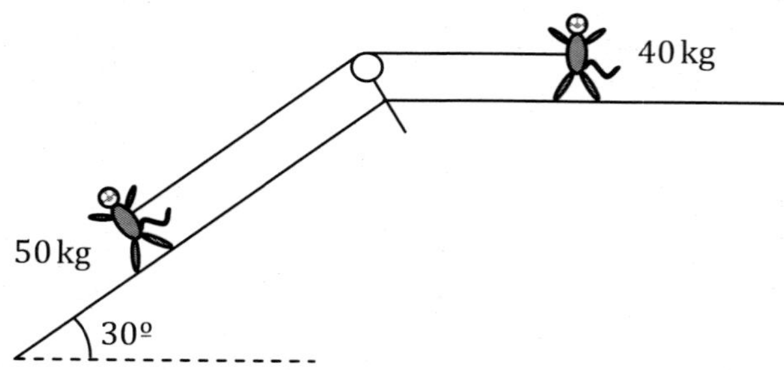

(B) Sum the x- and y-components of the forces acting on each monkey. Rewrite each sum in symbols in terms of the forces labeled in your FBD's.

(C) Sum the torques acting on the pulley in symbols in terms of the forces labeled in your extended FBD.

(D) Solve for the acceleration of the system.

Want help? Check the hints section at the back of the book.

Answer: $\frac{5}{2}$ m/s²

34 ROLLING WITHOUT SLIPPING

Relevant Terminology

Kinetic energy – work that can be done by changing speed. Kinetic energy is considered to be energy of motion because all moving objects have kinetic energy.
Translational kinetic energy – The kinetic energy that results from the motion of the center of mass of an object. This is the type of kinetic energy that we considered in Chapter 22.
Rotational kinetic energy – The kinetic energy that results from rotation.

Essential Concepts

Rotating objects can have two different kinds of kinetic energy:

- **Translational** kinetic energy (KE_t) from the motion of the center of mass.
- **Rotational** kinetic energy (KE_r) from the rotation.

Rolling objects have **both** kinds of kinetic energy because in addition to its rotation the center of mass of the object is moving.

Kinetic Energy Equations

Translational kinetic energy (KE_t) is proportional to speed (v) squared:

$$KE_{t0} = \frac{1}{2}mv_0^2 \quad , \quad KE_t = \frac{1}{2}mv^2$$

Rotational kinetic energy (KE_r) is proportional to angular speed (ω) squared:

$$KE_{r0} = \frac{1}{2}I\omega_0^2 \quad , \quad KE_r = \frac{1}{2}I\omega^2$$

The **total kinetic energy** (KE) includes both translational and rotational kinetic energy:

$$KE_0 = KE_{t0} + KE_{r0} = \frac{1}{2}mv_0^2 + \frac{1}{2}I\omega_0^2 \quad , \quad KE = KE_t + KE_r = \frac{1}{2}mv^2 + \frac{1}{2}I\omega^2$$

We use the above equations in the context of **conservation of energy** (Chapter 22).

$$PE_0 + KE_0 + W_{nc} = PE + KE$$

Recall a few of the equations relevant to conservation of energy:

- Gravitational potential energy is proportional to height (h). For a non-astronomical change in altitude:

$$PE_{g0} = mgh_0 \quad , \quad PE_g = mgh$$

- Spring potential energy is proportional to the square of the spring's displacement from equilibrium (x), where k is the spring constant:

$$PE_{s0} = \frac{1}{2}kx_0^2 \quad , \quad PE_s = \frac{1}{2}kx^2$$

Symbols and SI Units

Symbol	Name	SI Units
PE_0	initial potential energy	J
PE	final potential energy	J
KE_0	initial (total) kinetic energy	J
KE	final (total) kinetic energy	J
KE_{t0}	initial translational kinetic energy	J
KE_t	final translational kinetic energy	J
KE_{r0}	initial rotational kinetic energy	J
KE_r	final rotational kinetic energy	J
W_{nc}	nonconservative work	J
m	mass	kg
g	gravitational acceleration	m/s^2
h_0	initial height (relative to the reference height)	m
h	final height (relative to the reference height)	m
v_0	initial speed	m/s
v	final speed	m/s
I	moment of inertia	$kg \cdot m^2$
ω_0	initial angular speed	rad/s
ω	final angular speed	rad/s
k	spring constant	N/m (or kg/s^2)
x_0	initial displacement of a spring from equilibrium	m
x	final displacement of a spring from equilibrium	m

Strategy for Rolling without Slipping Problems

To solve a problem where an object rolls without slipping, follow these steps:

1. Apply the law of conservation of energy **unless** the problem asks for (or gives you) the acceleration. (If you need to find acceleration, see Step 10.) It may be helpful to review the strategy for conservation of energy in Chapter 22.
2. Draw a diagram of the path. Label the initial position (i), final position (f), and the **reference height** (RH).
3. Is there a spring involved in the problem? If so, also mark these positions in your diagram: equilibrium (EQ), fully compressed (FC), and fully stretched (FS).
4. Write out the law of **conservation of energy** in symbols:
$$PE_0 + KE_0 + W_{nc} = PE + KE$$
5. Rewrite each term of the conservation of energy equation in symbols as follows:
   - Initial potential energy may include both gravitational and spring potential energy: $PE_0 = mgh_0 + \frac{1}{2}kx_0^2$.
   - At the initial position (i), is the object moving? If so, then replace KE_0 with $\frac{1}{2}mv_0^2 + \frac{1}{2}I\omega_0^2$. Otherwise, if at the initial position (i) the object is at rest, then KE_0 equals zero.
   - For an object that rolls without slipping, there is friction (otherwise the object would slip), but $W_{nc} = 0$ because when an object rolls (as opposed to an object that slides) the friction force does **not** subtract energy from the system. See the note on the following page.
   - Final potential energy may include both gravitational and spring potential energy: $PE_0 = mgh + \frac{1}{2}kx^2$.
   - At the final position (f), is the object moving? If so, then replace KE with $\frac{1}{2}mv^2 + \frac{1}{2}I\omega^2$. Otherwise, if at the final position (f) the object is at rest, then KE equals zero.
6. It may help to think of the conservation of energy equation as looking like this:
$$mgh_0 + \frac{1}{2}kx_0^2 + \frac{1}{2}mv_0^2 + \frac{1}{2}I\omega_0^2 = mgh + \frac{1}{2}kx^2 + \frac{1}{2}mv^2 + \frac{1}{2}I\omega^2$$
In practice, one or more of these terms will be zero (as we learned in Chapter 22).
7. Substitute the correct expression for moment of inertia from Chapter 32 for I.
8. Rewrite ω_0 and ω using the equations $\omega_0 = \frac{v_0}{R}$ and $\omega = \frac{v}{R}$. (We used this equation in Chapters 16 and 28.)
9. Use algebra to solve for the desired unknown.
10. If you need to find acceleration (or if you're given acceleration), sum both the components of the forces and the torques using the strategy from Chapter 33.

Note Regarding Friction

Although there must be friction in order for an object to roll without slipping, $W_{nc} = 0$ since friction doesn't subtract mechanical energy from the system. To see this, compare the sliding block to the rolling ball shown below. For the sliding block, friction decelerates the block, converting mechanical energy into heat and internal energy. In contrast, for the rolling ball, a friction force opposite to velocity actually creates a forward torque which would **not** oppose the angular velocity to create angular deceleration. The difference is that $W_{nc} = 0$ for a ball that rolls perfectly without slipping, whereas W_{nc} is nonzero for a block that slides with friction.

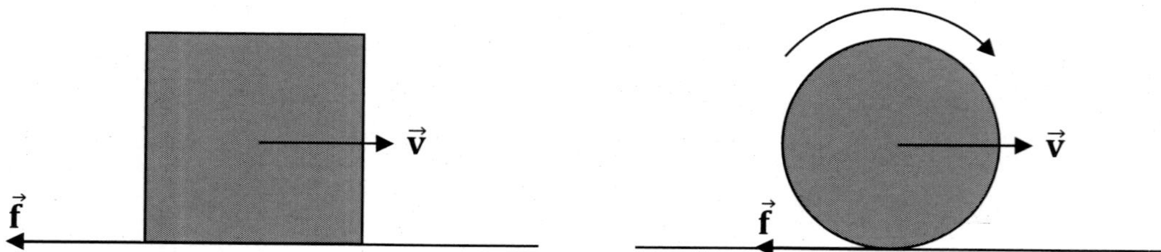

You can see this firsthand with a hockey puck or other object with a similar shape (as long as it rolls well):

- When the hockey puck slides along concrete or any surface with significant friction, the hockey puck comes to rest quickly.
- When you change the orientation of the hockey puck such that it rolls, it travels much further in comparison.

This simple experiment demonstrates that friction doesn't have nearly the significant effect on rolling as it has on sliding. In the case of perfect, idealized rolling without slipping, $W_{nc} = 0$, whereas W_{nc} is significant in the case of sliding.

The main effect of friction on rolling without slipping is actually indirect. Friction helps a round object roll, which enhances the inertia of the object. An object has translational inertia, which resists translational acceleration. A rolling object also has **rotational inertia** (moment of inertia), which resists angular acceleration. If a round object rolls down an incline, for example, gravity must overcome both types of inertia, and it is essentially this additional inertia that results in a lower acceleration compared to sliding without friction.

A round object that rolls **up** an incline without slipping is even more instructive. With enough friction to roll without slipping, the object actually rolls **farther** up the incline than it would without friction. That's because gravity must overcome two kinds of inertia (the translational inertia is overcome by the sum of the forces and the moment of inertia is overcome by the sum of the torques).

Example: A solid sphere rolls without slipping down an incline starting from rest. The solid sphere reaches the bottom of the incline after rolling 56 m along the incline. Determine the speed of the solid sphere as it reaches the bottom of the incline.

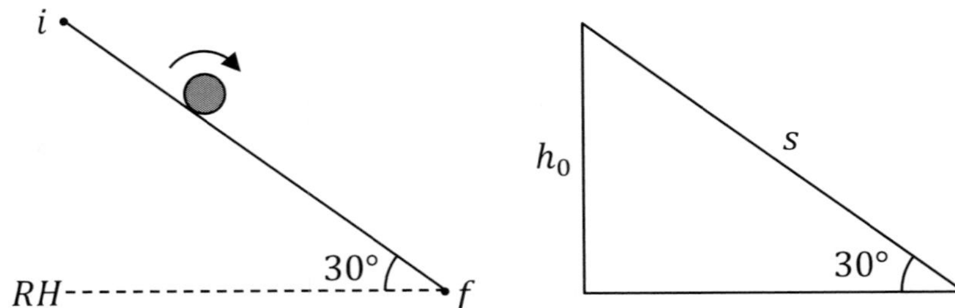

Draw the path and label the initial position (i), final position (f), and reference height (RH). The initial position (i) is where it starts, while the final position (f) is just before the sphere reaches the bottom of the incline (the final speed is **not** zero). We choose the reference height (RH) to be at the bottom of the incline. Write out conservation of energy:

$$PE_0 + KE_0 + W_{nc} = PE + KE$$

Let's analyze this term by term (we'll ignore the spring equations since they don't apply):

- $PE_0 = mgh_0$ since i is **not** at the same height as RH.
- $KE_0 = 0$ since the solid sphere is at rest at i.
- $W_{nc} = 0$ for rolling without slipping problems (even though there is friction – see the note on the previous page).
- $PE = 0$ since f is at the same height as RH.
- $KE = \frac{1}{2}mv^2 + \frac{1}{2}I\omega^2$ since the solid sphere is both translating and rotating at f.

Make the above substitutions into the conservation of energy equation:

$$mgh_0 = \frac{1}{2}mv^2 + \frac{1}{2}I\omega^2$$

Look up the equation for the moment of a inertia of a solid sphere (Chapter 32):

$$mgh_0 = \frac{1}{2}mv^2 + \frac{1}{2}\left(\frac{2}{5}mR^2\right)\omega^2$$

Divide both sides by the mass m. Use the equation $\omega = \frac{v}{R}$ (after which R^2 will cancel).

$$gh_0 = \frac{1}{2}v^2 + \frac{1}{2}\left(\frac{2}{5}R^2\right)\left(\frac{v}{R}\right)^2 = \frac{1}{2}v^2 + \frac{1}{5}v^2 = \left(\frac{1}{2} + \frac{1}{5}\right)v^2 = \left(\frac{5}{10} + \frac{2}{10}\right)v^2 = \frac{7}{10}v^2$$

The problem gave us the total distance ($s = 56$ m) traveled down the incline, but we need the initial height (h_0). We can solve for h_0 using trig. See the diagram above.

$$\sin 30° = \frac{h_0}{s} \quad \text{such that} \quad h_0 = s \sin 30° = \frac{s}{2} = \frac{56}{2} = 28 \text{ m}$$

Plug the initial height into the equation we arrived at from conservation of energy:

$$v = \sqrt{\frac{10}{7}gh_0} = \sqrt{\frac{10}{7}(9.81)(28)} \approx \sqrt{\frac{10}{7}(10)(28)} = \sqrt{400} = 20 \text{ m/s}$$

Example: Determine the acceleration of a solid sphere as it rolls without slipping down a 30° incline.

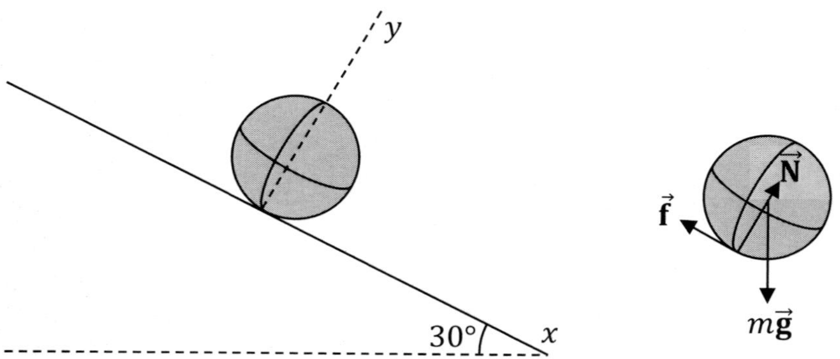

Since the problem asks for acceleration, we should sum the components of the forces and the torques. Draw and label an extended FBD for the solid sphere. This means to draw the shape of the object and draw each force where it effectively acts on the object (Chapter 33). Weight pulls downward, normal force pushes perpendicular to the surface, and friction pushes up the incline. (Although the problem didn't explicitly mention friction, it's implied since the solid sphere "rolls without slipping.") As we learned in a similar example in Chapter 14, the x-component of weight has a sine since x is opposite to 30°, while the y-component of weight has a cosine since y is adjacent to 30°. See page 135. Friction is the only force that exerts a torque (since the other forces point to or from the axis of rotation). The moment of inertia of a solid sphere is $I = \frac{2}{5}mR^2$ (Chapter 32).

$$\sum F_x = ma_x \quad , \quad \sum F_y = ma_y \quad , \quad \sum \tau = I\alpha$$

$$mg\sin 30° - f = ma_x \quad , \quad N - mg\cos 30° = 0 \quad , \quad Rf\sin 90° = \frac{2}{5}mR^2\alpha$$

Use the substitution $a_x = R\alpha$:

$$Rf\sin 90° = \frac{2}{5}mRa_x$$

$$f = \frac{2}{5}ma_x$$

Plug this equation in for friction in the x-sum:

$$mg\sin 30° - f = ma_x$$

$$mg\sin 30° - \frac{2}{5}ma_x = ma_x$$

$$g\left(\frac{1}{2}\right) = \frac{2}{5}a_x + a_x = \left(\frac{2}{5} + 1\right)a_x = \left(\frac{2}{5} + \frac{5}{5}\right)a_x = \frac{7}{5}a_x$$

$$a_x = \frac{5}{7}g\left(\frac{1}{2}\right) = \frac{5}{14}g = \frac{5(9.81)}{14} \approx \frac{5(10)}{14} = \frac{25}{7}\ \text{m/s}^2$$

201. A hollow sphere rolls without slipping down an incline from rest. The hollow sphere reaches the bottom of the incline after descending a height of 75 m. Determine the speed of the hollow sphere as it reaches the bottom of the incline.

Want help? Check the hints section at the back of the book.

Answer: 30 m/s

202. A donut for which $I = \frac{3mR^2}{4}$ rolls without slipping up a 30° incline with an initial speed of 40 m/s. How far does the donut travel up the incline?

Want help? Check the hints section at the back of the book.

Answer: 280 m

35 CONSERVATION OF ANGULAR MOMENTUM

Relevant Terminology

Momentum – mass times velocity.
Angular velocity – a combination of angular speed and direction, where angular speed is the instantaneous rate at which the angle changes (as measured from the center of the circle).
Moment of Inertia – the natural tendency of a rigid body to maintain constant angular momentum.
Angular momentum – moment of inertia times angular velocity.

Essential Concepts

Objects have inertia, which is a natural tendency to maintain constant momentum. A net external force is needed for an object to change its momentum. Similarly, rigid bodies (those that don't change shape during rotation) have moment of inertia (or rotational inertia), which is a natural tendency to maintain constant angular momentum. A net external torque is needed for an object to change its angular momentum.

More mass further from the axis of rotation results in a larger moment of inertia, which makes the rigid body more resistant to changes to its angular momentum.

The angular momentum of a system is conserved when the net torque acting on the system equals zero. In practice, students tend to have trouble telling whether or not the net torque acting on a system is zero. It may be easier to study the variety of examples and problems that you can find that involve conservation of angular momentum to help develop a feel for the kinds of problems where the law of conservation of angular momentum should be applied.

Angular Momentum Equations

Angular momentum ($\vec{\boldsymbol{L}}$) equals moment of inertia (I) times angular velocity ($\vec{\boldsymbol{\omega}}$):

$$\vec{\boldsymbol{L}} = I\vec{\boldsymbol{\omega}}$$

Conservation of angular momentum can be expressed in the form:

$$I_{10}\vec{\boldsymbol{\omega}}_{10} + I_{20}\vec{\boldsymbol{\omega}}_{20} = I_1\vec{\boldsymbol{\omega}}_1 + I_2\vec{\boldsymbol{\omega}}_2$$

To calculate how much **kinetic energy** the system gains or loses, compare the final kinetic energy ($KE = \frac{1}{2}I_1\omega_1^2 + \frac{1}{2}I_2\omega_2^2$) to the total kinetic energy ($KE_0 = \frac{1}{2}I_{10}\omega_{10}^2 + \frac{1}{2}I_{20}\omega_{20}^2$).

$$\%\text{ change} = \frac{|KE - KE_0|}{KE_0} \times 100\%$$

Symbols and SI Units

Symbol	Name	SI Units
$\vec{\boldsymbol{L}}$	angular momentum	kg·m²/s
I	moment of inertia	kg·m²
$\vec{\boldsymbol{\omega}}$	angular velocity	rad/s
ω	angular speed	rad/s
I_{10}	initial moment of inertia of object 1	kg·m²
I_{20}	initial moment of inertia of object 2	kg·m²
I_1	final moment of inertia of object 1	kg·m²
I_2	final moment of inertia of object 2	kg·m²
$\vec{\boldsymbol{\omega}}_{10}$	initial angular velocity of object 1	rad/s
$\vec{\boldsymbol{\omega}}_{20}$	initial angular velocity of object 2	rad/s
$\vec{\boldsymbol{\omega}}_1$	final angular velocity of object 1	rad/s
$\vec{\boldsymbol{\omega}}_2$	final angular velocity of object 2	rad/s
m_1	mass of object 1	kg
m_2	mass of object 2	kg
R_1	radius of object 1	m
R_2	radius of object 2	m

Notes Regarding Units

The SI units of angular momentum are $\text{kg}\cdot\text{m}^2/\text{s}$. This follows from the equation for angular momentum:

$$\vec{\boldsymbol{L}} = I\vec{\boldsymbol{\omega}}$$

The SI units of moment of inertia ($\text{kg}\cdot\text{m}^2$) times the SI units of angular velocity (rad/s) make the SI units of angular momentum ($\text{kg}\cdot\text{m}^2/\text{s}$). (As usual, the radians disappear in the context of meters, as a radian represents a fraction of the way around a circle. For example, in the arc length equation, $s = R\theta$, a meter equals a meter times a radian.)

Angular Vectors

An arrow above a symbol, as in $\vec{\boldsymbol{\omega}}$, reminds you that a quantity is a vector: It includes direction. For an object moving in a circle, clockwise rotations and counterclockwise rotations will involve a different sign for $\vec{\boldsymbol{\omega}}$.

More complex rotations (compared to an object rotating in a circle in a plane) involve vector products (Chapter 29):

$$\vec{\boldsymbol{v}} = \vec{\boldsymbol{\omega}} \times \vec{\boldsymbol{r}} \quad , \quad \vec{\boldsymbol{L}} = \vec{\boldsymbol{r}} \times \vec{\boldsymbol{p}} \quad , \quad \vec{\boldsymbol{\tau}} = \vec{\boldsymbol{r}} \times \vec{\boldsymbol{F}}$$

$$\sum \vec{\boldsymbol{\tau}} = \frac{d\vec{\boldsymbol{L}}}{dt}$$

Strategy for Applying the Law of Conservation of Angular Momentum

To apply the law of conservation of angular momentum, follow these steps:

1. The total angular momentum of a system is conserved if the net torque ($\sum \tau$) acting on the system equals zero.
2. Declare your choice of the positive sense of rotation. For example, if you declare clockwise rotations to be positive, any object rotating counterclockwise would then have a negative angular velocity ($\vec{\boldsymbol{\omega}}$).
3. Express **conservation of angular momentum** for the system by writing the following equation.
$$I_{10}\vec{\boldsymbol{\omega}}_{10} + I_{20}\vec{\boldsymbol{\omega}}_{20} = I_1\vec{\boldsymbol{\omega}}_1 + I_2\vec{\boldsymbol{\omega}}_2$$
(If there is just one object in the system, like a spinning ice skater, then you only need one term on each side of the equation instead of two.)
4. Substitute the correct expression for **moment of inertia** from Chapter 32 for I_{10}, I_{20}, I_1, and I_2.
5. Is there a **pointlike** object in the problem? (An object is "pointlike" if it is small in size compared to the radius of the circular motion that it is making.) If the problem involves a pointlike object and you're solving for (or you're given) its speed v (as opposed to its angular speed ω), it may be helpful to use the equation $\omega = \frac{v}{R}$ to relate its speed to its angular speed. (If you recall from Chapter 32 that the moment of inertia of a pointlike object is $I = mR^2$, you can write the angular momentum for a **pointlike** object either as $L = mR^2\omega$ or as $L = mvR$.)
6. If the problem asks you to determine the percentage of **kinetic energy** lost or gained in the process, first find $KE = \frac{1}{2}I_1\omega_1^2 + \frac{1}{2}I_2\omega_2^2$ and $KE_0 = \frac{1}{2}I_{10}\omega_{10}^2 + \frac{1}{2}I_{20}\omega_{20}^2$:
$$\%\text{ change} = \frac{|KE - KE_0|}{KE_0} \times 100\%$$

Example: A monkey is performing her exercise in the ice skating competition of the Winter Monkolympics. In one part of her routine, she spins while standing tall, with her arms straight up in the air and her legs straight down, maintaining excellent balance by positioning her center of gravity above the point of contact between her skate and the ice. Her angular speed is 12.0 rev/s during this part of her routine. Then she quickly stretches her arms horizontally outward, extends one of her legs horizontally backward, and leans her torso and head horizontally forward, effectively increasing her moment of inertia by a factor of 3 – still maintaining her balance exquisitely. What is her final angular speed?

This problem involves a redistribution of mass without any net torques acting on the ice skater. Since the net torque is zero, angular momentum is conserved for the system. There is just one object in the system: the monkey. So there is just one term on each side:

$$I_0\vec{\boldsymbol{\omega}}_0 = I\vec{\boldsymbol{\omega}}$$

We don't need to worry about what expression to use for moment of inertia since the problem tells us that it increases by a factor of 3. That is, $I = 3I_0$.

$$I_0\vec{\boldsymbol{\omega}}_0 = 3I_0\vec{\boldsymbol{\omega}}$$

$$\vec{\boldsymbol{\omega}} = \frac{\vec{\boldsymbol{\omega}}_0}{3}$$

$$\omega = \frac{12}{3} = 4.0 \text{ rev/s}$$

Example: A 4.0-kg solid disc with a radius of 50 cm is spinning with an initial angular speed of 24 rev/s on horizontal frictionless ice. A 2.0-kg thin ring with a radius of 50 cm is gently lowered onto the spinning solid disc. There is friction between the ring and the disc. What is the final angular speed of the system?

No net torque is acting on the system, so we conserve angular momentum for the system:

$$I_{10}\vec{\boldsymbol{\omega}}_{10} + I_{20}\vec{\boldsymbol{\omega}}_{20} = I_1\vec{\boldsymbol{\omega}}_1 + I_2\vec{\boldsymbol{\omega}}_2$$

From Chapter 32, the moment of inertia of the solid disc is $\frac{1}{2}m_1R^2$ and the moment of inertia of the thin ring is m_2R^2. The thin ring isn't spinning initially so $\vec{\boldsymbol{\omega}}_{20} = 0$. The solid disc and thin ring spin together afterward so $\vec{\boldsymbol{\omega}}_1 = \vec{\boldsymbol{\omega}}_2 = \vec{\boldsymbol{\omega}}$. Convert the radius from centimeters to meters: $R = 50 \text{ cm} = 0.50 \text{ m} = \frac{1}{2}\text{m}$.

$$\frac{1}{2}m_1R^2\omega_{10} + m_2R^2(0) = \left(\frac{1}{2}m_1R^2 + m_2R^2\right)\omega = \left(\frac{1}{2}m_1 + m_2\right)R^2\omega$$

$$\frac{1}{2}(4)\left(\frac{1}{2}\right)^2(24) = \left[\frac{1}{2}(4) + 2\right]\left(\frac{1}{2}\right)^2\omega$$

$$12 = 4\left(\frac{1}{4}\right)\omega = \omega$$

$$\omega = 12 \text{ rev/s}$$

Example: A horizontal record (a solid disc) is held in place by a small vertical axle. A hamster climbs onto the record. The system is initially at rest. There is friction between the record and hamster, but not between the record and the surface upon which it rests. (Also neglect any friction with the axle.) The 250-g hamster runs 40 cm/s (relative to the table) in a circle with a radius of 6.0 cm centered about the axle. The 500-g record has a diameter of 18 cm. What is the angular speed of the record while the hamster runs?

Since the net torque acting on the system (the record plus the hamster) equals zero, we may apply the law of conservation of angular momentum:

$$I_{10}\vec{\boldsymbol{\omega}}_{10} + I_{20}\vec{\boldsymbol{\omega}}_{20} = I_1\vec{\boldsymbol{\omega}}_1 + I_2\vec{\boldsymbol{\omega}}_2$$

From Chapter 32, the moment of inertia of the solid disc is $\frac{1}{2}m_1R_1^2$ and the moment of inertia of the pointlike hamster is $m_2R_2^2$. The record and hamster are both initially at rest: $\vec{\boldsymbol{\omega}}_{10} = \vec{\boldsymbol{\omega}}_{20} = 0$. Convert the radii from centimeters to meters:

$$R_1 = \frac{D_1}{2} = \frac{18 \text{ cm}}{2} = 9.0 \text{ cm} = 0.09 \text{ m} = \frac{9}{100}\text{m}$$

$$R_2 = 6.0 \text{ cm} = 0.06 \text{ m} = \frac{6}{100}\text{m} = \frac{3}{50}\text{m}$$

Convert the masses from grams to kilograms:

$$m_1 = 500 \text{ g} = 0.500 \text{ kg} = \frac{1}{2} \text{ kg}$$

$$m_2 = 250 \text{ g} = 0.250 \text{ kg} = \frac{1}{4} \text{ kg}$$

Also convert the speed from cm/s to m/s:

$$v_2 = 40 \text{ cm/s} = 0.40 \text{ m/s} = \frac{2}{5} \text{ m/s}$$

Use the equation $\omega = \frac{v}{R}$ to solve for the final angular speed of the hamster:

$$\omega_2 = \frac{v_2}{R_2} = \frac{\frac{2}{5}}{\frac{3}{50}} = \frac{2}{5} \div \frac{3}{50} = \frac{2}{5} \times \frac{50}{3} = \frac{20}{3} \text{ rad/s}$$

Now we're prepared to use the conservation of angular momentum equation.

$$\frac{1}{2}m_1R^2(0) + m_2R^2(0) = \frac{1}{2}m_1R_1^2\omega_1 + m_2R_2^2\omega_2$$

$$0 = \frac{1}{2}\left(\frac{1}{2}\right)\left(\frac{9}{100}\right)^2 \omega_1 + \left(\frac{1}{4}\right)\left(\frac{3}{50}\right)^2\left(\frac{20}{3}\right)$$

$$0 = \frac{81}{40{,}000}\omega_1 + \frac{3}{500}$$

$$\omega_1 = -\frac{3}{500}\frac{40{,}000}{81} = -\frac{80}{27} \text{ rad/s}$$

The minus sign signifies that the record rotates opposite to the motion of the hamster.

203. A 30-kg monkey is placed at rest at the center of a merry-go-round. The merry-go-round is a large solid disc which has a mass of 120 kg and a diameter of 16 m. A gorilla spins the merry-go-round at a rate of $\frac{1}{4}$ rev/s and lets go. As the merry-go-round spins, the monkey walks outward until he reaches the edge. Find the angular speed of the merry-go-round when the monkey reaches the edge. Neglect any resistive forces like friction with the axle.

Want help? Check the hints section at the back of the book.

Answer: $\frac{1}{6}$ rev/s

204. A monkey is SO frustrated with his slow internet connection that he picks up his laptop, slams it against a brick wall, and then jumps high into the air and stomps on it. At that exact moment, the earth suddenly contracts until it has one-third of its initial radius. How long will a 'day' be now?

Want help? Check the hints section at the back of the book.

Answers: $\frac{8}{3}$ hr

205. As illustrated below, a 200-g pointlike object on a frictionless table is connected to a light, inextensible cord that passes through a hole in the table. A monkey underneath the table is pulling on the cord to create tension. The monkey is initially pulling the cord such that the pointlike object slides with an initial speed of 5.0 m/s in a circle with a 50-cm diameter centered about the hole. The monkey increases the tension until the pointlike object slides in a circle with a 20-cm diameter. Determine the final speed of the pointlike object.

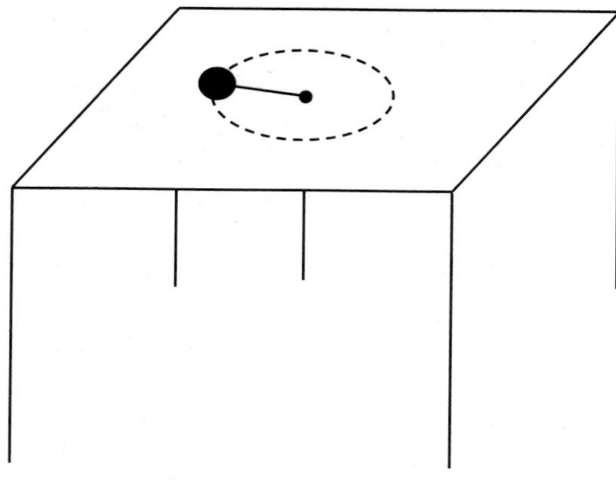

Want help? Check the hints section at the back of the book.

Answers: $\frac{25}{2}$ m/s

HINTS, INTERMEDIATE ANSWERS, AND EXPLANATIONS

How to Use This Section Effectively

Think of hints and intermediate answers as training wheels. They help you proceed with your solution. When you stray from the right path, the hints help you get back on track. The answers also help to build your confidence.

However, if you want to succeed in a physics course, you must eventually learn to rely less and less on the hints and intermediate answers. Make your best effort to solve the problem on your own before checking for hints, answers, and explanations. When you need a hint, try to find just the hint that you need to get over your current hurdle. Refrain from reading additional hints until you get further into the solution.

When you make a mistake, think about what you did wrong and what you should have done differently. Try to learn from your mistake so that you don't repeat the mistake in other solutions.

It's natural for students to check hints and intermediate answers repeatedly in the early chapters. However, at some stage, you would like to be able to consult this section less frequently. When you can solve more problems without help, you know that you're really beginning to master physics.

Would You Prefer to See Full Solutions?

Full solutions are like a security blanket: Having them makes students feel better. But full solutions are also dangerous: Too many students rely too heavily on the full solutions, or simply read through the solutions instead of working through the solutions on their own. Students who struggle through their solutions and improve their solutions only as needed tend to earn better grades in physics (though comparing solutions **after** solving a problem is always helpful).

It's a challenge to get just the right amount of help. In the ideal case, you would think your way through every solution on your own, seek just the help you need to make continued progress with your solution, and wait until you've solved the problem as best you can before consulting full solutions or reading every explanation.

With this in mind, full solutions to all problems are contained in a separate book. This workbook contains hints, intermediate answers, explanations, and several directions to help walk you through the steps of every solution, which should be enough to help most students figure out how to solve all of the problems. However, if you need the security of seeing **full solutions to all problems**, look for the book 100 Instructive Calculus-based Physics Examples with ISBN 978-1-941691-17-5. The solution to every problem in this workbook can be found in that book.

How to Cover up Hints that You Don't Want to See too Soon

There is a simple and effective way to cover up hints and answers that you don't want to see too soon:

- Fold a blank sheet of paper in half and place it in the hints and answers section. This will also help you bookmark this handy section of the book.
- Place the folded sheet of paper just below your current reading position. The folded sheet of paper will block out the text below.
- When you want to see the next hint or intermediate answer, just drop the folded sheet of paper down slowly, just enough to reveal the next line.
- This way, you won't reveal more hints or answers than you need.

You learn more when you force yourself to struggle through the problem. Consult the hints and answers when you really need them, but try it yourself first. After you read a hint, try it out and think it through as best you can before consulting another hint. Similarly, when checking intermediate answers to build confidence, try not to see the next answer before you have a chance to try it on your own first.

Chapter 1: Algebra Essentials

1. Apply the rule $x^m x^n = x^{m+n}$.
   - Add the exponents together.
   - $x^2 x^3 x^4 = x^{2+3+4}$
   - The answer is x^9.

---

2. Apply the rules $x^m x^n = x^{m+n}$ and $\frac{x^m}{x^n} = x^{m-n}$.
   - When multiplying, add the exponents together. When dividing, subtract them.
   - $x^4 x^5 = x^{4+5}$ and $\frac{x^9}{x^6} = x^{9-6}$.
   - The answer is x^3.

---

3. Apply the rule $\frac{x^m}{x^n} = x^{m-n}$.
   - Subtract the exponents.
   - When you subtract a negative number, the two minus signs make a plus sign.
   - $\frac{x^{-7}}{x^{-8}} = x^{-7-(-8)} = x^{-7+8}$.
   - The answer is x. (It's the same thing as x^1.)

---

4. Apply the rules $x^m x^n = x^{m+n}$ and $\frac{x^m}{x^n} = x^{m-n}$.
   - When multiplying, add the exponents together. When dividing, subtract them.
   - $x^3 x^{-6} = x^{3-6}$, $x^{-2}x^5 = x^{-2+5}$, and $\frac{x^{-3}}{x^3} = x^{-3-3}$.
   - The answer is x^{-6}.

---

5. Apply the rules $x^{1/2} = \sqrt{x}$, $x^1 = x$, and $x^m x^n = x^{m+n}$.
   - Add the exponents together.
   - $x\sqrt{x} = x^1 x^{1/2} = x^{1+1/2}$.
   - Add fractions with a **common denominator**: $1 + \frac{1}{2} = \frac{2}{2} + \frac{1}{2} = \frac{2+1}{2} = \frac{3}{2}$.
   - The answer is $x^{3/2}$.

---

6. Apply the rules $x^{1/2} = \sqrt{x}$ and $\frac{x^m}{x^n} = x^{m-n}$.
   - Subtract the exponents.
   - $\frac{x^{3/2}}{\sqrt{x}} = \frac{x^{3/2}}{x^{1/2}} = x^{3/2-1/2}$.
   - Note that $\frac{3}{2} - \frac{1}{2} = 1$.
   - The answer is x. (It's the same thing as x^1.)

---

7. Apply the rule $x^{-1/2} = \frac{1}{x^{1/2}} = \frac{1}{\sqrt{x}}$.
   - $(4x)^{-1/2} = \frac{1}{\sqrt{4x}}$.
   - **Rationalize the denominator.** Multiply the numerator and denominator by $\sqrt{4x}$.
   - $\frac{1}{\sqrt{4x}} = \frac{1}{\sqrt{4x}}\frac{\sqrt{4x}}{\sqrt{4x}} = \frac{\sqrt{4x}}{4x}$.
   - Apply the rule $\sqrt{ax} = \sqrt{a}\sqrt{x}$.
   - $\sqrt{4x} = \sqrt{4}\sqrt{x} = 2\sqrt{x}$. Note that $\frac{2\sqrt{x}}{4x} = \frac{2}{4}\frac{\sqrt{x}}{x} = \frac{\sqrt{x}}{2x}$.
   - The answer is $\frac{\sqrt{x}}{2x}$.

---

8. Make a **common denominator**.
   - Multiply $\frac{2}{x^2}$ by $\frac{x}{x}$.
   - $\frac{2}{x^2} = \frac{2}{x^2}\frac{x}{x} = \frac{2x}{x^3}$. Now subtract the fractions.
   - The answer is $\frac{2x-3}{x^3}$.

---

9. Make a **common denominator**.
   - Multiply $\sqrt{x}$ by $\frac{\sqrt{x}}{\sqrt{x}}$.

- $\sqrt{x} = \frac{\sqrt{x}}{1}\frac{\sqrt{x}}{\sqrt{x}} = \frac{x}{\sqrt{x}}$. Now add the fractions.
- You should get $\frac{x+1}{\sqrt{x}}$. It's the same thing as $\frac{1+x}{\sqrt{x}}$.
- **Rationalize the denominator.** Multiply the numerator and denominator by $\sqrt{x}$.
- $\frac{x+1}{\sqrt{x}} = \frac{x+1}{\sqrt{x}}\frac{\sqrt{x}}{\sqrt{x}} = \frac{(x+1)\sqrt{x}}{x}$.
- Distribute the $\sqrt{x}$.
- $(x+1)\sqrt{x} = x\sqrt{x} + \sqrt{x}$.
- The answer is $\frac{x\sqrt{x}+\sqrt{x}}{x}$.

---

10. Distribute the $6x$.
    - $6x(x+9) = 6x(x) + 6x(9)$.
    - The answer is $6x^2 + 54x$.

---

11. Distribute the $-3x^2$.
    - Two minus signs make a plus sign: $-3x^2(-2x^4) = 3x^2(2x^4)$.
    - $-3x^2(5x^6 - 2x^4) = -3x^2(5x^6) + 3x^2(2x^4)$.
    - Apply the rule $x^m x^n = x^{m+n}$.
    - The answer is $-15x^8 + 6x^6$.

---

12. Distribute the $\sqrt{x}$.
    - $\sqrt{x}(x + \sqrt{x}) = \sqrt{x}(x) + \sqrt{x}\sqrt{x}$.
    - Apply the rules $x^{1/2} = \sqrt{x}$, $\sqrt{x}\sqrt{x} = x$, $x^m x^n = x^{m+n}$, and $x^1 = x$.
    - The answer is $x^{3/2} + x$.

---

13. Think of this problem as $3x(x+5) + 2(x+5)$. Distribute twice.
    - $(3x+2)(x+5) = 3x(x) + 3x(5) + 2(x) + 2(5)$.
    - This gives you $3x^2 + 15x + 2x + 10$.
    - Combine like terms: $15x + 2x = 17x$.
    - The answer is $3x^2 + 17x + 10$.

---

14. Rewrite $(6x-4)^2$ as $(6x-4)(6x-4)$.
    - $(6x-4)(6x-4) = 6x(6x) + 6x(-4) - 4(6x) - 4(-4)$.
    - Two negative numbers multiplied together make a positive number.
    - The intermediate answer is $36x^2 - 24x - 24x + 16$.
    - Combine like terms: $-24x - 24x = -48x$.
    - The answer is $36x^2 - 48x + 16$.

---

15. Factor out an x.
    - Note that $x^3 = x^2 x$.
    - $x^3 - 4x = x^2 x - 4x$.
    - The answer is $x(x^2 - 4)$. You can check your answer by distributing.

---

16. Factor out $4x^6$. Apply the rule $\frac{x^m}{x^n} = x^{m-n}$.
    - $\frac{8x^9}{4x^6} = 2x^3$ and $\frac{12x^6}{4x^6} = 3$. (Note that $x^0 = 1$.)
    - $8x^9 + 12x^6 = 4x^6(2x^3) + 4x^6(3)$.
    - The answer is $4x^6(2x^3 + 3)$. You can check your answer by distributing.

---

17. Factor out $9x^3$. Apply the rule $\frac{x^m}{x^n} = x^{m-n}$.
    - $\frac{45x^7}{9x^3} = 5x^4$, $\frac{18x^5}{9x^3} = 2x^2$, and $\frac{27x^3}{9x^3} = 3$.
    - $45x^7 - 18x^5 + 27x^3 = 9x^3(5x^4) - 9x^3(2x^2) + 9x^3(3)$.
    - The answer is $9x^3(5x^4 - 2x^2 + 3)$. You can check your answer by distributing.

---

18. Think about the **perfect squares**: $1^2 = 1, 2^2 = 4, 3^2 = 9, 4^2 = 16, 5^2 = 25, 6^2 = 36$, etc.
    - What is the largest perfect square that factors into 18? It's 9.
    - Write 18 as $(2)(9)$.
    - Apply the rule $\sqrt{ax} = \sqrt{a}\sqrt{x}$.
    - $\sqrt{18} = \sqrt{(2)(9)} = \sqrt{2}\sqrt{9}$
    - The answer is $3\sqrt{2}$. (Note that $\sqrt{9} = 3$.)

---

19. Think about the **perfect squares**: $1^2 = 1, 2^2 = 4, 3^2 = 9, 4^2 = 16, 5^2 = 25, 6^2 = 36$, etc.
    - What is the largest perfect square that factors into 108? It's 36.
    - Write 108 as $(3)(36)$.
    - Apply the rule $\sqrt{ax} = \sqrt{a}\sqrt{x}$.
    - $\sqrt{108} = \sqrt{(3)(36)} = \sqrt{3}\sqrt{36}$
    - The answer is $6\sqrt{3}$. (Note that $\sqrt{36} = 6$.)

---

20. This quadratic equation is already in standard form.
    - Identify the constants a, b, and c. Include the minus signs.
    - The constants are $a = 2$, $b = -2$, and $c = -40$.
    - (It would also be okay to divide the equation by 2. Then you would get $a = 1$, $b = -1$, and $c = -20$. However, if you do this, your intermediate answers will be different from those that follow.)
    - The squareroot works out to $\sqrt{324} = 18$.

- The two solutions are $\frac{2+18}{4}$ and $\frac{2-18}{4}$.
- These solutions simplify to 5 and –4.

---

21. Reorder the terms in **standard form**.
    - Put y^2 first, then y, followed by the constant.
    - Identify the constants a, b, and c. Include the minus signs.
    - The constants are $a = 2$, $b = 3$, and $c = -27$.
    - The squareroot works out to $\sqrt{225} = 15$.
    - The two solutions are $\frac{-3+15}{4}$ and $\frac{-3-15}{4}$.
    - These solutions simplify to 3 and $-\frac{9}{2}$.

---

22. Bring all three terms to the same side of the equation.
    - Use algebra to put t^2 first, then t, followed by the constant.
    - The constants are $a = 2$, $b = 6$, and $c = -8$.
    - (It would also be okay to use $a = -2$, $b = -6$, and $c = 8$, but don't mix and match.)
    - The squareroot works out to $\sqrt{100} = 10$.
    - The two solutions are $\frac{-6+10}{4}$ and $\frac{-6-10}{4}$.
    - These solutions simplify to 1 and –4.

---

23. Bring all six terms to the same side of the equation.
    - **Combine like terms**: Combine the two x^2 terms, the two x terms, and the two constants.
    - Use algebra to put x^2 first, then x, followed by the constant.
    - You should get $0 = 2x^2 - 17x + 8$.
    - (It would also be okay to have $-2x^2 + 17x - 8 = 0$.)
    - The constants are $a = 2$, $b = -17$, and $c = 8$.
    - (It would also be okay to use $a = -2$, $b = 17$, and $c = -8$, but don't mix and match.)
    - The squareroot works out to $\sqrt{225} = 15$.
    - The two solutions are $\frac{17+15}{4}$ and $\frac{17-15}{4}$.
    - These solutions simplify to 8 and $\frac{1}{2}$.

---

24. If you solve for y in the top equation, you get $y = \frac{18-3x}{2}$.
    - If you substitute this into the bottom equation, you get $8x - 5\left(\frac{18-3x}{2}\right) = 17$.
    - Distribute the 5 to get $8x - 45 + \frac{15x}{2} = 17$. When you distribute, the two minus signs make a plus. Note that $(5)(18)/2 = (5)(9) = 45$ and $(-5)(-3)/2 = 15/2$.

- Combine like terms to get $\frac{31x}{2} = 62$. (To combine the x terms, you will first need to make a common denominator.)
- Solve the previous equation to find that $x = 4$.
- Plug this into $y = \frac{18-3x}{2}$ to find that $y = 3$.

---

25. If you solve for y in the top equation, you get $y = \frac{10-3z}{4}$.
- If you substitute this into the bottom equation, you get $5\left(\frac{10-3z}{4}\right) - 2z = -22$.
- Distribute the 5 to get $\frac{25}{2} - \frac{15z}{4} - 2z = -22$. Note that (5)(10)/4 = 25/2 and 5(−3)/4 = −15/4.
- Combine like terms to get $-\frac{23z}{4} = -\frac{69}{2}$. (To combine like terms, you will first need to make a common denominator. Alternatively, you could multiply every term of the previous equation by 4.)
- Solve the previous equation to find that $z = 6$. (Note that the minus signs cancel.)
- Plug this into $y = \frac{10-3z}{4}$ to find that $y = -2$.

---

26. First solve for z in the bottom equation to get $z = 13 - 2x$.
- Plug this into both the top and middle equations to get:
- $-x - 4y = 18$
- $-12x + 5y = -49$
- (You will also need to distribute and combine like terms to arrive at the two equations above.)
- Solve the two equations above just like you solved problems 24 and 25.
- If you solve for x in the equation $-x - 4y = 18$, you get $x = -4y - 18$. (Add x to both sides and subtract 18 from both sides. If instead you add $4y$ to both sides, you will then need to multiply both sides by −1.)
- If you substitute $x = -4y - 18$ into the equation $-12x + 5y = -49$, you get $-12(-4y - 18) + 5y = -49$.
- Distribute the −12 to get $48y + 216 + 5y = -49$. When you distribute, two minuses make a plus.
- Combine like terms to get $53y = -265$.
- Solve the previous equation to find that $y = -5$.
- Plug this into $x = -4y - 18$ to find that $x = 2$.
- Plug this into $z = 13 - 2x$ to find that $z = 9$.

Chapter 2: Calculus Essentials

27. Take the derivative using the formula $\frac{d}{dx}(ax^b) = bax^{b-1}$.
    - Compare $6x^5$ to the general form ax^b.
    - Note that $a = 6$ and $b = 5$.
    - The answer is $30x^4$.

---

28. Take the derivative using the formula $\frac{d}{dt}(at^b) = bat^{b-1}$.
    - Compare t to the general form at^b.
    - Note that $a = 1$ and $b = 1$ (since $1t^1 = t$).
    - The answer is 1. (Recall from Chapter 1 that $t^0 = 1$.)

---

29. Take the derivative using the formula $\frac{d}{dx}(ax^b) = bax^{b-1}$.
    - Compare $2x^{-3}$ to the general form ax^b.
    - Note that $a = 2$ and $b = -3$.
    - Note that $b - 1 = -3 - 1 = -4$.
    - The answer is $-6x^{-4}$. The answer can also be expressed as $-\frac{6}{x^4}$.

---

30. Take the derivative using the formula $\frac{d}{dt}(at^b) = bat^{b-1}$.
    - Apply the rule $t^{-m} = \frac{1}{t^m}$.
    - Rewrite $\frac{3}{t^6}$ as $3t^{-6}$.
    - Compare $3t^{-6}$ to the general form at^b.
    - Note that $a = 3$ and $b = -6$.
    - Note that $b - 1 = -6 - 1 = -7$.
    - The answer is $-18t^{-7}$. The answer can also be expressed as $-\frac{18}{t^7}$.

---

31. Apply the rule $\frac{d}{dx}(y_1 + y_2 + y_3 + y_4) = \frac{dy_1}{dx} + \frac{dy_2}{dx} + \frac{dy_3}{dx} + \frac{dy_4}{dx}$.
    - Take the derivative of each term using the formula $\frac{d}{dx}(ax^b) = bax^{b-1}$.
    - The separate derivatives are:

    $$\frac{d}{dx}(3x^8) = 24x^7 \quad , \quad \frac{d}{dx}(-6x^5) = -30x^4 \quad , \quad \frac{d}{dx}(9x^2) = 18x \quad , \quad \frac{d}{dx}(-4) = 0$$

    - (Recall that the derivative of a constant is zero.)
    - Combine these derivatives together.
    - The answer is $24x^7 - 30x^4 + 18x$.

---

32. Take the derivative using the formula $\frac{d}{du}(au^b) = bau^{b-1}$.

- Apply the rule $\frac{1}{\sqrt{u}} = u^{-1/2}$.
- Compare $u^{-1/2}$ to the general form au^b.
- Note that $a = 1$ and $b = -\frac{1}{2}$.
- Note that $b - 1 = -\frac{1}{2} - 1 = -\frac{3}{2}$.
- The intermediate answer is $-\frac{1}{2}u^{-3/2}$. However, this answer is not in **standard form**.
- **Rationalize the denominator** in order to express the answer in standard form.
- Apply the rule $u^{-m} = \frac{1}{u^m}$.
- Rewrite $-\frac{1}{2}u^{-3/2}$ as $-\frac{1}{2u^{3/2}}$.
- Multiply the numerator and denominator by $u^{1/2}$.
- $-\frac{1}{2u^{3/2}}\frac{u^{1/2}}{u^{1/2}} = -\frac{u^{1/2}}{2u^2}$. Apply the rule $u^{1/2} = \sqrt{u}$.
- The final answer is $-\frac{\sqrt{u}}{2u^2}$. It's the same as $-\frac{u^{1/2}}{2u^2}$. (It's also equivalent to $-\frac{1}{2}u^{-3/2}$.)

---

33. First find the anti-derivative using the formula $\int ax^b\,dx = \frac{ax^{b+1}}{b+1}$.

- Compare $8x^3$ to the general form ax^b.
- Note that $a = 8$ and $b = 3$.
- The anti-derivative is $2x^4$.
- Evaluate $2x^4$ over the limits.
- $[2x^4]_{x=0}^{2} = 2(2)^4 - 2(0)^4$.
- The answer is 32.

---

34. First find the anti-derivative using the formula $\int at^b\,dt = \frac{at^{b+1}}{b+1}$.

- Compare t^4 to the general form at^b.
- Note that $a = 1$ and $b = 4$.
- The anti-derivative is $\frac{t^5}{5}$.
- Evaluate $\frac{t^5}{5}$ over the limits.
- $\left[\frac{t^5}{5}\right]_{t=-5}^{5} = \frac{(5)^5}{5} - \frac{(-5)^5}{5}$.
- Note that $-(-5)^5 = -(-3125) = 3125$.
- The answer is 1250.

---

35. First find the anti-derivative using the formula $\int ax^b\,dx = \frac{ax^{b+1}}{b+1}$.

- Compare $900x^{-3}$ to the general form ax^b.

- Note that $a = 900$ and $b = -3$.
- Note that $b + 1 = -3 + 1 = -2$.
- The anti-derivative is $-450x^{-2}$.
- Evaluate $-450x^{-2}$ over the limits.
- $[-450x^{-2}]_{x=3}^{5} = -450(5)^{-2} - [-450(3)^{-2}]$.
- The two minus signs make a plus sign: $-450(5)^{-2} + 450(3)^{-2}$.
- Note that $5^{-2} = \frac{1}{5^2} = \frac{1}{25}$ and $3^{-2} = \frac{1}{3^2} = \frac{1}{9}$.
- The answer is 32.

---

36. First find the anti-derivative using the formula $\int au^b \, du = \frac{au^{b+1}}{b+1}$.
    - Apply the rule $u^{-m} = \frac{1}{u^m}$.
    - Rewrite $\frac{1}{u^4}$ as u^{-4}.
    - Compare $81u^{-4}$ to the general form au^b.
    - Note that $a = 81$ and $b = -4$.
    - Note that $b + 1 = -4 + 1 = -3$.
    - The anti-derivative is $-27u^{-3}$.
    - Evaluate $-27u^{-3}$ over the limits.
    - $[-27u^{-3}]_{u=1}^{3} = -27(3)^{-3} - [-27(1)^{-3}]$.
    - The two minus signs make a plus sign: $-27(3)^{-3} + 27(1)^{-3}$.
    - Note that $3^{-3} = \frac{1}{3^3} = \frac{1}{27}$ and $1^{-3} = \frac{1}{1^3} = \frac{1}{1} = 1$.
    - The answer is 26.

---

37. First find the anti-derivative using the formula $\int ax^b \, dx = \frac{ax^{b+1}}{b+1}$.
    - Apply the rule $\frac{1}{\sqrt{x}} = x^{-1/2}$.
    - Rewrite $\frac{3}{\sqrt{x}}$ as $3x^{-1/2}$.
    - Compare $3x^{-1/2}$ to the general form ax^b.
    - Note that $a = 3$ and $b = -\frac{1}{2}$.
    - Note that $b + 1 = -\frac{1}{2} + 1 = \frac{1}{2}$.
    - Note that $\frac{a}{b+1} = \frac{3}{1/2} = 3(2) = 6$. (To divide by a fraction, multiply by its **reciprocal**.)
    - The anti-derivative is $6x^{1/2}$.
    - Apply the rule $\sqrt{x} = x^{1/2}$ to rewrite $6x^{1/2}$ as $6\sqrt{x}$.
    - Evaluate $6\sqrt{x}$ over the limits.
    - $\left[6\sqrt{x}\right]_{x=4}^{9} = 6\sqrt{9} - 6\sqrt{4}$. Note that $\sqrt{9} = 3$ and $\sqrt{4} = 2$.
    - The answer is 6.

---

38. First find the anti-derivative using the formula $\int at^b\,dt = \frac{at^{b+1}}{b+1}$.

- Compare $14t^{3/4}$ to the general form at^b.
- Note that $a = 14$ and $b = \frac{3}{4}$.
- Note that $b + 1 = \frac{3}{4} + 1 = \frac{3}{4} + \frac{4}{4} = \frac{7}{4}$. (Get a **common denominator**.)
- Note that $\frac{a}{b+1} = \frac{14}{7/4} = \frac{14(4)}{7} = 8$. (To divide by a fraction, multiply by its **reciprocal**.)
- The anti-derivative is $8t^{7/4}$.
- Evaluate $8t^{7/4}$ over the limits.
- $\left[8t^{7/4}\right]_{t=1}^{16} = 8(16)^{7/4} - 8(1)^{7/4}$.
- Note that $(16)^{7/4} = \left(16^{1/4}\right)^7 = \left(\sqrt[4]{16}\right)^7 = 2^7 = 128$. Also, $(1)^{7/4} = 1$.
- Try entering 16^(7/4) on your calculator to see that $(16)^{7/4} = 128$.
- The answer is 1016.

---

39. Apply the rule $\int (y_1 + y_2)\,dx = \int y_1\,dx + \int y_2\,dx$.

- Find the anti-derivative of each term using the formula $\int ax^b\,dx = \frac{ax^{b+1}}{b+1}$.
- The separate anti-derivatives are:

$$\int 12x^3\,dx = 3x^4 \quad , \quad \int 9x^2\,dx = 3x^3$$

- Combine these anti-derivatives together to make $3x^4 - 3x^3$.
- Evaluate $3x^4 - 3x^3$ over the limits.
- $[3x^4 - 3x^3]_{x=2}^{4} = [3(4)^4 - 3(4)^3] - [3(2)^4 - 3(2)^3]$.
- Distribute the minus sign. Two minus signs make a plus sign.
- You should get $3(4)^4 - 3(4)^3 - 3(2)^4 + 3(2)^3$.
- The final answer is 552.

---

40. Apply the rule $\int (y_1 + y_2 + y_3)\,dx = \int y_1\,dx + \int y_2\,dx + \int y_3\,dx$.

- Find the anti-derivative of each term using the formula $\int ax^b\,dx = \frac{ax^{b+1}}{b+1}$.
- The separate anti-derivatives are:

$$\int 12x^5\,dx = 2x^6 \quad , \quad \int 24x^3\,dx = 6x^4 \quad , \quad \int 48x\,dx = 24x^2$$

- Combine these anti-derivatives together to make $2x^6 + 6x^4 - 24x^2$.
- Evaluate $2x^6 + 6x^4 - 24x^2$ over the limits.
- $[2x^6 + 6x^4 - 24x^2]_{x=-1}^{2} = [2(2)^6 + 6(2)^4 - 24(2)^2] - [2(-1)^6 + 6(-1)^4 - 24(-1)^2]$.
- Distribute the minus sign. Two minus signs make a plus sign.
- You should get $2(2)^6 + 6(2)^4 - 24(2)^2 - 2(-1)^6 - 6(-1)^4 + 24(-1)^2$.
- The final answer is 144.

Chapter 3: One-dimensional Uniform Acceleration

41. Two knowns are given as numbers. The word "rest" gives you the third known.
    - The three knowns are $\Delta x = 90$ m, $t = 6.0$ s, and $v_{x0} = 0$.
    - Use the equation $\Delta x = v_{x0}t + \frac{1}{2}a_x t^2$.
    - After plugging in numbers and simplifying, you should get $90 = 18a_x$.
    - Solve this equation to find that $a_x = 5.0 \text{ m/s}^2$.

---

42. All three knowns are given as numbers. The units should help identify the knowns.
    - The three knowns are $v_{x0} = 15$ m/s, $a_x = -4.0 \text{ m/s}^2$, and $t = 6.0$ s.
    - Use the equation $v_x = v_{x0} + a_x t$.
    - After plugging in numbers and simplifying, the answer is $v_x = -9.0$ m/s.

---

43. All three knowns are given as numbers. The units should help identify the knowns.
    - The three knowns are $v_{x0} = 10$ m/s, $v_x = 30$ m/s, and $a_x = 8.0 \text{ m/s}^2$.
    - Use the equation $v_x^2 = v_{x0}^2 + 2a_x \Delta x$.
    - After plugging in numbers and simplifying, you should get $800 = 16\Delta x$.
    - Solve this equation to find that $\Delta x = 50$ m.

---

44. Ignore the 500 g. The mass of the banana **doesn't** affect the answer.
    - Two of the knowns are given as numbers. The word "rest" gives you the third known.
    - The three knowns are $\Delta y = -36$ m, $t = 4.0$ s, and $v_{y0} = 0$.
    - Δy is negative because the banana finishes **below** the starting point.
    - Use the equation $\Delta y = v_{y0}t + \frac{1}{2}a_y t^2$.
    - After plugging in numbers and simplifying, you should get $-36 = 8a_y$.
    - Solve this equation to find that $a_y = -\frac{9}{2} \text{ m/s}^2$.

---

45. Two of the knowns are given as numbers. You should know the third known since the banana falls near earth's surface.
    - The three knowns are $v_{y0} = 20$ m/s, $\Delta y = -60$ m, and $a_y = -9.81 \text{ m/s}^2$.
    - Δy is negative because the banana finishes **below** the starting point.
    - a_y is negative for free fall problems (since we choose $+y$ to be up).
    - Use the equation $\Delta y = v_{y0}t + \frac{1}{2}a_y t^2$. Round -9.81 to -10 if not using a calculator.
    - After plugging in numbers and simplifying, you should get a quadratic equation. Put the quadratic equation in **standard form**: $5t^2 - 20t - 60 \approx 0$. If you're using a calculator, you should use $4.905t^2$ instead of $5t^2$.

- The constants are $a \approx 5$, $b = -20$, and $c = -60$. (Using a calculator, $a = 4.905$.)
- (It would also be okay to divide by 5 and use $a \approx 1$, $b = -4$, and $c = -12$.)
- The two solutions to the quadratic are –2 s and 6 s.
- The final answer is $t \approx 6.0$ s. If using a calculator, you should get $t = 6.1$ s.

---

46. The 'trick' to this problem is to work with only **half the trip**.
    - Work with the trip up. The time will be 0.5 s.
    - You should know one number since the monkey is near earth's surface.
    - If you work with the trip up, the final speed (at the top) will be zero.
    - The three knowns are $t = 0.5$ s, $a_y = -9.81$ m/s^2, and $v_y = 0$.
    - Use the equation $v_y = v_{y0} + a_y t$.
    - The final answer is $v_{y0} \approx 5.0$ m/s. If using a calculator, you should get 4.9 m/s.

Chapter 4: One-dimensional Motion with Calculus

47. Choose the appropriate equations from page 48.
    - Velocity equals the **derivative** of position with respect to time.
    - Use the equation $v_x = \frac{dx}{dt}$.
    - The derivative is $v_x = 12t^2 - 16t$.
    - (The third term vanishes: Recall that the derivative of a constant equals zero.)
    - Plug in the given value for time: $t = 5.0$ s. (Don't ignore the square in t^2).
    - The velocity at the specified time is $v_x = 220$ m/s.
    - Acceleration equals the **derivative** of velocity with respect to time.
    - Use the equation $a_x = \frac{dv_x}{dt}$.
    - Take a derivative of $v_x = 12t^2 - 16t$ with respect to time.
    - The derivative is $a_x = 24t - 16$.
    - Plug in the given value for time: $t = 7.0$ s.. (It's not 5.0 s.)
    - The acceleration at the specified time is $a_x = 152$ m/s^2.

---

48. Choose the appropriate equations from page 48.
    - Net displacement is the **integral** of velocity.
    - Use the equation $\Delta x = \int_{t=t_0}^{t} v_x \, dt$.
    - Perform the definite integral $\int_{t=2}^{8} \sqrt{2t} \, dt$.
    - Apply the rules $\sqrt{ax} = \sqrt{a}\sqrt{x}$ and $x^{1/2} = \sqrt{x}$.
    - Rewrite $\sqrt{2t}$ as $2^{1/2} t^{1/2}$.
    - Perform the definite integral $\int_{t=2}^{8} 2^{1/2} t^{1/2} \, dt$.

- The anti-derivative is $\frac{2^{3/2}t^{3/2}}{3}$.
- Note that $2^{1/2}\left(\frac{t^{3/2}}{3/2}\right) = 2^{1/2}\left(\frac{2t^{3/2}}{3}\right) = \frac{2^{3/2}t^{3/2}}{3}$. Note that $2^{1/2}(2) = 2^{1/2}(2^1) = 2^{3/2}$.
- Evaluate $\frac{2^{3/2}t^{3/2}}{3}$ over the limits.
- $\left[\frac{2^{3/2}t^{3/2}}{3}\right]_{t=2}^{8} = \frac{2^{3/2}(8)^{3/2}}{3} - \frac{2^{3/2}(2)^{3/2}}{3}$.
- Apply the rule $(x^m)^n = x^{mn}$ to write $(8)^{3/2} = (2^3)^{3/2} = 2^{9/2}$.
- Apply the rule $x^m x^n = x^{m+n}$ to write $2^{3/2}(2)^{9/2} = 2^6$ and $2^{3/2}(2)^{3/2} = 2^3$.
- The net displacement for the specified time interval is $\Delta x = \frac{56}{3}$ m.
- Acceleration equals the **derivative** of velocity with respect to time.
- Use the equation $a_x = \frac{dv_x}{dt}$.
- Take a derivative of $v_x = 2^{1/2}t^{1/2}$ with respect to time. (Recall that $\sqrt{2t} = 2^{1/2}t^{1/2}$.)
- The derivative is $a_x = 2^{-1/2}t^{-1/2} = \frac{1}{\sqrt{2t}}$. Note that $\frac{2^{1/2}}{2} = 2^{1/2-1} = 2^{-1/2}$.
- Plug in the given value for time: $t = 8.0$ s.
- The acceleration at the specified time is $a_x = \frac{1}{4}$ m/s^2.

---

49. Choose the appropriate equations from page 48.
    - Velocity equals initial velocity plus the **integral** of acceleration.
    - Use the equation $v_x = v_{x0} + \int_{t=t_0}^{t} a_x\, dt$.
    - Evaluate $18 + \int_{t=0}^{3} 24t^2\, dt$.
    - The anti-derivative is $8t^3$.
    - Evaluate $8t^3$ over the limits. Remember to add this to 18.
    - $18 + [8t^3]_{t=0}^{3} = 18 + 8(3)^3 - 8(0)^3$.
    - The velocity at the specified time equals $v_x = 234$ m/s.
    - Net displacement is the **integral** of velocity.
    - Use the equation $\Delta x = \int_{t=t_0}^{t} v_x\, dt$.
    - The velocity function is $18 + 8t^3$ (because $v_x = v_{x0} + \int_{t=t_0}^{t} a_x\, dt$).
    - (Recall that we found the expression $8t^3$ in the previous part of the solution.)
    - Perform the definite integral $\int_{t=0}^{3}(18 + 8t^3)\, dt$. Integrate each term separately.
    - The anti-derivative is $18t + 2t^4$.
    - Evaluate $18t + 2t^4$ over the limits.
    - $[18t + 2t^4]_{t=0}^{3} = 18(3) + 2(3)^4 - 18(0) - 2(0)^4$.
    - The net displacement for the specified time interval is $\Delta x = 216$ m.

Chapter 5: Geometry Essentials

50. Use the formulas for the perimeter and area of a rectangle.
    - The knowns are $L = 6$ m and $W = 4$ m.
    - Use the equations $P = 2L + 2W$ and $A = LW$.
    - The perimeter is $P = 20$ m and the area is $A = 24$ m^2.

---

51. You need to find the hypotenuse of the triangle before you can find the perimeter.
    - The knowns are $a = 3$ m and $b = 4$ m. (Think "b" for "base.")
    - Use the Pythagorean theorem: $a^2 + b^2 = c^2$.
    - Plug in the given values to find that $c = 5$ m.
    - Add up the lengths of the sides to find the perimeter of the triangle.
    - Use the formula $P = a + b + c$. The perimeter is $P = 12$ m.
    - The base is $b = 4$ m and the height is $h = 3$ m.
    - Use the formula $A = \frac{1}{2}bh$. The area is $A = 6$ m^2.

---

52. First solve for the length of an edge.
    - The known is $A = 36$ m^2.
    - Use the formula $A = L^2$ to find that $L = 6$ m.
    - Set $W = L$ in the formula for the perimeter of a rectangle ($W = L$ for a square).
    - Use the formula $P = 4L$ to get $P = 24$ m.

---

53. Use the Pythagorean theorem.
    - The knowns are $a = 5$ m and $b = 12$ m. (Think "b" for "base.")
    - Use the equation $a^2 + b^2 = c^2$.
    - Plug in the given values to find that $c = 13$ m.

---

54. Use the Pythagorean theorem.
    - The knowns are $b = \sqrt{3}$ m and $c = 2$ m. (Think "b" for "base.")
    - Use the equation $a^2 + b^2 = c^2$.
    - Unlike the previous problem, you're given the hypotenuse. Solve for a.
    - Plug in the given values to find that $a = 1$ m.

---

55. The diagonal divides the rectangle into two right triangles. Work with one triangle.
    - The knowns are $a = 6$ m and $b = 8$ m. (Think "b" for "base.")
    - Use the Pythagorean theorem: $a^2 + b^2 = c^2$.
    - Plug in the given values to find that $c = 10$ m.

---

56. You're given the diameter.
- The known is $D = 6$ m.
- Use the equations $R = \frac{D}{2}$, $C = 2\pi R$, and $A = \pi R^2$.
- The answers are $R = 3$ m, $C = 6\pi$ m, and $A = 9\pi$ m^2.

---

57. First solve for the radius.
- The known is $A = 16\pi$ m^2.
- Use the equation $A = \pi R^2$.
- Solve for the radius: $R = 4$ m.
- Use the equation $C = 2\pi R$.
- The circumference is $C = 8\pi$ m.

Chapter 6: Motion Graphs

58. (A) **Read** the graph.
- For the total distance traveled, find the distance traveled for each segment.
- You should get 80 m backwards, 40 m forward, and 20 m backwards.
- Add up the absolute values of these distances: $TDT = |d_1| + |d_2| + |d_3|$.
- The total distance traveled is $TDT = 140$ m.
- Read the initial and final values: $x_i = 20$ m and $x_f = -40$ m.
- Use the formula $ND = x_f - x_i$.
- The net displacement for the trip is $ND = -60$ m.

(B) Find the **slope** of the first line segment.
- Read off the endpoints: (0, 20 m) and (20 s, −60 m).
- Use the slope formula: $slope = \frac{x_2 - x_1}{t_2 - t_1}$.
- The velocity is $v_x = -4.0$ m/s during the first 20 s.

---

59. (A) Find the **slope** of the second line segment.
- Read off the endpoints: (20 s, 0) and (30 s, −60 m/s).
- Use the slope formula: $slope = \frac{v_{2x} - v_{1x}}{t_2 - t_1}$.
- The acceleration is $a_x = -6.0$ m/s^2 during the second segment.

(B) Find the **area** between each solid line segment and the dashed t-axis.
- Make a triangle for the first 20 s, a triangle for the next 10 s, and a rectangle for the last 20 s.
- Find these areas using the formulas $A = \frac{1}{2}bh$ and $A = LW$.
- The "areas" (they don't actually have units of ordinary "area") are $A_1 = 200$ m,

$A_2 = -300$ m, and $A_3 = -1200$ m.

- Add these areas together to find the net displacement.
- The net displacement is $ND = -1300$ m. Note that -1300 m $= -1.3$ km.

---

60. (A) Find the **area** between each solid line segment and the dashed t-axis.

- Make a rectangle for the first 20 s, a triangle from 20 s to approximately 26.5 s, and a triangle from approximately 26.5 s to 50 s.
- Find these areas using the formulas $A = \frac{1}{2}bh$ and $A = LW$.
- The "areas" (they don't actually have units of ordinary "area") are $A_1 = -800$ m/s, $A_2 = -130$ m/s, and $A_3 = 235$ m/s.
- Add these areas together to get -695 m/s. Area is **not** the answer.
- Use the equation $v_x = v_{x0} + area$. (The initial velocity is given in the paragraph.)
- The answer is $v_x = -545$ m/s.
- (If you get an answer between -500 and -600, that's probably close enough.)

(B) **Read** the graph.

- Where does the graph cross the t-axis?
- It first crosses at 26.7 s. It crosses again at the very end, at 50.0 s.
- (If you get a number between 26 s and 27 s, that's probably close enough. How did we get 26.7 s? The equation for the line is $x = 6t - 160$. Set $x = 0$ and solve for t.)

Chapter 7: Two Objects in Motion

61. Check that your equation of constraint is correct: $\Delta x_1 - \Delta x_2 = d$. The distances will really "add" together, but since they are headed in opposite directions, a minus sign is needed to make them "add."

- The second monkey has zero acceleration since his velocity is constant.
- The knowns are $v_{10} = 0$, $a_1 = \frac{1}{8}$ m/s^2, $v_{20} = -15$ m/s, $a_2 = 0$, and $d = 1600$ m.
- Use the equations $\Delta x_1 = v_{10}t + \frac{1}{2}a_1 t^2$ and $\Delta x_2 = v_{20}t + \frac{1}{2}a_2 t^2$.
- After plugging in numbers, you should have $\Delta x_1 = \frac{1}{16}t^2$ and $\Delta x_2 = -15t$.
- Substitute these expressions into $\Delta x_1 - \Delta x_2 = d$.
- You will get a quadratic equation. Put it in **standard form**: $\frac{1}{16}t^2 + 15t - 1600 = 0$.
- The constants are $a = \frac{1}{16}$, $b = 15$, and $c = -1600$.
- Use the quadratic formula to find that $t = 80$ s.
- The answers are $\Delta x_1 = 400$ m and $\Delta x_2 = -1200$ m.

---

62. Check that your equation of constraint is correct: $t_1 = t_2 + \Delta t$. The thief runs for 2.0 s more than his uncle runs, so t_1 (for the thief) is 2.0 s larger than t_2 (for the uncle).

- The thief has zero acceleration since his velocity is constant.
- The knowns are $v_{10} = 9.0$ m/s, $a_1 = 0$, $v_{20} = 0$, $a_2 = 4.0$ m/s^2, and $\Delta t = 2.0$ s.
- Use the equations $\Delta x = v_{10}t_1 + \frac{1}{2}a_1 t_1^2$ and $\Delta x = v_{20}t_2 + \frac{1}{2}a_2 t_2^2$.
- After plugging in numbers, you should have $\Delta x = 9t_1$ and $\Delta x = 2t_2^2$.
- Substitute $t_1 = t_2 + \Delta t$ into the equation $\Delta x = 9t_1$.
- Set the two Δx's equal to each other.
- You will get a quadratic equation. Put it in **standard form**: $2t_2^2 - 9t_2 - 18 = 0$.
- The constants are $a = 2$, $b = -9$, and $c = -18$.
- Use the quadratic formula to find that $t_2 = 6.0$ s. Therefore $t_1 = 8.0$ s.
- The answer is $\Delta x = 72$ m.

---

63. Check that your equation of constraint is correct: $-\Delta y_1 + \Delta y_2 = d$. The distances will really "add" together, but since the net displacements are opposite (one finished below while the other above where it starts), a minus sign is needed to make them "add."

- The monkey has zero acceleration since his velocity is constant.
- The knowns are $v_{10} = -5.0$ m/s, $a_1 = 0$, $v_{20} = 40$ m/s, $a_2 = -9.81$ m/s^2, and $d = 90$ m. The sign of v_{10} is negative because the monkey is traveling downward. The sign of a_2 is negative because it's freely falling with $+y$ upward.
- Use the equations $\Delta y_1 = v_{10}t + \frac{1}{2}a_1 t^2$ and $\Delta y_2 = v_{20}t + \frac{1}{2}a_2 t^2$.
- After plugging in numbers, you should have $\Delta y_1 = -5t$ and $\Delta y_2 \approx 40t - 5t^2$.
- Substitute these expressions into $-\Delta y_1 + \Delta y_2 = d$.
- You will get a quadratic equation. Put it in **standard form**: $-5t^2 + 45t - 90 \approx 0$. If you have $5t^2 - 45t + 90 \approx 0$ that's the same. You may also divide by 5.
- The constants are $a \approx -5$, $b = 45$, and $c = -90$ (or $a \approx 5$, $b = -45$, and $c = 90$). (You may also divide all three constants by 5.)
- Use the quadratic formula to find that $t \approx 3.0$ s (if you round a_y to ≈ -10 m/s^2).
- The answers are $\Delta y_1 \approx -15$ m and $\Delta y_2 \approx 75$ m.
- With a calculator, $a = -4.905$ in the quadratic, $t = 2.95$ s, and $\Delta y_2 = 75.3$ m.

Chapter 8: Net and Average Values

64. Note that west and east are opposite. In the answers below, east is positive.
- Add the given distances to find the total distance traveled: $TDT = |d_1| + |d_2| + |d_3|$.
- For net displacement, the second displacement is negative: $ND = d_1 - d_2 + d_3$.
- The answers are $TDT = 120$ m and $ND = -20$ m (if you choose east to be positive). The minus sign means that the net displacement is 20 m to the west.

---

65. Note that west and east are opposite. In the answers below, east is positive.
- Add the times together: $TT = t_1 + t_2$.
- The total time is $TT = 10.0$ s.
- Add the given distances to find the total distance traveled: $TDT = |d_1| + |d_2|$.
- For net displacement, one displacement is negative: $ND = -d_1 + d_2$.
- The answers are $TDT = 300$ m and $ND = 60$ m to the east.
- Use the formulas $\begin{matrix}\text{ave.}\\\text{spd.}\end{matrix} = \frac{TDT}{TT}$ and $\begin{matrix}\text{ave.}\\\text{vel.}\end{matrix} = \frac{ND}{TT}$.
- The answers are $\begin{matrix}\text{ave.}\\\text{spd.}\end{matrix} = 30$ m/s and $\begin{matrix}\text{ave.}\\\text{vel.}\end{matrix} = 6.0$ m/s to the east.

---

66. Note that east and south are **perpendicular**. This impacts the net displacement.
- Add the times together: $TT = t_1 + t_2$.
- The total time is $TT = 20.0$ s.
- Add the given distances to find the total distance traveled: $TDT = |d_1| + |d_2|$.
- For net displacement, apply the Pythagorean theorem: $ND = \sqrt{d_1^2 + d_2^2}$.
- The answers are $TDT = 140$ m and $ND = 100$ m.
- Use the formulas $\begin{matrix}\text{ave.}\\\text{spd.}\end{matrix} = \frac{TDT}{TT}$ and $\begin{matrix}\text{ave.}\\\text{vel.}\end{matrix} = \frac{ND}{TT}$.
- The answers are $\begin{matrix}\text{ave.}\\\text{spd.}\end{matrix} = 7.0$ m/s and $\begin{matrix}\text{ave.}\\\text{vel.}\end{matrix} = 5.0$ m/s.

---

67. Examine the definition of **average acceleration**. The solution is simpler than it seems.
- Add the times together: $TT = t_1 + t_2 + t_3$.
- The total time is $TT = 20.0$ s.
- What is the initial velocity? What is the final velocity?
- These are given: $v_i = 5.0$ m/s to the east and $v_f = 0$ (the gorilla comes to "rest").
- Use the formula $\begin{matrix}\text{ave.}\\\text{accel.}\end{matrix} = \frac{v_f - v_i}{TT}$.
- The answer is $\begin{matrix}\text{ave.}\\\text{accel.}\end{matrix} = -\frac{1}{4}$ m/s^2. The minus sign means $\frac{1}{4}$ m/s^2 to the "west." The average acceleration is opposite to the motion because it is **deceleration**.

Chapter 9: Trigonometry Essentials

68. First use the Pythagorean theorem to find the hypotenuse.
    - The knowns are $a = 10$ and $b = 24$. (Think "b" for "base.")
    - Use the Pythagorean theorem: $a^2 + b^2 = c^2$.
    - Plug in the given values to find that $c = 26$.
    - Identify the opposite to θ, adjacent to θ, and hypotenuse.
    - In relation to θ, the sides are: $opp. = 10$, $adj. = 24$, and $hyp. = 26$.
    - Use the formulas $\sin\theta = \frac{\text{opp.}}{\text{hyp.}}$, $\cos\theta = \frac{\text{adj.}}{\text{hyp.}}$, and $\tan\theta = \frac{\text{opp.}}{\text{adj.}}$.
    - The answers are $\sin\theta = \frac{5}{13}$, $\cos\theta = \frac{12}{13}$, and $\tan\theta = \frac{5}{12}$. (Note, for example: $\frac{10}{26} = \frac{5}{13}$.)

---

69. First use the Pythagorean theorem to find the missing side.
    - The knowns are $c = 8$ and $b = 4$. (Think "b" for "base.")
    - Use the Pythagorean theorem: $a^2 + b^2 = c^2$.
    - Plug in the given values to find that $a = 4\sqrt{3}$.
    - Identify the opposite to θ, adjacent to θ, and hypotenuse.
    - In relation to θ, the sides are: $opp. = 4$, $adj. = 4\sqrt{3}$, and $hyp. = 8$.
    - Use the formulas $\sin\theta = \frac{\text{opp.}}{\text{hyp.}}$, $\cos\theta = \frac{\text{adj.}}{\text{hyp.}}$, and $\tan\theta = \frac{\text{opp.}}{\text{adj.}}$.
    - The answers are $\sin\theta = \frac{1}{2}$, $\cos\theta = \frac{\sqrt{3}}{2}$, and $\tan\theta = \frac{\sqrt{3}}{3}$ $\left(\text{or } \frac{1}{\sqrt{3}} \text{ since } \frac{1}{\sqrt{3}} = \frac{1}{\sqrt{3}}\frac{\sqrt{3}}{\sqrt{3}} = \frac{\sqrt{3}}{3}\right)$.

---

70. Simply read the table on page 85. You can build fluency through practice.

(A) $\sin 60° = \frac{\sqrt{3}}{2}$ (B) $\cos 45° = \frac{\sqrt{2}}{2}$ (C) $\tan 30° = \frac{\sqrt{3}}{3}$ (D) $\sin 45° = \frac{\sqrt{2}}{2}$

(E) $\cos 30° = \frac{\sqrt{3}}{2}$ (F) $\tan 60° = \sqrt{3}$ (G) $\sin 90° = 1$ (H) $\cos 90° = 0$

(I) $\tan 45° = 1$ (J) $\sin 30° = \frac{1}{2}$ (K) $\cos 60° = \frac{1}{2}$ (L) $\tan 90° =$ undef.

(M) $\sin 0° = 0$ (N) $\cos 0° = 1$ (O) $\tan 0° = 0$ (P) $\sin 60° = \frac{\sqrt{3}}{2}$

(Q) $\cos 30° = \frac{\sqrt{3}}{2}$ (R) $\tan 45° = 1$

---

71. First find the reference angle. Evaluate the trig function at the **reference angle**. Next determine the **sign** of trig function in the given Quadrant. Combine these two answers. **Note**: If you use a calculator, make sure that your mode is set to **degrees** (**not** radians).

(A) The reference angle is 30°. The sign is positive. The answer is: $\sin 150° = \frac{1}{2}$.

(B) The reference angle is 60°. The sign is negative. The answer is: $\cos 240° = -\frac{1}{2}$.

(C) The reference angle is 60°. The sign is negative. The answer is: $\tan 300° = -\sqrt{3}$.

(D) The reference angle is 45°. The sign is negative. The answer is: $\sin 315° = -\frac{\sqrt{2}}{2}$.
(E) The reference angle is 45°. The sign is negative. The answer is: $\cos 135° = -\frac{\sqrt{2}}{2}$.
(F) The reference angle is 30°. The sign is positive. The answer is: $\tan 210° = \frac{\sqrt{3}}{3}$.
(G) The reference angle is 60°. The sign is negative. The answer is: $\sin 240° = -\frac{\sqrt{3}}{2}$.
(H) The reference angle is 0°. The sign is negative. The answer is: $\cos 180° = -1$.
(I) The reference angle is 45°. The sign is positive. The answer is: $\tan 225° = 1$.
(J) The reference angle is 0°. The answer is: $\sin 180° = 0$.
(K) The reference angle is 90°. The answer is: $\cos 270° = 0$.
(L) The reference angle is 60°. The sign is negative. The answer is: $\tan 300° = -\sqrt{3}$.
(M) The reference angle is 60°. The sign is negative. The answer is: $\sin 300° = -\frac{\sqrt{3}}{2}$.
(N) The reference angle is 30°. The sign is positive. The answer is: $\cos 330° = \frac{\sqrt{3}}{2}$.
(O) The reference angle is 0°. The answer is: $\tan 180° = 0$.
(P) The reference angle is 90°. The sign is negative. The answer is: $\sin 270° = -1$.
(Q) The reference angle is 45°. The sign is negative. The answer is: $\cos 225° = -\frac{\sqrt{2}}{2}$.
(R) The reference angle is 30°. The sign is negative. The answer is: $\tan 150° = -\frac{\sqrt{3}}{3}$.

---

72. First find the reference angle using the chart on page 85. Next look at the sign of the argument in order to determine which Quadrants the answers lie in. Find the correspondding answers using the equations in Step 3 of page 90.

Note: If you use a calculator, make sure that your mode is set to **degrees** (**not** radians). (Most calculators only give one of the two answers, and you must apply your trig skills to find the alternate answer. Also note that sometimes you need to add 360° to a calculator's answer to check your answer. For example, if a calculator gives you −60°, add 360° to get 300°. The two angles −60° and 300° are really the same.)

(A) The reference angle is 60°. The angles lie in Quadrants III and IV.
The answers are: $\sin^{-1}\left(-\frac{\sqrt{3}}{2}\right) = 240°$ or $300°$.
(B) The reference angle is 45°. The angles lie in Quadrants I and IV.
The answers are: $\cos^{-1}\left(\frac{\sqrt{2}}{2}\right) = 45°$ or $315°$.
(C) The reference angle is 45°. The angles lie in Quadrants II and IV.
The answers are: $\tan^{-1}(-1) = 135°$ or $315°$.
(D) The reference angle is 30°. The angles lie in Quadrants I and II.
The answers are: $\sin^{-1}\left(\frac{1}{2}\right) = 30°$ or $150°$.
(E) The reference angle is 30°. The angles lie in Quadrants II and III.
The answers are: $\cos^{-1}\left(-\frac{\sqrt{3}}{2}\right) = 150°$ or $210°$.

(F) The reference angle is 60°. The angles lie in Quadrants II and IV.
The answers are: $\tan^{-1}(-\sqrt{3}) = 120°$ or $300°$.

(G) The reference angle is 45°. The angles lie in Quadrants I and II.
The answers are: $\sin^{-1}\left(\frac{\sqrt{2}}{2}\right) = 45°$ or $135°$.

(H) The reference angle is 60°. The angles lie in Quadrants I and IV.
The answers are: $\cos^{-1}\left(\frac{1}{2}\right) = 60°$ or $300°$.

(I) The reference angle is 60°. The angles lie in Quadrants I and III.
The answers are: $\tan^{-1}(\sqrt{3}) = 60°$ or $240°$.

(J) This is one of the special angles. The answer is: $\sin^{-1}(1) = 90°$.

(K) The reference angle is 30°. The angles lie in Quadrants I and IV.
The answers are: $\cos^{-1}\left(\frac{\sqrt{3}}{2}\right) = 30°$ or $330°$.

(L) The reference angle is 45°. The angles lie in Quadrants I and III.
The answers are: $\tan^{-1}(1) = 45°$ or $225°$.

(M) The reference angle is 60°. The angles lie in Quadrants I and II.
The answers are: $\sin^{-1}\left(\frac{\sqrt{3}}{2}\right) = 60°$ or $120°$.

(N) This is one of the special angles. The answer is: $\cos^{-1}(-1) = 180°$.

(O) This is one of the special angles. The answers are: $\tan^{-1}(0) = 0°$ or $180°$.

(P) This is one of the special angles. The answers are: $\sin^{-1}(0) = 0°$ or $180°$.

---

73. First find the anti-derivative.

- The anti-derivative is $-\cos\theta$.
- Evaluate the anti-derivative over the limits.
- $[-\cos\theta]_{\theta=180°}^{360°} = -\cos(360°) - [-\cos(180°)] = -\cos(360°) + \cos(180°)$.
- The answer is -2.

---

74. First find the anti-derivative.

- The anti-derivative is $\sin\theta$.
- Evaluate the anti-derivative over the limits.
- $[\sin\theta]_{\theta=120°}^{240°} = \sin(240°) - \sin(120°)$.
- The answer is $-\sqrt{3}$. Note that $-\frac{\sqrt{3}}{2} - \frac{\sqrt{3}}{2} = -\sqrt{3}$.

Chapter 10: Vector Addition

75. Check the signs of your components: M_x, M_y, and B_x are negative.
   (A) Use the component equations, such as $M_x = M\cos\theta_M$.
   Answers: $M_x = -18$ ☺, $M_y = -18\sqrt{3}$ ☺, $B_x = -9$ ☺, $B_y = 9\sqrt{3}$ ☺.
   (B) Add respective components together. For example, $R_x = M_x + B_x$.
   Answers: $R_x = -27$ ☺, $R_y = -9\sqrt{3}$ ☺ (or -15.6 ☺).
   (C) Use the Pythagorean theorem, $R = \sqrt{R_x^2 + R_y^2}$, and inverse tangent, $\theta_R = \tan^{-1}\left(\frac{R_y}{R_x}\right)$.
   Note that $\left(\sqrt{3}\right)^2 = 3$. Thus, $\left(9\sqrt{3}\right)^2 = 9^2(3) = 243$. Note that $\sqrt{972} = 18\sqrt{3}$.
   The reference angle is 30°. Since $R_x < 0$ and $R_y < 0$, the answer lies in Quadrant III.
   Answers: $R = 18\sqrt{3}$ ☺ (or 31.2 ☺), $\theta_R = 210°$. ($\theta_{III} = 180° + \theta_{ref}$.)

---

76. Check the signs of your components: Ψ_x, Φ_x, and Φ_y are negative.
   (A) Use the component equations, such as $\Psi_x = \Psi\cos\theta_\Psi$. There are **six** equations.
   Answers: $\Psi_x = -3$ N, $\Psi_y = 3$ N, $\Phi_x = -6$ N, $\Phi_y = -6$ N, $\Omega_x = 12$ N, $\Omega_y = 0$.
   (B) Add respective components together. For example, $R_x = \Psi_x + \Phi_x + \Omega_x$.
   Answers: $R_x = 3$ N, $R_y = -3$ N.
   (C) Use the Pythagorean theorem, $R = \sqrt{R_x^2 + R_y^2}$, and inverse tangent, $\theta_R = \tan^{-1}\left(\frac{R_y}{R_x}\right)$.
   Note that $(-3)^2 = +9$. Note that $\sqrt{18} = \sqrt{(9)(2)} = \sqrt{9}\sqrt{2} = 3\sqrt{2}$.
   The reference angle is 45°. Since $R_x > 0$ and $R_y < 0$, the answer lies in Quadrant IV.
   Answers: $R = 3\sqrt{2}$ N (or 4.24 N), $\theta_R = 315°$. ($\theta_{IV} = 360° - \theta_{ref}$.)

---

77. Check the signs of your components: F_y, T_x, and T_y are negative.
   (A) Use the component equations, such as $T_x = T\cos\theta_T$.
   Answers: $F_x = 0$, $F_y = -16$ m, $T_x = -8$ m, $T_y = -8$ m.
   (B) **Subtract** respective components. For example, $B_x = F_x - T_x$. Note that $\vec{\mathbf{F}}$ is **first**.
   Answers: $B_x = 8$ m, $B_y = -8$ m.
   (C) Use the Pythagorean theorem, $B = \sqrt{B_x^2 + B_y^2}$, and inverse tangent, $\theta_B = \tan^{-1}\left(\frac{B_y}{B_x}\right)$.
   Note that $(-8)^2 = +64$. Note that $\sqrt{128} = \sqrt{(64)(2)} = \sqrt{64}\sqrt{2} = 8\sqrt{2}$.
   The reference angle is 45°. Since $B_x > 0$ and $B_y < 0$, the answer lies in Quadrant IV.
   Answers: $B = 8\sqrt{2}$ m (or 11.3 m), $\theta_B = 315°$. ($\theta_{IV} = 360° - \theta_{ref}$.)

---

78. Check the signs of your components: M_x, M_y, and S_x are negative.
   (A) Use the component equations, such as $M_x = M\cos\theta_M$.
   Answers: $M_x = -2\sqrt{3}$ N, $M_y = -2$ N, $S_x = -\frac{3\sqrt{3}}{2}$ N, $S_y = \frac{3}{2}$ N.

(B) Use the equations $P_x = 3M_x - 2S_x$ and $P_y = 3M_y - 2S_y$.

Answers: $P_x = -3\sqrt{3}$ N (or -5.20 N), $P_y = -9$ N.

(C) Use the Pythagorean theorem, $P = \sqrt{P_x^2 + P_y^2}$, and inverse tangent, $\theta_P = \tan^{-1}\left(\frac{P_y}{P_x}\right)$.

Note that $\left(\sqrt{3}\right)^2 = 3$. Thus, $\left(3\sqrt{3}\right)^2 = 3^2(3) = 27$. Note that $\sqrt{108} = 6\sqrt{3}$.

The reference angle is 60°. Since $P_x < 0$ and $P_y < 0$, the answer lies in Quadrant III.

Answers: $P = 6\sqrt{3}$ N (or 10.4 N), $\theta_P = 240°$. ($\theta_{III} = 180° + \theta_{ref}$.)

---

79. Identify the components of the given vector.
   - Compare $\vec{\mathbf{J}} = \sqrt{3}\,\hat{\mathbf{x}} - \hat{\mathbf{y}}$ to $\vec{\mathbf{J}} = J_x\hat{\mathbf{x}} + J_y\hat{\mathbf{y}}$ in order to determine the components.
   - The components are $J_x = \sqrt{3}$ and $J_y = -1$.
   - (Note that $-\hat{\mathbf{y}}$ is the same thing as $-1\hat{\mathbf{y}}$.)
   - Use the equation $J = \sqrt{J_x^2 + J_y^2}$.
   - The magnitude of the vector is $J = 2$.
   - Use the equation $\theta_J = \tan^{-1}\left(\frac{J_y}{J_x}\right)$.
   - The reference angle is 30°. The answer lies in Quadrant IV since $J_x > 0$ and $J_y < 0$.
   - The direction of the vector is $\theta_J = 330°$. ($\theta_{IV} = 360° - \theta_{ref}$.)

---

80. Identify the given values.
   - The known values are $K = 4$ and $\theta_K = 120°$.
   - Use trig to find the components of the vector: $K_x = K\cos\theta_K$ and $K_y = K\sin\theta_K$.
   - The components of the vector are $K_x = -2$ and $K_y = 2\sqrt{3}$.
   - Substitute these values into the equation $\vec{\mathbf{K}} = K_x\hat{\mathbf{x}} + K_y\hat{\mathbf{y}}$.
   - The answer is $\vec{\mathbf{K}} = -2\,\hat{\mathbf{x}} + 2\sqrt{3}\,\hat{\mathbf{y}}$.

---

81. **Combine like terms** together.
   - Combine the $\hat{\mathbf{x}}$-terms together to make $8\sqrt{3}\,\hat{\mathbf{x}}$.
   - Combine the $\hat{\mathbf{y}}$-terms together to make $-24\,\hat{\mathbf{y}}$.
   - The resultant vector is $\vec{\mathbf{R}} = 8\sqrt{3}\,\hat{\mathbf{x}} - 24\,\hat{\mathbf{y}}$.
   - Use the equation $R = \sqrt{R_x^2 + R_y^2}$.
   - The magnitude of the resultant vector is $R = 16\sqrt{3}$.
   - Use the equation $\theta_R = \tan^{-1}\left(\frac{R_y}{R_x}\right)$.
   - The reference angle is 60°. The answer lies in Quadrant IV since $R_x > 0$ and $R_y < 0$.
   - The direction of the resultant vector is $\theta_R = 300°$. ($\theta_{IV} = 360° - \theta_{ref}$.)

---

82. **Combine like terms** together.

- First multiply $\vec{\mathbf{L}}$ by 3 to make $3\vec{\mathbf{L}} = 21\sqrt{2}\,\hat{\mathbf{x}} + 9\sqrt{2}\,\hat{\mathbf{y}}$.
- Next multiply $\vec{\mathbf{R}}$ by 4 to make $4\vec{\mathbf{R}} = 8\sqrt{2}\,\hat{\mathbf{x}} - 4\sqrt{2}\,\hat{\mathbf{y}}$.
- Substitute these into $\vec{\mathbf{Q}} = 3\vec{\mathbf{L}} - 4\vec{\mathbf{R}}$. **Distribute** the minus sign.
- Note that $-4\vec{\mathbf{R}} = -(8\sqrt{2}\,\hat{\mathbf{x}} - 4\sqrt{2}\,\hat{\mathbf{y}}) = -8\sqrt{2}\,\hat{\mathbf{x}} + 4\sqrt{2}\,\hat{\mathbf{y}}$.
- Combine the $\hat{\mathbf{x}}$-terms together to make $13\sqrt{2}\,\hat{\mathbf{x}}$.
- Combine the $\hat{\mathbf{y}}$-terms together to make $13\sqrt{2}\,\hat{\mathbf{y}}$.
- The vector $\vec{\mathbf{Q}}$ equals $\vec{\mathbf{Q}} = 13\sqrt{2}\,\hat{\mathbf{x}} + 13\sqrt{2}\,\hat{\mathbf{y}}$.
- Use the equation $Q = \sqrt{Q_x^2 + Q_y^2}$.
- The magnitude of $\vec{\mathbf{Q}}$ equals $Q = 26$.
- Use the equation $\theta_Q = \tan^{-1}\left(\frac{Q_y}{Q_x}\right)$.
- The reference angle is $45°$. The answer lies in Quadrant I since $Q_x > 0$ and $Q_y > 0$.
- The direction of $\vec{\mathbf{Q}}$ is $\theta_Q = 45°$. ($\theta_I = \theta_{ref}$.)

Chapter 11: Projectile Motion

83. Check your signs: Δy and a_y are negative.

- Begin with the trig equations. For example, $v_{x0} = v_0 \cos\theta_0$.
- The components of the initial velocity are $v_{x0} = 20\sqrt{3}$ m/s and $v_{y0} = 20$ m/s.
- What are the four known symbols? They are v_{x0}, v_{y0}, Δy, and a_y.
- $\Delta y < 0$ because the banana finishes **below** the starting point. $a_y < 0$ in free fall.
- You know: $v_{x0} = 20\sqrt{3}$ m/s, $v_{y0} = 20$ m/s, $\Delta y = -60$ m, and $a_y = -9.81$ m/s^2.
- Use the equation $\Delta y = v_{y0}t + \frac{1}{2}a_y t^2$ to find the time. You should get $5t^2 - 20t - 60 = 0$. Use the **quadratic formula** with $a \approx 5$, $b = -20$, and $c = -60$.
- The time is $t \approx 6.0$ s. (It works out to 6.1 s if you don't round a_y to ≈ -10 m/s^2.)
- Use the equation $\Delta x = v_{x0}t$. The answer is $\Delta x \approx 120\sqrt{3}$ m if you round a_y to ≈ -10 m/s^2. (If you use a calculator and don't round, $\Delta x = 211$ m.)

---

84. Check your signs: Δy and a_y are negative.

- Begin with the trig equations. For example, $v_{x0} = v_0 \cos\theta_0$.
- What is the launch angle?
- For a **horizontal** launch, the launch angle is $\theta_0 = 0°$ (since θ_0 is the angle that the initial velocity makes with the horizontal).
- The components of the initial velocity are $v_{x0} = 40$ m/s and $v_{y0} = 0$.
- What are the four known symbols? They are v_{x0}, v_{y0}, Δy, and a_y.

- $\Delta y < 0$ because the banana finishes **below** the starting point. $a_y < 0$ in free fall.
- You know: $v_{x0} = 40$ m/s, $v_{y0} = 0$, $\Delta y = -80$ m, and $a_y = -9.81$ m/s^2.
- Use the equation $v_y^2 = v_{y0}^2 + 2a_y\Delta y$ to find v_y. However, v_y is **not** the answer.
- Check your intermediate answer: $v_y \approx -40$ m/s (if you round a_y to ≈ -10 m/s^2).
- Note that $\sqrt{1600} = \pm 40$ m/s. Note that $(-40)^2 = 1600$. Choose the negative root.
- v_y is **negative** because it is heading downward.
- Use the equation $v = \sqrt{v_{x0}^2 + v_y^2}$. Be sure to use v_y (and not v_{y0}).
- The final speed is $v \approx 40\sqrt{2}$ m/s (or 56 m/s if you don't round a_y).
- Use the equation $\theta_v = \tan^{-1}\left(\frac{v_y}{v_{x0}}\right)$. The answer lies in Quadrant IV since $v_y < 0$.
- The direction of the final velocity is $\theta_v = 315°$. (The reference angle is 45°.)

---

85. Check your signs: Δy, v_{y0}, and a_y are negative.

- Begin with the trig equations. For example, $v_{x0} = v_0 \cos\theta_0$.
- The launch angle θ_0 lies in **Quadrant IV** because the textbook is thrown **downward**.
- The launch angle is $\theta_0 = 330°$. (Using $-30°$ with the minus sign is fine, too.)
- The components of the initial velocity are $v_{x0} = 10\sqrt{3}$ m/s and $v_{y0} = -10$ m/s.
- $v_{y0} < 0$ because the textbook is moving downward initially.
- What are the four known symbols? They are v_{x0}, v_{y0}, Δy, and a_y.
- $\Delta y < 0$ because the textbook finishes **below** the starting point. $a_y < 0$ in free fall.
- You know: $v_{x0} = 10\sqrt{3}$ m/s, $v_{y0} = -10$ m/s, $\Delta y = -75$ m, and $a_y = -9.81$ m/s^2.
- Use the equation $\Delta y = v_{y0}t + \frac{1}{2}a_y t^2$ to find the time.
- You will get a quadratic equation.
- The constants are $a \approx 5$, $b = 10$, and $c = -75$ (or $a \approx -5$, $b = -10$, and $c = 75$). (You may also divide all three constants by 5.)
- The time is $t = 3.0$ s.
- Use the equation $\Delta x = v_{x0}t$.
- The answer is $\Delta x \approx 30\sqrt{3}$ m. (If you use a calculator and don't round, $\Delta x = 52$ m.)

Chapter 12: Two-dimensional Motion with Calculus

86. Choose the appropriate equations from page 119.
    - Velocity equals the **derivative** of position with respect to time.
    - Use the equation $\vec{\boldsymbol{v}} = \frac{d\vec{\mathbf{r}}}{dt}$.
    - The derivative is $\vec{\boldsymbol{v}} = 8t^3\,\hat{\mathbf{x}} - 12t^2\,\hat{\mathbf{y}}$.
    - Plug in the given value for time: $t = 3.0$ s.
    - The velocity at the specified time is $\vec{\boldsymbol{v}} = 216\,\hat{\mathbf{x}} - 108\,\hat{\mathbf{y}}$.
    - Acceleration equals the **derivative** of velocity with respect to time.
    - Use the equation $\vec{\mathbf{a}} = \frac{d\vec{\boldsymbol{v}}}{dt}$.
    - Take a derivative of $\vec{\boldsymbol{v}} = 8t^3\,\hat{\mathbf{x}} - 12t^2\,\hat{\mathbf{y}}$ with respect to time.
    - The derivative is $\vec{\boldsymbol{a}} = 24t^2\,\hat{\mathbf{x}} - 24t\,\hat{\mathbf{y}}$.
    - Plug in the given value for time: $t = 5.0$ s..
    - The acceleration at the specified time is $\vec{\boldsymbol{a}} = 600\,\hat{\mathbf{x}} - 120\,\hat{\mathbf{y}}$.

---

87. Choose the appropriate equations from page 119.
    - Net displacement is the **integral** of velocity.
    - Use the equation $\Delta\vec{\mathbf{r}} = \int_{t=t_0}^{t} \vec{\boldsymbol{v}}\,dt$.
    - Perform the definite integral $\int_{t=0}^{3} (8t^3\,\hat{\mathbf{x}} - 9t^2\,\hat{\mathbf{y}})\,dt$.
    - The anti-derivative is $2t^4\,\hat{\mathbf{x}} - 3t^3\,\hat{\mathbf{y}}$.
    - Evaluate $2t^4\,\hat{\mathbf{x}} - 3t^3\,\hat{\mathbf{y}}$ over the limits.
    - $[2t^4\,\hat{\mathbf{x}} - 3t^3\,\hat{\mathbf{y}}]_{t=0}^{3} = [2(3)^4\,\hat{\mathbf{x}} - 3(3)^3\,\hat{\mathbf{y}}] - [2(0)^4\,\hat{\mathbf{x}} - 3(0)^3\,\hat{\mathbf{y}}]$.
    - The net displacement for the specified time interval is $\Delta\vec{\mathbf{r}} = 162\,\hat{\mathbf{x}} - 81\,\hat{\mathbf{y}}$.
    - Acceleration equals the **derivative** of velocity with respect to time.
    - Use the equation $\vec{\mathbf{a}} = \frac{d\vec{\boldsymbol{v}}}{dt}$.
    - Take a derivative of $\vec{\boldsymbol{v}} = 8t^3\,\hat{\mathbf{x}} - 9t^2\,\hat{\mathbf{y}}$ with respect to time.
    - The derivative is $\vec{\boldsymbol{a}} = 24t^2\,\hat{\mathbf{x}} - 18t\,\hat{\mathbf{y}}$.
    - Plug in the given value for time: $t = 6.0$ s..
    - The acceleration at the specified time is $\vec{\boldsymbol{a}} = 864\,\hat{\mathbf{x}} - 108\,\hat{\mathbf{y}}$.

GET A DIFFERENT ANSWER?

If you get a different answer and can't find your mistake even after consulting the hints and explanations, what should you do?

Please contact the author, Dr. McMullen.

How? Visit one of the author's blogs (see below). Either use the Contact Me option, or click on one of the author's articles and post a comment on the article.

www.monkeyphysicsblog.wordpress.com
www.improveyourmathfluency.com
www.chrismcmullen.wordpress.com

Why?

- If there happens to be a mistake (although much effort was put into perfecting the answer key), the correction will benefit other students like yourself in the future.
- If it turns out not to be a mistake, **you may learn something** from Dr. McMullen's reply to your message.

99.99% of students who walk into Dr. McMullen's office believing that they found a mistake with an answer discover one of two things:

- They made a mistake that they didn't realize they were making and learned from it.
- They discovered that their answer was actually the same. This is actually fairly common. For example, the answer key might say $t = \frac{\sqrt{3}}{3}$ s. A student solves the problem and gets $t = \frac{1}{\sqrt{3}}$ s. These are actually the same: Try it on your calculator and you will see that both equal about 0.57735. Here's why: $\frac{1}{\sqrt{3}} = \frac{1}{\sqrt{3}}\frac{\sqrt{3}}{\sqrt{3}} = \frac{\sqrt{3}}{3}$.

Two experienced physics teachers solved every problem in this book to check the answers, and dozens of students used this book and provided feedback before it was published. Every effort was made to ensure that the final answer given to every problem is correct.

But all humans, even those who are experts in their fields and who routinely aced exams back when they were students, make an occasional mistake. So if you believe you found a mistake, you should report it just in case. Dr. McMullen will appreciate your time.

Chapter 13: Newton's Laws of Motion

88. Review Newton's three laws of motion and the related terminology.

(A) What is the definition of acceleration?

- Acceleration is the instantaneous rate at which velocity changes.
- The answer is: "changing velocity."

(B) What is Newton's first law?

- Use the concept of **inertia** to answer this question.
- The banana was initially moving horizontally. It retains this horizontal component to its velocity because the banana has inertia.
- The answer is: "on the X." The banana's horizontal velocity matches the train's velocity. The banana and train travel the same horizontal distance.

(C) What is Newton's first law?

- Use the concept of **inertia** to answer this question.
- The banana was initially moving horizontally. It retains this horizontal component to its velocity because the banana has inertia.
- Whereas the banana's horizontal component of velocity is constant, the train's velocity is decreasing because the train is decelerating.
- The answer is: "east of the X." The banana's horizontal velocity exceeds the velocity of the train. (The train is slowing down. The banana is not.)

(D) What is Newton's third law?

- According to Newton's third law, the forces are **equal** in magnitude.
- The answer is: "the same." ("Equal and opposite" would also be okay.)

(E) What is Newton's second law.

- We already reasoned that the forces are equal in part (D).
- The net force on either person is mass times acceleration (Newton's second law).
- Since the forces are equal, more mass implies less acceleration.
- The football player has more mass, so he accelerates less during the collision.
- The answer is: "less."

(F) What is Newton's first law?

- Objects have inertia, which is a natural tendency to maintain constant velocity.
- If velocity is constant, **acceleration** is zero.
- Therefore, objects have a natural tendency to have **zero** acceleration.
- The answer is: "zero."

(G) Which of Newton's laws involves a push backwards?

- When the gun exerts a force on the bullet, the bullet exerts a force back on the gun that is equal in magnitude and **opposite** in direction.
- The shooter experiences this opposite force (push back) due to Newton's **third** law.
- The answer is: "third."

(H) What is Newton's first law?

- Use the concept of **inertia** to answer this question.
- This question involves the same reasoning as part (B). Compare these questions.
- The cannonball was initially moving horizontally with the same velocity as the ship. It retains this horizontal component to its velocity because it has inertia.
- The answer is: "in the cannon." The cannonball's horizontal velocity matches the ship's velocity. The cannonball and ship travel the same horizontal distance.

(I) What is Newton's third law?

- According to Newton's third law, the forces are **equal** in magnitude.
- The answer is: "25 N."

---

89. Review Newton's three laws of motion and the related terminology.

(A) What happens to mass and weight when the monkey visits the moon?

- **Mass** remains the **same**, while weight is reduced by a factor of 6.
- Which quantity is 24 kg? Look at the units to figure this out.
- The SI unit of mass is the kilogram (kg). The SI unit of weight is the Newton (N).
- Based on the units, the mass of the monkey is 24 kg.
- The monkey has the same mass on the moon.
- What is the equation for weight?
- Use the equation $W_e = mg_e$ to find the monkey's weight on the earth.
- The monkey's weight on earth is $W_e \approx 240$ N (or 235 N if you don't round gravity).
- The monkey weighs 6 times less on the moon.
- Divide W_e by 6 to find the monkey's weight on the moon.
- The answer is $W_m \approx 40$ N (or 39 N you don't round earth's gravity).
- The answers are: "24 kg" and "40 N." (The **mass** is 24 kg, the **weight** is $\approx$40 N.)

(B) What is Newton's first law?

- **Inertia** is the natural tendency of an object to maintain constant **momentum**.
- This means that an object resists changes to its momentum.
- The answer is: "momentum." ("Velocity" is also a good answer.)

(C) What is Newton's third law? What is Newton's second law?

- According to Newton's third law, the forces are **equal** in magnitude.
- The net force on either animal is mass times acceleration (Newton's second law).
- Since the forces are equal, more mass implies less acceleration.
- The gorilla has more mass, so the gorilla accelerates less during the collision.
- The answers are: "the same" and "less."

(D) Which quantity is defined as mass (m) times velocity ($\vec{v}$)?

- Consider the equation $\vec{\mathbf{p}} = m\vec{v}$. What is $\vec{\mathbf{p}}$?
- $\vec{\mathbf{p}}$ represents momentum. Momentum equals mass times velocity.
- The answer is: "momentum."

(E) Which quantity is defined as mass (m) times acceleration ($\vec{\mathbf{a}}$)?

- Consider the equation $\sum \vec{F} = m\vec{\mathbf{a}}$. What is $\sum \vec{F}$?
- $\sum \vec{F}$ represents net force. Net force equals mass times acceleration.
- This is Newton's second law.
- The answer is: "net force."

(F) What is Newton's third law?

- According to Newton's third law, the forces are **equal** in magnitude.
- The answer is: "equal to." (It's more precise to say "equal and opposite to.")

(G) What is the definition of acceleration?

- Acceleration is the instantaneous rate at which velocity changes.
- How can the monkey change his velocity if his speed is constant?
- Velocity is a combination of speed and direction.
- Since the monkey's speed is constant, he must change the direction of his velocity.
- The answer is: "change direction." (It would be okay to say "run in a circle.")

(H) What is Newton's second law?

- According to Newton's second law, net force equals mass times acceleration.
- If the net force is zero, the acceleration must be zero also.
- What is the definition of acceleration?
- Acceleration is the instantaneous rate at which velocity changes.
- Zero acceleration implies constant velocity (that is, velocity doesn't change).
- The answers are: "acceleration" and "velocity."

(I) Where does the mass of an object factor into the free fall equations?

- It doesn't. Recall the equations from Chapter 3. Mass isn't in those equations.
- In a perfect vacuum, all objects fall with the same acceleration, **regardless of mass**.
- The banana and feather both have an acceleration of $a_y = -9.81$ m/s^2.
- Since they have the same acceleration and descend the same height, and since they are released simultaneously, they reach the ground at the same time.
- The answer is: "at the same time."

(J) What is Newton's third law?

- Apply Newton's third law to the banana.
- If the monkey exerts a force on the banana by throwing it, what will happen?
- According to Newton's third law, the banana exerts a force back on the monkey that is equal in magnitude and opposite in direction to the force that the monkey exerts on the banana.
- The monkey should throw the banana directly away from his house so that the equal and opposite reaction will push the monkey toward his house.
- The answer is: "throwing the banana directly away from his house."

90. Review Newton's three laws of motion and the related terminology.

(A) What is Newton's first law?

- Objects have **inertia**, which is a natural tendency to travel with constant velocity.
- This means to travel with constant speed in a **straight line**.
- After leaving the metal arc, there is no longer a force pushing on the golf ball that will change the direction of its velocity. After leaving the metal arc, the golf ball's path is determined by its inertia.
- The golf ball will continue in a straight line tangent to the metal arc (path C).
- The answer is: "C."

(B) What is Newton's first law?

- The box of bananas has **inertia** – a natural tendency to maintain constant velocity.
- Initially, the box of bananas has a horizontal velocity equal to the plane's velocity.
- When the box of bananas is released, it acquires a vertical component of velocity to go along with its initial horizontal velocity.
- Since we neglect air resistance unless otherwise stated, the horizontal component of the box's velocity will remain constant due to the box's inertia. The box's vertical component of velocity will increase due to gravity.
- Which path begins horizontally (due to the box's inertia), and tilts more and more vertically as time goes on? That is path D.
- The answer would be the same if a gorilla stood at the top of a tall building and threw the box of bananas horizontally. The box is a **projectile**, following the path of projectile motion (Chapter 11), whether released from a plane or thrown by a gorilla.
- The answer is: "D."

(C) Where does the mass of an object factor into the free fall equations?

- It doesn't. Recall the equations from Chapter 3. Mass isn't in those equations.
- Neglecting air resistance, all objects fall with the same acceleration, **regardless of mass**. (Even allowing for air resistance, a banana and box of bananas released from a couple of meters above the ground will strike the ground at nearly the same time.)
- The banana and box of bananas both have an acceleration of $a_y = -9.81 \text{ m/s}^2$.
- Since they have the same acceleration and descend the same height, and since they are released simultaneously, they reach the ground at the same time.
- The answer is: "They strike the ground at the same time."

(D) What determines how much time either bullet will travel through the air?

- Gravity! Note that gravity affects both bullets the same way: Both bullets have the **same acceleration**, $a_y = -9.81 \text{ m/s}^2$, since both bullets are in **free fall**.
- Both bullets have $v_{y0} = 0$ (the shot bullet has a nonzero v_{x0}, but its v_{y0} is zero).
- The y-equations of projectile motion (Chapter 11) are identical to the equations of one-dimensional uniform acceleration (Chapter 3). So t will be the same for each.
- The answer is: "They strike the ground at the same time."

(E) What is Newton's first law?

- Objects have **inertia**, which is a natural tendency to travel with constant velocity.
- The necklace resists changes to its velocity. The necklace resists acceleration.
- When the car speeds up, the necklace resists this increase in velocity. The necklace leans **backward** when the car speeds up (assuming the car isn't in reverse).
- When the car slows down, the necklace resists this decrease in velocity. The necklace leans **forward** when the car slows down (assuming the car isn't in reverse).
- When the car travels with constant velocity along a level road, the necklace has nothing to resist, so it leans **straight down**.
- When the car rounds a turn to the left with constant speed, the necklace wants to keep going in a straight line due to its inertia, so it leans to the driver's **right**. The necklace leans outward.

Chapter 14: Applications of Newton's Second Law

91. Check your FBD's. You need three FBD's: one for each object.

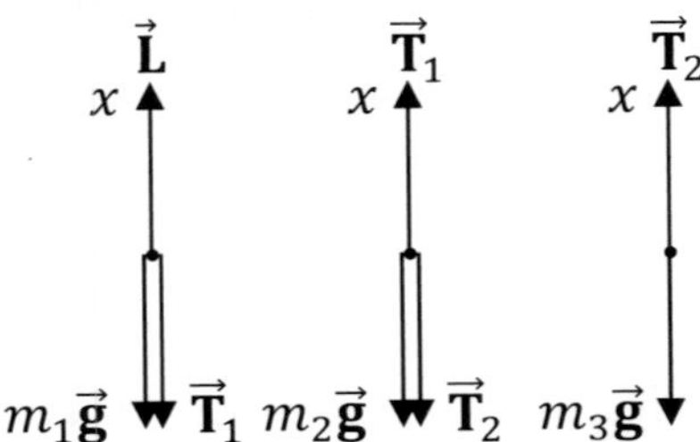

- Each object has weight ($m_1\vec{\mathbf{g}}$, $m_2\vec{\mathbf{g}}$, and $m_3\vec{\mathbf{g}}$) pulling straight down.
- A lift force ($\vec{\mathbf{L}}$) pulls upward on the helicopter.
- There are two pairs of tension ($\vec{\mathbf{T}}_1$ and $\vec{\mathbf{T}}_2$) forces: one pair in each rope.
- Label $+x$ up for each object (each object accelerates upward).
- Write out the three sums for Newton's second law (for example, $\sum F_{1x} = m_1 a_x$):
$$L - m_1 g - T_1 = m_1 a_x \quad , \quad T_1 - m_2 g - T_2 = m_2 a_x \quad , \quad T_2 - m_3 g = m_3 a_x$$
- Add all three equations together in order to cancel the tension forces.
$$L - m_1 g - m_2 g - m_3 g = m_1 a_x + m_2 a_x + m_3 a_x$$
- Plug in numbers and solve for acceleration.
- The answer is $a_x \approx 5.0 \text{ m/s}^2$. (If you don't round gravity, $a_x = 5.2 \text{ m/s}^2$.)
- Solve for tension in the equations from Newton's second law (the sums).
$$T_1 = L - m_1 g - m_1 a_x \quad , \quad T_2 = m_3 a_x + m_3 g$$
- Plug numbers into these equations.
- The tensions are $T_1 = 3000$ N and $T_2 = 750$ N.

92. Check your FBD's. You need two FBD's: one for each object.

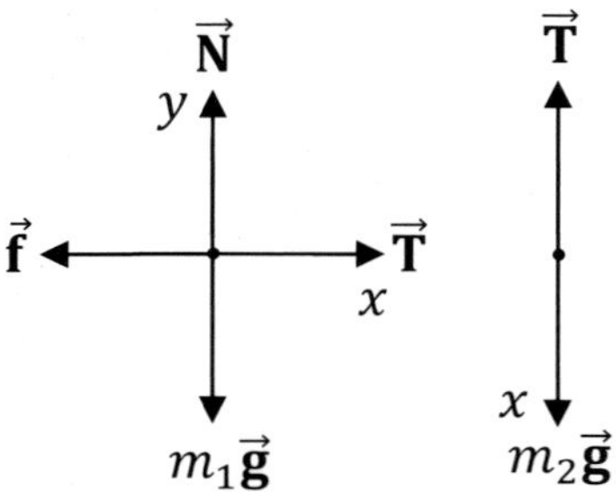

- Each object has weight ($m_1\vec{\mathbf{g}}$ and $m_2\vec{\mathbf{g}}$) pulling straight down.
- A normal force ($\vec{\mathbf{N}}$) supports the box (perpendicular to the surface).
- Friction ($\vec{\mathbf{f}}$) acts opposite to the velocity of the box of bananas.
- There is a pair of tension ($\vec{\mathbf{T}}$) forces in the connecting cord.
- Label $+x$ in the direction that each object accelerates. (Think of the pulley as bending the x-axis.) Label $+y$ perpendicular to x.
- Write out the sums: $\sum F_{1x} = m_1 a_x$, $\sum F_{1y} = m_1 a_y$, and $\sum F_{2x} = m_2 a_x$.

$$T - f = m_1 a_x \quad , \quad N - m_1 g = 0 \quad , \quad m_2 g - T = m_2 a_x$$

- $a_y = 0$ because the box doesn't accelerate vertically. The box has a_x, not a_y.
- Solve for normal force in the y-sum. Plug in numbers.
- Normal force is $N \approx 200$ N (or 196 N if you don't round g to 10 m/s^2).
- What is the equation for friction? Use the equation $f = \mu N$. Plug in numbers.
- Friction force is $f \approx 100$ N (or 98 N if you don't round g to 10 m/s^2.)
- Add the x-equations together in order to cancel the tension forces.

$$m_2 g - f = m_1 a_x + m_2 a_x$$

- Plug in numbers and solve for acceleration.
- The answer is $a_x \approx 4.0$ m/s^2. (If you don't round gravity, $a_x = 3.9$ m/s^2.)
- Solve for tension in one of the original equations (from the x-sums).

$$T = f + m_1 a_x$$

- Plug numbers into this equation.
- The tension is $T \approx 180$ N. (If you don't round gravity, $T = 177$ N.)

---

93. Check your FBD's. You need two FBD's: one for each object.

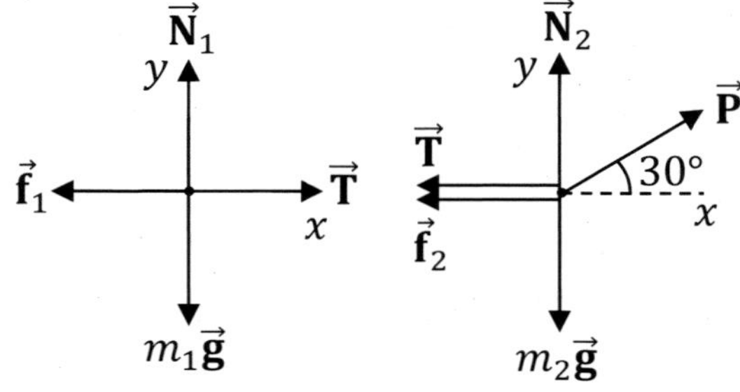

- Each object has weight ($m_1\vec{\mathbf{g}}$ and $m_2\vec{\mathbf{g}}$) pulling straight down.
- Two different normal forces ($\vec{\mathbf{N}}_1$ and $\vec{\mathbf{N}}_2$) support the boxes.
- Two different friction forces ($\vec{\mathbf{f}}_1$ and $\vec{\mathbf{f}}_2$) pull backwards on the two boxes.
- There is a pair of tension ($\vec{\mathbf{T}}$) forces in the connecting cord.
- The monkey's pull ($\vec{\mathbf{P}}$) acts on the right box.
- Write out the sums: $\sum F_{1x} = m_1 a_x$, $\sum F_{1y} = m_1 a_y$, $\sum F_{2x} = m_2 a_x$, and $\sum F_{2y} = m_2 a_y$.

$$T - f_1 = m_1 a_x \quad , \quad N_1 - m_1 g = 0 \quad , \quad P\cos 30° - T - f_2 = m_2 a_x \quad , \quad N_2 + P\sin 30° - m_2 g = 0$$

- Since $\vec{\mathbf{P}}$ doesn't lie on an axis, it goes in both the $\sum F_{2x}$ and $\sum F_{2y}$ sums with trig.
- The **components** of $\vec{\mathbf{P}}$ are $P_x = P\cos 30°$ and $P_y = P\sin 30°$.
- Solve for the normal forces in the y-sums. Plug in numbers.
- The normal forces are $N_1 \approx 100\sqrt{3}$ N and $N_2 \approx 150\sqrt{3}$ N (if you round g).
- Use the equation $f = \mu N$. Plug in numbers.
- The friction forces are $f_1 \approx 60$ N and $f_2 \approx 90$ N (if you round g).
- Add the x-equations together in order to cancel the tension forces.

$$P\cos 30° - f_1 - f_2 = m_1 a_x + m_2 a_x$$

- Plug in numbers and solve for acceleration.
- The answer is $a_x \approx \frac{5\sqrt{3}}{2}$ m/s^2. (If you don't round gravity, $a_x = 4.4$ m/s^2.)
- Solve for tension in one of the original equations (from the x-sums).

$$T = f_1 + m_1 a_x$$

- Plug numbers into this equation. The tension is $T = 135$ N.

---

94. Check your FBD. You just need one for this problem.

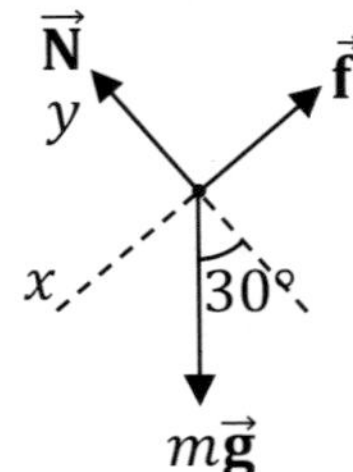

- Weight ($m\vec{\mathbf{g}}$) pulls straight down.
- Normal force ($\vec{\mathbf{N}}$) pushes perpendicular to the surface.
- Friction ($\vec{\mathbf{f}}$) acts opposite to the velocity of the box of bananas.
- Label $+x$ in the direction that the box accelerates: **down the incline**. Label $+y$ perpendicular to x: along the normal.
- Write out the sums for Newton's second law: $\sum F_x = ma_x$ and $\sum F_y = ma_y$.

$$mg\sin 30° - f = ma_x \quad , \quad N - mg\cos 30° = 0$$

- The example on page 135 explains which sums $mg\sin 30°$ and $mg\cos 30°$ go in.
- $a_y = 0$ because the box doesn't accelerate perpendicular to the incline.

- Solve for normal force in the y-sum. Normal force equals $N = \frac{mg\sqrt{3}}{2}$.
- Use the equation $f = \mu N$. Plug in the expression for normal force.
- Friction force is $f = \frac{3mg}{10}$.
- Plug the expression for friction into the equation from the x-sum.

$$\frac{mg}{2} - \frac{3mg}{10} = ma_x$$

- Mass cancels out. Plug in numbers and solve for acceleration.
- The answer is $a_x = 2.0 \text{ m/s}^2$.

---

95. Check your FBD's. You need two FBD's: one for each object.

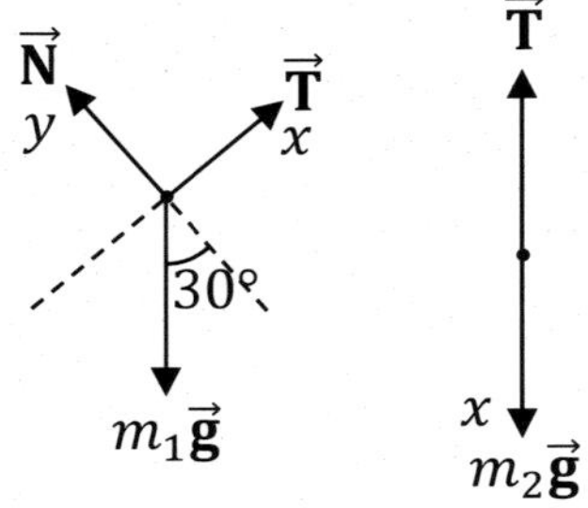

- Each object has weight ($m_1\vec{\mathbf{g}}$ and $m_2\vec{\mathbf{g}}$) pulling straight down.
- Normal force ($\vec{\mathbf{N}}$) pushes perpendicular to the surface.
- There is a pair of tension ($\vec{\mathbf{T}}$) forces in the connecting cord.
- Label $+x$ in the direction that the each object accelerates: **up the incline** for the box and straight down for the monkey. Label $+y$ perpendicular to x: along the normal.
- Write out the sums: $\sum F_{1x} = m_1 a_x$, $\sum F_{1y} = m_1 a_y$, and $\sum F_{2x} = m_2 a_x$.

$$T - m_1 g \sin 30° = m_1 a_x \quad , \quad N - m_1 g \cos 30° = 0 \quad , \quad m_2 g - T = m_2 a_x$$

- The example on page 135 explains which sums $m_1 g \sin 30°$ and $m_1 g \cos 30°$ go in.
- Add the x-equations together in order to cancel the tension forces.

$$m_2 g - m_1 g \sin 30° = m_1 a_x + m_2 a_x$$

- Plug in numbers and solve for acceleration.
- The answer is $a_x \approx 4.0 \text{ m/s}^2$. (If you don't round gravity, $a_x = 3.9 \text{ m/s}^2$.)

---

96. Check your FBD's. You need two FBD's: one for each object.

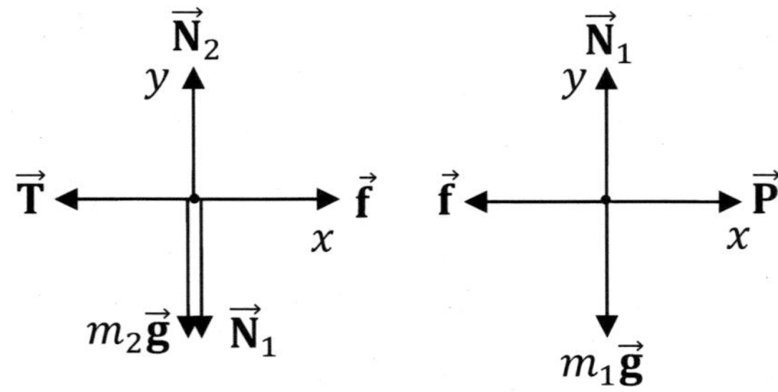

- We call the top box object 1 (right figure) and the bottom box object 2 (left figure).
- Each object has weight ($m_1\vec{\mathbf{g}}$ and $m_2\vec{\mathbf{g}}$) pulling straight down.
- The monkey's pull ($\vec{\mathbf{P}}$) acts on the top box.
- Tension ($\vec{\mathbf{T}}$) in the cord pulls the bottom box to the left.
- Friction force ($\vec{\mathbf{f}}$) pulls to the left on the top box (resisting the 200-N pull).
- There is an equal and opposite friction force ($\vec{\mathbf{f}}$) pulling the bottom box to the right due to **<u>Newton's third law</u>**. This friction force creates tension in the cord.
- Two different normal forces ($\vec{\mathbf{N}}_1$ and $\vec{\mathbf{N}}_2$) support the boxes.
- There is also an equal and opposite normal force $\vec{\mathbf{N}}_1$ pushing downward on the bottom box. This follows from **<u>Newton's third law</u>**. Since the bottom box pushes upward on the top box with a force $\vec{\mathbf{N}}_1$, the top box must push downward on the bottom box with an equal and opposite force.
- Write out the sums: $\sum F_{1x} = m_1 a_{1x}$, $\sum F_{1y} = m_1 a_y$, $\sum F_{2x} = m_2 a_{2x}$, and $\sum F_{2y} = m_2 a_y$.

  $P - f = m_1 a_{1x}$, $N_1 - m_1 g = 0$, $f - T = 0$, $N_2 - N_1 - m_2 g = 0$
- Solve for N_1 in the $\sum F_{1y}$ sum. Plug in numbers.
- The normal force is $N_1 \approx 400$ N. (It's 392 N if you don't round gravity.)
- Use the equation $f = \mu N$. Plug in numbers.
- The friction force is $f \approx 100$ N. (It's 98 N if you don't round gravity.)
- Now you can solve for acceleration in the $\sum F_{1x}$ sum. Plug in numbers.
- The answer is $a_{1x} \approx \frac{5}{2}$ m/s^2. (If you don't round gravity, $a_{1x} = 2.6$ m/s^2.)
- Solve for tension in the $\sum F_{2x}$ sum. Plug in numbers. Note that $a_{2x} = 0$.
- The tension is $T = 100$ N. (It's 98 N if you don't round gravity.)

Chapter 15: Hooke's Law

97. Check your FBD. You just need one for this problem.

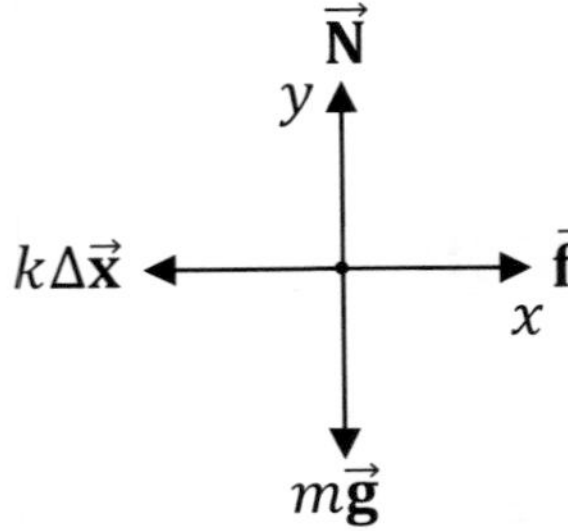

- Weight ($m\vec{\mathbf{g}}$) pulls straight down.
- Normal force ($\vec{\mathbf{N}}$) pushes perpendicular to the surface.
- The spring exerts a restoring force ($k\Delta\vec{\mathbf{x}}$) **<u>towards equilibrium</u>** (to the left).
- Friction force ($\vec{\mathbf{f}}$) pulls to the right (opposite to the velocity).

- Write out the sums for Newton's second law: $\sum F_x = ma_x$ and $\sum F_y = ma_y$.

$$-k\Delta x + f = ma_x \quad , \quad N - mg = 0$$

- Solve for normal force in the y-sum.
- Normal force equals $N \approx 30$ N. (It's 29 N if you don't round gravity.)
- Use the equation $f = \mu N$. Plug in numbers.
- Friction force is $f \approx \frac{15}{2}$ N. (It's 7.4 N if you don't round gravity.)
- Now you can solve for acceleration in the x-sum.
- The answer is $a_x \approx -5.0 \text{ m/s}^2$ (or -5.1 m/s^2 if you don't round gravity). The minus sign means that the box is accelerating to the **left** (since we chose $+x$ to be right).

---

98. Check your FBD. You just need one for this problem.

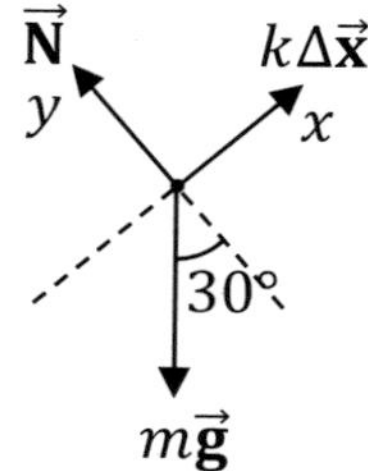

- Weight ($m\vec{\mathbf{g}}$) pulls straight down.
- Normal force ($\vec{\mathbf{N}}$) pushes perpendicular to the surface.
- The spring exerts a restoring force ($k\Delta\vec{\mathbf{x}}$) towards its original equilibrium position (up the incline).
- Label $+x$ along the incline. Label $+y$ along the normal (perpendicular to x).
- Write out the sums for Newton's second law: $\sum F_x = ma_x$ and $\sum F_y = ma_y$.

$$k\Delta x - mg \sin 30° = 0 \quad , \quad N - mg \cos 30° = 0$$

- The example on page 135 explains which sums $mg \sin 30°$ and $mg \cos 30°$ go in.
- The acceleration is zero (the box isn't moving at all).
- Solve for the spring constant in the x-sum. Plug in numbers.
- The answer is $k \approx 45$ N/m. (If you don't round gravity, $k = 44$ N/m.)

Chapter 16: Uniform Circular Motion

99. Check your knowns. Which symbols are given in the problem?

- The knowns are $D = 12$ m, $\omega = \frac{1}{2}$ rad/s, and $t = 4.0$ min.
- Before you do anything else, find the radius and convert the time to **seconds**.
- The radius is $R = 6.0$ m and the time is $t = 240$ s.

(A) Use the equation $v = R\omega$.

- The speed is $v = 3.0$ m/s.

(B) Use the equation $a_c = \frac{v^2}{R}$.

- The acceleration is $a_c = \frac{3}{2}$ m/s^2.

(C) Use the equation $s = vt$.

- The time must be in **seconds** (don't use minutes).
- The total distance traveled is $s = 720$ m.

(D) Use the equation $\omega = \frac{2\pi}{T}$.

- Solve for the period. Multiply both sides by T. Divide both sides by ω.
- The period is $T = 4\pi$ s. If you use a calculator, it is approximately $T \approx 13$ s.

(E) Use the equation $s = R\theta$.

- Solve for θ. Convert from radians to revolutions: 2π rad $= 1$ rev.
- The monkey completes $\theta = \frac{60}{\pi}$ revolutions.

---

100. Check your knowns. Which symbols are given in the problem?

- The knowns are $v = 4.0$ m/s, $T = 8\pi$ s, and $\theta = \frac{20}{\pi}$ rev.
- Before you do anything else, convert θ from revolutions to **radians**.
- The angle is $\theta = 40$ rad.

(A) Use the equation $\omega = \frac{2\pi}{T}$.

- The angular speed is $\omega = \frac{1}{4}$ rad/s.

(B) Use the equation $v = R\omega$.

- Solve for the radius.
- The radius is $R = 16$ m.

(C) Use the equation $a_c = \frac{v^2}{R}$.

- The acceleration is $a_c = 1.0$ m/s^2.

(D) Use the equation $s = R\theta$.

- The angle θ must be in **radians** (don't use revolutions).
- The monkey travels $s = 640$ m.

(E) Use the equation $f = \frac{1}{T}$.

- The frequency is $f = \frac{1}{8\pi}$ Hz.

Chapter 17: Uniform Circular Motion with Newton's Second Law

101. Check your FBD. You just need one for this problem.

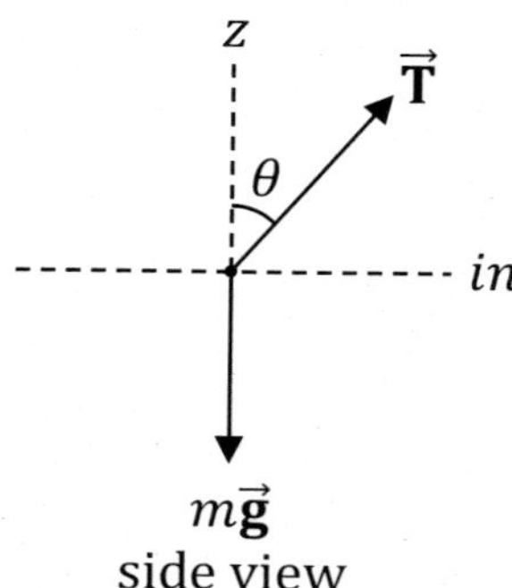

- Weight ($m\vec{\mathbf{g}}$) pulls straight down.
- There is tension ($\vec{\mathbf{T}}$) along the chain.
- Label *in* toward the **center** of the horizontal circle: to the right. Label $+z$ perpendicular to the plane of the circle: straight up.
- Write out the sums for Newton's second law: $\sum F_{in} = ma_c$ and $\sum F_z = 0$.
$$T \sin 30° = ma_c \quad , \quad T \cos 30° - mg = 0$$
- The inward component of $\vec{\mathbf{T}}$ is opposite to 30°, so the inward component of $\vec{\mathbf{T}}$ has a sine. The z-component of $\vec{\mathbf{T}}$ is adjacent to 30°, so it has a cosine.
- Solve for tension in the z-sum equation.
- The mass equals the mass of the monkey **plus** the seat: $m = 25 \text{ kg} + 5 \text{ kg} = 30 \text{ kg}$.
- The tension is $T \approx 200\sqrt{3}$ N. (It's 340 N if you don't round gravity.)
- Solve for acceleration in the inward sum.
- The radius of the circle equals the radius of the disc plus the inward component of the chain. See the diagram below.
- The radius is $R = R_{disc} + L \sin 30° = \frac{D_{disc}}{2} + \frac{L}{2} = \frac{48\sqrt{3}}{2} + \frac{32\sqrt{3}}{2} = 40\sqrt{3}$ m.

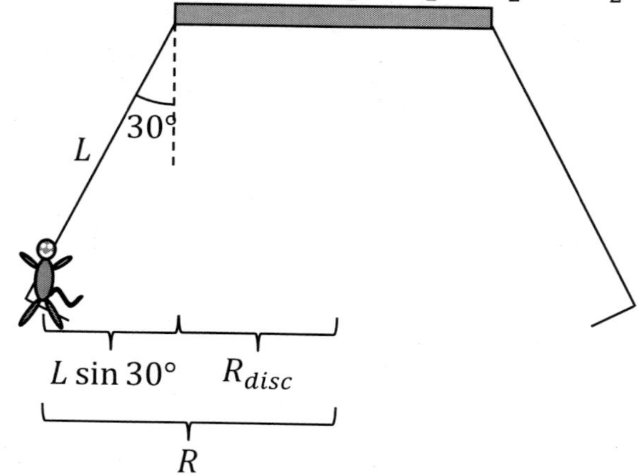

- The acceleration is $a_c \approx \frac{10\sqrt{3}}{3}$ m/s^2. (It's 5.67 m/s^2 if you don't round gravity.)
- Use the equation $a_c = \frac{v^2}{R}$. Solve for speed. Multiply both sides by R. Divide by a_c.
- The speed is $v = 20$ m/s.

---

102. Check your FBD. You just need one for this problem.

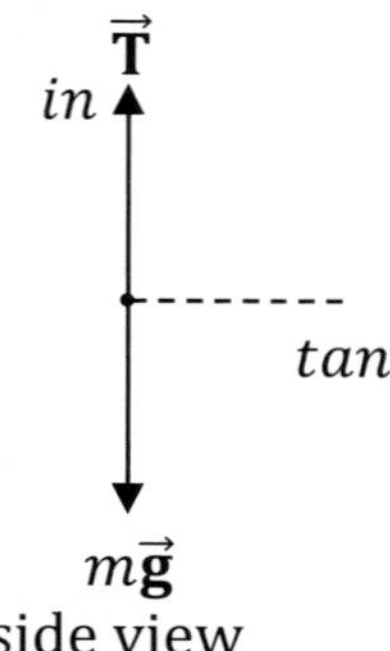

- Note that this is a **vertical** circle, unlike the horizontal circle in the previous problem.
- Draw the FBD for the mouse when it is at the **bottom** of its arc because part (D) specifically asks about the bottom of the arc.
- Weight ($m\vec{\mathbf{g}}$) pulls straight down.
- There is tension ($\vec{\mathbf{T}}$) in the tail (straight up when the mouse is at the bottom).
- Label *in* toward the **center** of the vertical circle: straight up.
- Write out the sum for Newton's second law: $\sum F_{in} = ma_c$.

$$T - mg = ma_c$$

- Tension points inward, so it is positive. Weight points outward, so it is negative.
- Convert the radius from centimeters (cm) to meters (m): 100 cm = 1 m.
- The radius is $R = 2.00$ m.
- Use the equation $a_c = \frac{v^2}{R}$. Express R in meters (not centimeters).
- The acceleration is $a_c = 8.0$ m/s^2.
- Convert the mass from grams (g) to kilograms (kg).
- The mass is $m = \frac{1}{2}$ kg (which is the same as 0.500 kg).
- Plug the acceleration into the equation for tension (above).
- The tension is $T \approx 9.0$ N. (It's 8.9 N if you don't round gravity.)

---

103. Check your FBD (on the next page). You just need one for this problem.
- Weight ($m\vec{\mathbf{g}}$) pulls straight down.
- Normal force ($\vec{\mathbf{N}}$) is perpendicular to the wall: it is horizontal (to the right).
- Friction ($\vec{\mathbf{f}}_s$) is along the surface: it is up. Friction prevents the monkey from falling.
- Label *in* toward the **center** of the circle: to the right. Label z perpendicular to the plane of the circle: straight up.

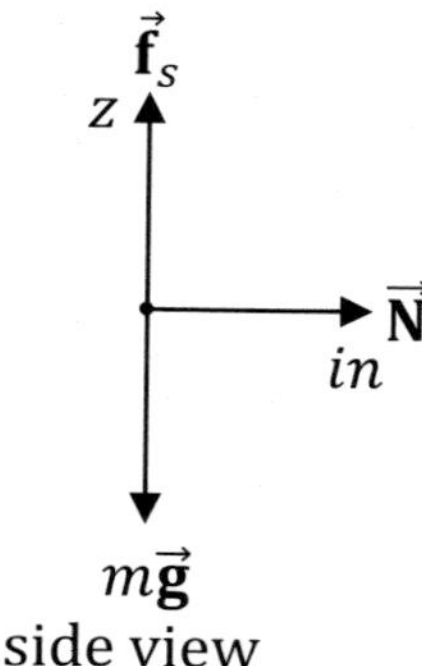

- Write out the sums for Newton's second law: $\sum F_{in} = ma_c$ and $\sum F_z = 0$.
$$N = ma_c \quad , \quad f_s - mg = 0$$
- Normal force points **inward**, so it goes in the $\sum F_{in}$ sum.
- Solve for the friction force in the z-sum.
- The friction force equals $f_s \approx 500$ N. (It's 491 N if you don't round gravity.)
- Recall that the force (f_s) of **static** friction is less than or equal to the coefficient (μ_s) of static friction times normal force (N): $f_s \leq \mu_s N$.
- Substitute the expression for normal force from the inward sum into the friction inequality: $f_s \leq \mu_s m a_c$.
- Plug in numbers to solve for the minimum acceleration.
- The acceleration is $a_c \geq 40$ m/s^2. (It's 39 m/s^2 if you don't round gravity.)
- Use the equation $a_c = \frac{v^2}{R}$. Solve for speed. Multiply both sides by R. Divide by a_c.
- The speed is $v \geq 20$ m/s. Therefore, the minimum speed needed is 20 m/s.

---

104. Check your FBD (on the next page). You just need one for this problem.
- Weight ($m\vec{\mathbf{g}}$) pulls straight down.
- Normal force ($\vec{\mathbf{N}}$) is **perpendicular** to the incline.
- Label in toward the **center** of the circle: to the right. Label z perpendicular to the plane of the circle: straight up. See the diagram on the next page.
- Note the difference in coordinates for this incline problem compared to previous incline problems (Chapter 14). In Chapter 14, when the object was sliding down the incline, we chose $+x$ to be along the incline. In this problem, the racecar **isn't** sliding up or down the incline. The racecar is traveling in a **horizontal** circle. In circular motion problems, the acceleration is inward, so we label in toward the center of the circle. Therefore, in is horizontal (to the right for the position shown).
- Write out the sums for Newton's second law: $\sum F_{in} = ma_c$ and $\sum F_z = 0$.
$$N \sin 30° = ma_c \quad , \quad N \cos 30° - mg = 0$$
- The inward component of $\vec{\mathbf{N}}$ is opposite to 30°, so the inward component of $\vec{\mathbf{N}}$ has a sine. The z-component of $\vec{\mathbf{N}}$ is adjacent to 30°, so it has a cosine.

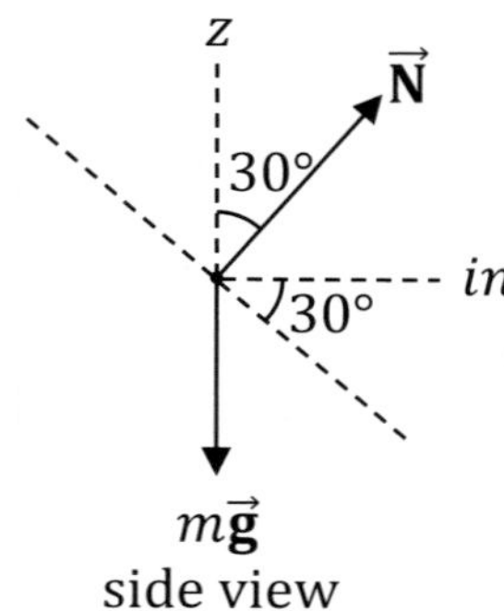

- In the z-sum, bring mg to the right side of the equation.
- After bringing mg to the right, divide the two equations.
- You should get $\tan 30° = \frac{a_c}{g}$.
- Solve for the acceleration. Plug in numbers.
- Divide the given diameter by two in order to find the radius.
- The radius is $R = 90\sqrt{3}$ m.
- The acceleration is $a_c \approx \frac{10\sqrt{3}}{3}$ m/s^2. (It's 5.77 m/s^2 if you don't round gravity.)
- Use the equation $a_c = \frac{v^2}{R}$. Solve for speed. Multiply both sides by R. Divide by a_c.
- The speed is $v = 30$ m/s.

Chapter 18: Newton's Law of Gravity

105. Make a list of the knowns and the desired unknown.
- The given knowns are $R = 6.0 \times 10^7$ m and $g = 20$ m/s^2.
- You should also know $G = 6.67 \times 10^{-11} \frac{\text{N}\cdot\text{m}^2}{\text{kg}^2} \approx \frac{2}{3} \times 10^{-10} \frac{\text{N}\cdot\text{m}^2}{\text{kg}^2}$.
- In part (A), you are solving for m_p.
- Use the equation $g = G\frac{m_p}{R^2}$.
- Solve for m_p. Multiply both sides by R^2. Divide both sides by G.
- You should get $m_p = \frac{gR^2}{G}$.
- Plug in numbers. Note that $(6.0 \times 10^7)^2 = 36 \times 10^{14}$. Note that $\frac{10^{14}}{10^{-10}} = 10^{24}$.
- The mass of the planet is $m_p = 1.08 \times 10^{27}$ kg. (It's the same as 1080×10^{24} kg.)
- In part (B), you are solving for F_g.
- Use the equation $F_g = G\frac{m_1 m_2}{R^2}$. Plug in numbers.
- The force is $F_g = 9.0 \times 10^{19}$ N. (It's the same as 0.090×10^{21} N.)

---

106. Study the ratio example at the bottom of page 168. These are **ratio** problems.
(A) Express the given numbers as ratios.

$$\frac{R_p}{R_e} = 5 \quad , \quad \frac{g_p}{g_e} = 3$$

- Write the equation for the surface gravity (g_p) of the planet using a subscript p for everything except G, and write a similar equation for the surface gravity (g_e) of earth using a subscript e for everything except G.
- Divide the two equations.

$$\frac{g_p}{g_e} = \frac{\frac{Gm_p}{R_p^2}}{\frac{Gm_e}{R_e^2}}$$

- Divide the fractions. This means to multiply by the reciprocal.

$$\frac{g_p}{g_e} = \frac{m_p}{m_e}\frac{R_e^2}{R_p^2}$$

- Solve for the ratio $\frac{m_p}{m_e}$.

$$\frac{m_p}{m_e} = \frac{g_p}{g_e}\left(\frac{R_p}{R_e}\right)^2$$

- Plug in the ratios for $\frac{R_p}{R_e}$ and $\frac{g_p}{g_e}$. (Find these numbers above).
- The answer is $m_p = 75m_e$. The planet's mass is 75× greater than earth's.

(B) Express the given numbers as ratios.

$$\frac{m_p}{m_e} = 6 \quad , \quad \frac{g_p}{g_e} = 96$$

- Write an equation for the surface gravity of each planet and divide the equations (multiplying by the reciprocal when dividing fractions), just like we did in part (A).

$$\frac{g_p}{g_e} = \frac{\frac{Gm_p}{R_p^2}}{\frac{Gm_e}{R_e^2}} = \frac{m_p}{m_e}\frac{R_e^2}{R_p^2}$$

- Solve for the ratio $\frac{R_p}{R_e}$. First isolate the ratio $\frac{R_p^2}{R_e^2}$ and then squareroot both sides.

$$\frac{R_p}{R_e} = \sqrt{\frac{m_p}{m_e}\frac{g_e}{g_p}}$$

- Plug in the ratios for $\frac{m_p}{m_e}$ and $\frac{g_p}{g_e}$. (Find these numbers above).
- Since $\frac{g_p}{g_e} = 96$, it follows that $\frac{g_e}{g_p} = \frac{1}{96}$. Plug in $\frac{1}{96}$ for $\frac{g_e}{g_p}$.
- The answer is $R_p = \frac{R_e}{4}$. The planet's radius is $\frac{1}{4}$× of earth's radius.

107. Study the example on page 169. The algebra of this problem is **identical** to that example.

- Set $g_m = g_k$.
- Write the equation for the surface gravity (g_m) of Mon using a subscript m for everything except G, and write a similar equation for the surface gravity (g_k) of Key using a subscript k for everything except G.
- Substitute these equations into $g_m = g_k$ and simplify.
$$\frac{m_m}{R_m^2} = \frac{m_k}{R_k^2}$$
- Cross multiply and take the squareroot of both sides. Recall that $\sqrt{x^2} = \pm x$. For example, $(-3)^2$ and 3^2 both equal 9, so $\sqrt{9}$ could equal -3 or $+3$.
$$\sqrt{m_m}\, R_k = \pm\sqrt{m_k}\, R_m$$
- The $\pm$ indicates that we should consider both possible signs.
- Relate R_m and R_k (which are the distances from the center of the planet to the point where the net gravitational field is zero – they are **not** the radii of the planet and moon) to the distance between Mon and Key.
$$R_m + R_k = d$$
- The distance d was given in the problem as $d = 9.0 \times 10^8$ m.
- Isolate R_k in the previous equation and substitute it into the equation with the squareroots. Distribute $\sqrt{m_m}$.
$$\sqrt{m_m}\, d - \sqrt{m_m}\, R_m = \pm\sqrt{m_k}\, R_m$$
- Bring $\sqrt{m_m}\, R_m$ to the right-hand side and factor R_m out. Solve for R_m.
$$R_m = \frac{d\sqrt{m_m}}{\sqrt{m_m} \pm \sqrt{m_k}}$$
- Plug in numbers. Only the positive sign provides an answer **between** Mon and Key.
- The answer is $R_m = 8.0 \times 10^8$ m from Planet Mon.
- (If you solve for R_k, the answer is $R_k = 1.0 \times 10^8$ m from Key.)

---

108. What are the R's for the equations for surface gravity in this problem?

- Apply the equation $g = G\frac{m_p}{R^2}$ to both Coco and Nut.
- Note that each R is one-half the distance between Coco and Nut.
- The value of R is $R = 2.0 \times 10^8$ m.
- Plug in numbers. Remember to square R in the formula.
- Note that $(2.0 \times 10^8)^2 = 4.0 \times 10^{16}$.
- The two gravitational accelerations are $g_c = \frac{1}{200}$ m/s^2 and $g_n = \frac{1}{100}$ m/s^2.
- Midway between the two planets, the gravitational fields will point in opposite directions. Subtract the two gravitational accelerations to find the net gravitational field at the midpoint. The answer is $g_{net} = \frac{1}{200}$ m/s$^2 = 0.0050$ m/s^2.

Chapter 19: Satellite Motion

109. Make a list of the knowns and the desired unknown.
- The given knowns are $R = 2.0 \times 10^7$ m and $m_p = 1.2 \times 10^{24}$ kg.
- You should also know $G = 6.67 \times 10^{-11} \frac{\text{N}\cdot\text{m}^2}{\text{kg}^2} \approx \frac{2}{3} \times 10^{-10} \frac{\text{N}\cdot\text{m}^2}{\text{kg}^2}$.
- In part (A), you are solving for v.
- Use the equation $v = \sqrt{G\frac{m_p}{R}}$.
- Plug in numbers. Note that $\sqrt{\frac{1.2 \times 10^7}{3}} = \sqrt{\frac{12 \times 10^6}{3}} = \sqrt{4 \times 10^6} = \sqrt{4}\sqrt{10^6}$.
- The orbital speed is $v = 2.0$ km/s. (It's the same as 2000 m/s or 2.0×10^3 m/s.)
- In part (B), you are solving for T.
- Use the equation $v = \frac{2\pi R}{T}$. Solve for T. Multiply both sides by T and divide by v.
- You should get $T = \frac{2\pi R}{v}$. Plug in numbers.
- The orbital period is $T = 2\pi \times 10^4$ s. (It's the same as 6.3×10^4 s.)

---

110. Make a list of the knowns and the desired unknown.
- The value given numerically is $m_p \approx 6.0 \times 10^{24}$ kg.
- You should also know $G = 6.67 \times 10^{-11} \frac{\text{N}\cdot\text{m}^2}{\text{kg}^2} \approx \frac{2}{3} \times 10^{-10} \frac{\text{N}\cdot\text{m}^2}{\text{kg}^2}$.
- The word "geosynchronous" should give you another known. What is it?
- A **geosynchronous** satellite has the same period as earth's rotation.
- What is the period of the earth's rotation about its axis?
- The period is 24 hours. Convert this to seconds.
- In SI units, the period is 86,400 s. (Note that 1 hr = 60 min. = 3600 s.)
- The unknown that you are solving for is R.
- Use the equation $T = 2\pi\sqrt{\frac{R^3}{Gm_p}}$.
- Solve for R. Square both sides. Multiply both sides by Gm_p.
- You should get $R = \sqrt[3]{\frac{Gm_pT^2}{4\pi^2}}$ where $\sqrt[3]{\ }$ means to take the **cube root**.
- Plug in numbers. You should get $\sqrt[3]{\frac{746496}{\pi^2}} \times 10^{18}$.
- Note: If your calculator doesn't have a cube root function, raise the number to the power of one-third: $\left(\frac{Gm_pT^2}{4\pi^2}\right)^{1/3}$. Be sure to enclose the $\frac{1}{3}$ in parentheses: ^(1/3).
- The orbital radius is $R = 72 \times \sqrt[3]{\frac{2}{\pi^2}} \times 10^6$ m or $R = 4.23 \times 10^7$ m.

Chapter 20: The Scalar Product

111. Identify the components of the given vectors.
- The components are: $A_x = 6$, $A_y = 1$, $A_z = 3$, $B_x = 2$, $B_y = 8$, and $B_z = 4$.
- Use the component form of the scalar product: $\vec{\mathbf{A}} \cdot \vec{\mathbf{B}} = A_x B_x + A_y B_y + A_z B_z$.
- The scalar product is $\vec{\mathbf{A}} \cdot \vec{\mathbf{B}} = 32$.

---

112. Identify the components of the given vectors.
- The components are: $A_x = 2$, $A_y = -1$, $A_z = 6$, $B_x = -3$, $B_y = 2$, and $B_z = 5$.
- Use the component form of the scalar product: $\vec{\mathbf{A}} \cdot \vec{\mathbf{B}} = A_x B_x + A_y B_y + A_z B_z$.
- The scalar product is $\vec{\mathbf{A}} \cdot \vec{\mathbf{B}} = 22$.

---

113. Identify the components of the given vectors.
- The components are: $A_x = 0$, $A_y = 3$, $A_z = 4$, $B_x = 9$, $B_y = -3$, and $B_z = -1$.
- (Observe that $\vec{\mathbf{A}}$ doesn't have an $\hat{\mathbf{x}}$: That's why $A_x = 0$.)
- Use the component form of the scalar product: $\vec{\mathbf{A}} \cdot \vec{\mathbf{B}} = A_x B_x + A_y B_y + A_z B_z$.
- The scalar product is $\vec{\mathbf{A}} \cdot \vec{\mathbf{B}} = -13$.

---

114. Identify the components of the given vectors.
- The components are: $A_x = 12$, $A_y = 4$, $A_z = -16$, $B_x = 2$, $B_y = -24$, and $B_z = -6$.
- Use the component form of the scalar product: $\vec{\mathbf{A}} \cdot \vec{\mathbf{B}} = A_x B_x + A_y B_y + A_z B_z$.
- The scalar product is $\vec{\mathbf{A}} \cdot \vec{\mathbf{B}} = 24$.

---

115. Use both forms of the scalar product together.
- Identify the components of the given vectors.
- The components are: $A_x = 0$, $A_y = 3$, $A_z = -3$, $B_x = 0$, $B_y = -6$, and $B_z = 6$.
- Use the equations $A = \sqrt{A_x^2 + A_y^2 + A_z^2}$ and $B = \sqrt{B_x^2 + B_y^2 + B_z^2}$. Plug in numbers.
- The magnitudes of the given vectors are $A = 3\sqrt{2}$ and $B = 6\sqrt{2}$.
- Set the two forms of the scalar product equal to each other.

$$A_x B_x + A_y B_y + A_z B_z = AB \cos\theta$$

- Plug numbers into the above equation.
- The left-hand side equals -36.
- The angle equals $180°$. Note that $\cos^{-1}(-1) = 180°$.

---

116. Use both forms of the scalar product together.
- Identify the components of the given vectors.
- The components are: $A_x = 3$, $A_y = -\sqrt{3}$, $A_z = 2$, $B_x = \sqrt{3}$, $B_y = -1$, and $B_z = 2\sqrt{3}$.

- Use the equations $A = \sqrt{A_x^2 + A_y^2 + A_z^2}$ and $B = \sqrt{B_x^2 + B_y^2 + B_z^2}$. Plug in numbers.
- The magnitudes of the given vectors are $A = 4$ and $B = 4$.
- Set the two forms of the scalar product equal to each other.

$$A_x B_x + A_y B_y + A_z B_z = AB \cos\theta$$

- Plug numbers into the above equation.
- The left-hand side equals $8\sqrt{3}$.
- The angle equals 30°. Note that $\cos^{-1}\left(\frac{\sqrt{3}}{2}\right) = 30°$.

Chapter 21: Work and Power

117. Which of the work equations should you use to solve this problem?
    - Use the equation $W_g = -mg\Delta h$ to find the work done "by gravity."
    - Since the box falls, Δh is negative: $\Delta h = -1.5$ m.
    - The two minus signs will make the answer positive.
    - The work done by gravity is $W_g \approx 750$ J. (It's 736 J if you don't round gravity.)

---

118. Which of the work equations should you use to solve this problem?
    - Use the equation $W_s = \pm\frac{1}{2}kx^2$ to find the work done "by the spring."
    - The work done "by the spring" is negative because the system travels **away** from the equilibrium position.
    - The work done by the spring is $W_s = -64$ J.

---

119. Which of the work equations should you use to solve this problem?
    - Use the equation $W_g = Gm_p m\left(\frac{1}{R} - \frac{1}{R_0}\right)$ for an **astronomical** change in altitude.
    - The rocket begins at $R_0 = 4.0 \times 10^6$ m.
    - Use the altitude equation from Chapter 19 to find R.
    - The altitude equation is $R = R_0 + h$.
    - The altitude is given: $h = 4.0 \times 10^6$ m.
    - The final R is $R = 8.0 \times 10^6$ m.
    - Plug numbers into the work equation.
    - The work done by gravity is $W_g = -1.25 \times 10^{10}$ J.

---

120. Which of the work equations should you use to solve this problem?
    - Use the equation $W_{net} = ma_x\Delta x$ to find the net work.
    - Use the equations of **uniform acceleration** (Chapter 3) to find a_x and Δx.
    - The three knowns for uniform acceleration are $v_{x0} = 0$, $v_x = 60$ m/s, and $t = 5.0$ s.

- Use the equation $v_x = v_{x0} + a_x t$ to find the acceleration.
- The acceleration is $a_x = 12 \text{ m/s}^2$.
- Use the equation $\Delta x = v_{x0}t + \frac{1}{2}a_x t^2$ to find the net displacement.
- The net displacement is $\Delta x = 150$ m.
- Plug numbers into the equation for the net work.
- The net work is $W_{net} = 360$ kJ. (It's the same as 360,000 J.)

---

121. Which of the work equations should you use to solve each part of this problem?

(A) Use the equation $W_g = -mg\Delta h$ to find the work done "by gravity."

- Since the box travels **down** the incline, Δh is negative.
- Draw a right triangle to relate s to Δh: $\sin 30° = \frac{-\Delta h}{s}$ where $s = 8.0$ m. $\Delta h = -4.0$ m.
- The two minus signs will make the answer positive.
- The work done by gravity is $W_g \approx 800$ J. (It's 785 J if you don't round gravity.)

(B) Use the equation $W_{nc} = -\mu\, N\, s$ to find the work done "by friction."

- The coefficient of friction is $\mu = \frac{\sqrt{3}}{4}$.
- Draw a FBD to find **normal force**. (Normal force does **not** equal weight.)

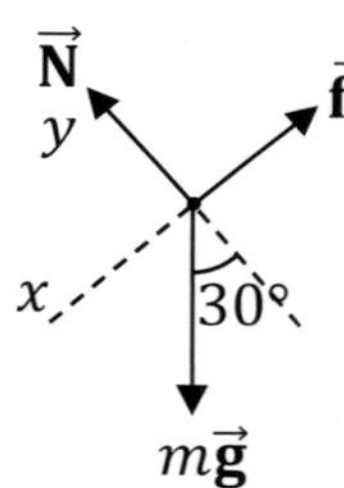

- Sum the y-components of the forces to solve for normal force: $\sum F_y = ma_y$.

$$N - mg\cos 30° = 0$$

- Note: $a_y = 0$ because the box doesn't accelerate perpendicular to the incline.
- Solve for normal force in the above equation.
- Normal force equals $N \approx 100\sqrt{3}$ N. (It's 170 N if you don't round gravity.)
- Plug numbers into the equation for nonconservative work.
- Don't forget to include the displacement: $s = 8.0$ m.
- The work done by friction is $W_{nc} = -600$ J (or -589 J if you don't round gravity).

(C) Use the equation $W = F\, s\cos\theta$ to find the work done "by the normal force."

- What is θ for the work done by normal force?
- $\theta = 90°$ because normal force is perpendicular to the surface.
- Since $\cos 90° = 0$, normal force doesn't do any work: $W_N = 0$.

(D) Use the equation $W_{net} = (\sum F_x)s$ to find the net work.

- Sum the x-components of the forces:

$$\sum F_x = mg \sin 30° - f$$

- Use the equation $f = \mu N$ to find the friction force.
- The friction force is $f = 75$ N.
- Plug in numbers to find a numerical value for $\sum F_x$.
- The net force is $\sum F_x = 25$ N.
- Plug $\sum F_x$ into the equation $W_{net} = (\sum F_x)s$.
- The net work is 200 J. (It's 196 J if you don't round gravity.)

---

122. Find the scalar product between the force and the displacement.
- Identify the components of $\vec{\mathbf{F}}$ and $\vec{\mathbf{s}}$.
- The components are: $F_x = 2$, $F_y = 6$, $F_z = 0$, $\Delta x = 3$, $\Delta y = -8$, and $\Delta z = 0$.
- Use the component form of the scalar product: $W = \vec{\mathbf{F}} \cdot \vec{\mathbf{s}} = F_x \Delta x + F_y \Delta y + F_z \Delta z$.
- The work done is $W = -42$ J.

---

123. Find the scalar product between the force and the displacement.
- Identify the components of $\vec{\mathbf{F}}$ and $\vec{\mathbf{s}}$.
- The components are: $F_x = 7$, $F_y = -9$, $F_z = 5$, $\Delta x = 6$, $\Delta y = 0$, and $\Delta z = -1$.
- (Observe that $\vec{\mathbf{s}}$ doesn't have a $\hat{\mathbf{y}}$: That's why $\Delta y = 0$.)
- Use the component form of the scalar product: $W = \vec{\mathbf{F}} \cdot \vec{\mathbf{s}} = F_x \Delta x + F_y \Delta y + F_z \Delta z$.
- The work done is $W = 37$ J.

---

124. Which of the work equations should you use to solve this problem?
- Use the equation $W = F s \cos\theta$ to find the work done "by the monkey's pull."
- The force to use is the monkey's pull: $F = 300$ N.
- The angle $\theta = 60°$ since the pull makes an angle of 60° with the displacement.
- The work done by the monkey's pull is $W = 1.8$ kJ. (It's the same as 1800 J.)

---

125. Which of the work equations should you use to solve each part of this problem?

(A) Use the equation $W = F s \cos\theta$ to find the work done "by the monkey's pull."
- The force to use in part (A) is the monkey's pull: $F = 160$ N.
- The angle $\theta = 30°$ since the pull makes an angle of 30° with the displacement.
- The work done by the monkey's pull is $W = 560\sqrt{3}$ J (or 970 J).

(B) Use the equation $W_{nc} = -\mu N s$ to find the work done "by friction."
- The coefficient of friction is $\mu = \frac{\sqrt{3}}{6}$.
- Draw a FBD to find **normal force**. (Normal force does **not** equal weight.)

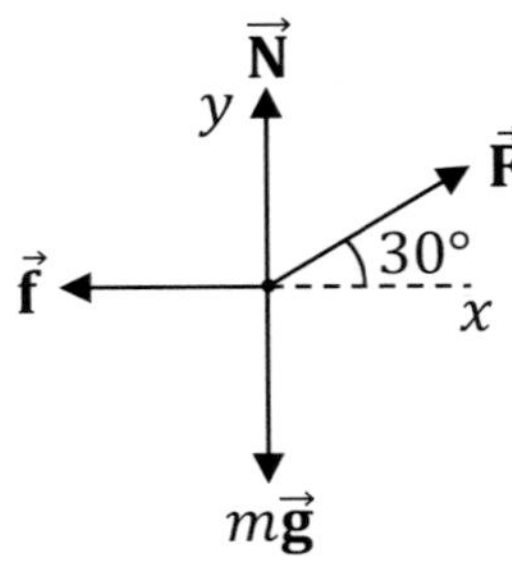

- Sum the y-components of the forces to solve for normal force: $\sum F_y = ma_y$.

$$N + F \sin 30° - mg = 0$$

- Note: $a_y = 0$ because the box doesn't accelerate up or down.
- Solve for normal force in the above equation. Plug in numbers, using $F = 160$ N.
- Normal force equals $N \approx 120$ N. (It's 116 N if you don't round gravity.)
- Plug numbers into the equation for nonconservative work.
- Don't forget to include the displacement: $s = 7.0$ m.
- The work done by friction is $W_{nc} \approx -140\sqrt{3}$ J (or -235 J if you don't round gravity).

(C) Use the equation $W = F\, s \cos\theta$ to find the work done "by the normal force."

- What is θ for the work done by normal force?
- $\theta = 90°$ because normal force is perpendicular to the surface.
- Since $\cos 90° = 0$, normal force doesn't do any work: $W_N = 0$.

---

126. Carry out the work integral: $W = \int_{x=x_0}^{x} F_x\, dx + \int_{y=y_0}^{y} F_y\, dy$.

- There is just a y-component of force in this problem: $W = \int_{y=y_0}^{y} F_y\, dy$.
- Get the integration limits from the y-coordinates of the initial and final points.
- Perform the integral $\int_{y=-1}^{2} 15y^4\, dy$.
- The anti-derivative is $3y^5$. Evaluate the anti-derivative over the limits.
- $[3y^5]_{y=-1}^{2} = 3(2)^5 - 3(-1)^5$. Note that $(-1)^5 = -1$ and $-3(-1) = +3$.
- The answer is $W = 99$ J.

---

127. Carry out the work integral: $W = \int_{x=x_0}^{x} F_x\, dx + \int_{y=y_0}^{y} F_y\, dy$.

- Get the integration limits from the (x, y)-coordinates of the initial and final points.
- Perform the integrals $\int_{x=2}^{4} x\, dx + \int_{y=3}^{6} y^2\, dy$.
- The anti-derivatives are $\frac{x^2}{2}$ and $\frac{y^3}{3}$. Evaluate the anti-derivatives over the limits.
- $\left[\frac{x^2}{2}\right]_{x=2}^{4} = \frac{(4)^2}{2} - \frac{(2)^2}{2}$ and $\left[\frac{y^3}{3}\right]_{y=3}^{6} = \frac{(6)^3}{3} - \frac{(3)^3}{3}$.
- The answer is $W = 69$ J.

---

128. Carry out the work integral: $W = \int_{x=x_0}^{x} F_x\,dx + \int_{y=y_0}^{y} F_y\,dy$.

- Get the integration limits from the (x, y)-coordinates of the initial and final points.
- Perform the integrals $\int_{x=0}^{4} xy\,dx + \int_{y=0}^{16} x^2\,dy$.
- Plug $y = x^2$ into xy so that the first integrand contains only x's.
- Plug $x = \sqrt{y}$ into x^2 so that the second integrand contains only y's.
- (Note that if you solve for x in the equation $y = x^2$, you get $x = \sqrt{y}$.)
- Perform the integrals $\int_{x=0}^{4} x^3\,dx + \int_{y=0}^{16} y\,dy$. (This is what you get after you make the substitutions $y = x^2$ and $x = \sqrt{y}$.)
- The anti-derivatives are $\frac{x^4}{4}$ and $\frac{y^2}{2}$. Evaluate the anti-derivatives over the limits.
- $\left[\frac{x^4}{4}\right]_{x=0}^{4} = \frac{(4)^4}{4} - \frac{(0)^4}{4}$ and $\left[\frac{y^2}{2}\right]_{y=0}^{16} = \frac{(16)^2}{2} - \frac{(0)^2}{2}$.
- The answer is $W = 192$ J.

---

129. (A) What is the formula for instantaneous power?

- Instantaneous power is the derivative of work with respect to time: $P = \frac{dW}{dt}$.
- First apply the rule $\sqrt{t} = t^{1/2}$.
- The derivative is $P = 36t^{-1/2} = \frac{36}{\sqrt{t}}$.
- Plug in the specified time.
- The instantaneous power at the specified time is $P = 12$ W.

(B) Which formula for average power should you use?

- Use the equation $\bar{P} = \frac{W}{t}$.
- Plug the final time into the given function ($W = 72\sqrt{t}$) to solve for the final work. (The initial work is zero when $t = 0$.)
- Divide the work by the time.
- The average power over the specified interval is $\bar{P} = 24$ W.

---

130. Find the scalar product between the force and the average velocity.

- Identify the components of $\vec{\mathbf{F}}$ and $\vec{\bar{v}}$.
- The components are: $F_x = 8$, $F_y = -2$, $F_z = 0$, $\bar{v}_x = 6$, $\bar{v}_y = -1$, and $\bar{v}_z = 0$.
- Use the component form of the scalar product: $W = \vec{\mathbf{F}} \cdot \vec{\bar{v}} = F_x\bar{v}_x + F_y\bar{v}_y + F_z\bar{v}_z$.
- The average power delivered by the monkey is $\bar{P} = 50$ W.

---

131. Which of the work and power equations should you use to solve this problem?

- First use the equation $W_{net} = ma_x \Delta x$ to find the net work.
- Use the equations of **uniform acceleration** (Chapter 3) to find a_x and Δx.
- The three knowns for uniform acceleration are $v_{x0} = 0$, $v_x = 60$ m/s, and $t = 5.0$ s.
- Use the equation $v_x = v_{x0} + a_x t$ to find the acceleration.
- The acceleration is $a_x = 12$ m/s^2.
- Use the equation $\Delta x = v_{x0} t + \frac{1}{2} a_x t^2$ to find the net displacement.
- The net displacement is $\Delta x = 150$ m.
- Plug numbers into the equation for the net work.
- The net work is $W_{net} = 360{,}000$ J.
- Now use the equation for average power: $\bar{P} = \frac{W}{t}$.
- The average power is $\bar{P} = 72$ kW. (This is the same as 72,000 W.)

Chapter 22: Conservation of Energy

132. Put i just after the throw and f just before impact. Put RH on the ground.

- $PE_{g0} = mgh_0$ since i is **not** at the same height as RH.
- $PE_{s0} = 0$ since at i there isn't a spring compressed or stretched from equilibrium.
- $PE_0 = PE_{g0} + PE_{s0} = mgh_0$.
- $KE_0 = \frac{1}{2} m v_0^2$ since the textbook is moving at i.
- $W_{nc} = 0$ since there are no frictional forces.
- $PE_g = 0$ since f is at the same height as RH.
- $PE_s = 0$ since at f there isn't a spring compressed or stretched from equilibrium.
- $PE = PE_g + PE_s = 0$.
- $KE = \frac{1}{2} m v^2$ since the textbook is moving at f.
- Write out conservation of energy: $PE_0 + KE_0 + W_{nc} = PE + KE$.
- Substitute the previous expressions into the conservation of energy equation.

$$mgh_0 + \frac{1}{2} m v_0^2 = \frac{1}{2} m v^2$$

- Mass cancels. Solve for the final speed.

$$v = \sqrt{v_0^2 + 2gh_0}$$

- The final speed just before impact is $v = 50$ m/s.

---

133. Put i at the bottom and f when it reaches its highest point. Put RH at the bottom.

- $PE_{g0} = 0$ since i is at the same height as RH.
- $PE_{s0} = 0$ since at i there isn't a spring compressed or stretched from equilibrium.
- $PE_0 = PE_{g0} + PE_{s0} = 0$.
- $KE_0 = \frac{1}{2}mv_0^2$ since the box is moving at i.
- $W_{nc} = 0$ since there are no frictional forces.
- $PE_g = mgh$ since f is **not** at the same height as RH.
- $PE_s = 0$ since at f there isn't a spring compressed or stretched from equilibrium.
- $PE = PE_g + PE_s = mgh$.
- $KE = 0$ since the box runs out of speed at f (otherwise it would rise higher).
- Write out conservation of energy: $PE_0 + KE_0 + W_{nc} = PE + KE$.
- Substitute the previous expressions into the conservation of energy equation.

$$\frac{1}{2}mv_0^2 = mgh$$

- Mass cancels. Solve for the final height.

$$h = \frac{v_0^2}{2g}$$

- The maximum height is $h \approx 80$ m. (It's 82 m if you don't round gravity.)

---

134. Put i at the top left, f_A at the bottom, and f_B at point B. Put RH at the bottom.

- $PE_{g0} = mgh_0$ since i is **not** at the same height as RH.
- $PE_{s0} = 0$ since at i there isn't a spring compressed or stretched from equilibrium.
- $PE_0 = PE_{g0} + PE_{s0} = mgh_0$.
- $KE_0 = 0$ since the banana is at **rest** at i.
- $W_{nc} = 0$ since there are no frictional forces.

(A) In part (A), the final position (f_A) is at point A (at the bottom of the arc).

- $PE_g = 0$ since f_A is at the same height as RH.
- $PE_s = 0$ since at f_A there isn't a spring compressed or stretched from equilibrium.
- $PE = PE_g + PE_s = 0$.
- $KE = \frac{1}{2}mv_A^2$ since the banana is moving at f_A. (That's where it moves fastest.)
- Write out conservation of energy: $PE_0 + KE_0 + W_{nc} = PE_A + KE_A$.
- Substitute the previous expressions into the conservation of energy equation.

$$mgh_0 = \frac{1}{2}mv_A^2$$

- Mass cancels. Solve for the speed at point A.

$$v_A = \sqrt{2gh_0}$$

- **Study the example on pages 213-214.** This problem is similar to that example.
- The above formula involves h_0 (the initial height), but we know L (the length of the pendulum). Draw a right triangle to relate h_0 to L (like we did on page 214).

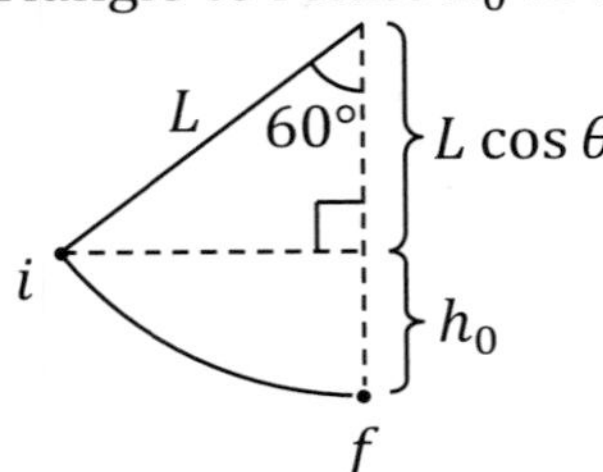

- Just like we did on **page 214**, apply trig and geometry to the figure above.

$$h_0 = L - L\cos 60°$$

- The length of the pendulum is given as $L = 10$ m. Solve for the initial height.
- The initial height is $h_0 = 5.0$ m.
- Plug the initial height into the previous equation for v_A.
- The speed at point A is $v_A \approx 10$ m/s. (It's 9.9 m/s if you don't round gravity.)

(B) In part (B), the final position (f_B) is at point B. (All of the initial values are the same.)

- $PE_g = mgh_B$ since f_B is **not** at the same height as RH.
- $PE_s = 0$ since at f_B there isn't a spring compressed or stretched from equilibrium.
- $PE = PE_g + PE_s = mgh_B$.
- $KE = \frac{1}{2}mv_B^2$ since the banana is moving at f_B. (Note: B isn't as high as i.)
- Write out conservation of energy: $PE_0 + KE_0 + W_{nc} = PE_B + KE_B$.
- Substitute the previous expressions into the conservation of energy equation.

$$mgh_0 = mgh_B + \frac{1}{2}mv_B^2$$

- Mass cancels. Solve for the speed at point B.

$$v_B = \sqrt{2g(h_0 - h_B)}$$

- In order to find h_B, draw a right triangle similar to the one we drew in part (A).
- Apply trig and geometry similar to what we did in part (A).

$$h_B = L - L\cos 30°$$

- The height at point B is $h_B = 10 - 5\sqrt{3}$ m.
- Plug the heights into the previous equation for v_B.
- The speed at B is $v_B \approx 10\sqrt{\sqrt{3} - 1}$ m/s. In decimals without rounding, it's 8.5 m/s.

---

135. Put i at the top, f_A at point A, and f_B at point B. Put RH at the bottom.

(A) In part (A), the final position (f_A) is at point A (at the bottom of the hill).

- $PE_{g0} = mgh_0$ since i is **not** at the same height as RH.

- $PE_{s0} = 0$ since at i there isn't a spring compressed or stretched from equilibrium.
- $PE_0 = PE_{g0} + PE_{s0} = mgh_0$.
- $KE_0 = 0$ since the box is at **rest** at i.
- $W_{nc} = 0$ in part (A) since there isn't friction between i and f_A.
- $PE_g = 0$ since f_A is at the same height as RH.
- $PE_s = 0$ since at f_A there isn't a spring compressed or stretched from equilibrium.
- $PE = PE_g + PE_s = 0$.
- $KE = \frac{1}{2}mv_A^2$ since the box is moving at f_A. (That's where it moves fastest.)
- Write out conservation of energy: $PE_0 + KE_0 + W_{nc} = PE_A + KE_A$.
- Substitute the previous expressions into the conservation of energy equation.

$$mgh_0 = \frac{1}{2}mv_A^2$$

- Mass cancels. Solve for the speed at point A.

$$v_A = \sqrt{2gh_0}$$

- The speed at point A is $v_A = 20$ m/s.

(B) In part (B), the final position (f_B) is at point B.

- $PE_{g0} = mgh_0$ since i is **not** at the same height as RH.
- $PE_{s0} = 0$ since at i there isn't a spring compressed or stretched from equilibrium.
- $PE_0 = PE_{g0} + PE_{s0} = mgh_0$.
- $KE_0 = 0$ since the box is at **rest** at i.
- $W_{nc} = -\mu Ns$ in part (B) because there is **friction** between f_A and f_B.
- $PE_g = 0$ since f_B is at the same height as RH.
- $PE_s = 0$ since at f_B there isn't a spring compressed or stretched from equilibrium.
- $PE = PE_g + PE_s = 0$.
- $KE = 0$ since the box is at **rest** at f_B.
- Write out conservation of energy: $PE_0 + KE_0 + W_{nc} = PE_B + KE_B$.
- Substitute the previous expressions into the conservation of energy equation.

$$mgh_0 - \mu Ns = 0$$

- Draw a FBD to find **normal force**.

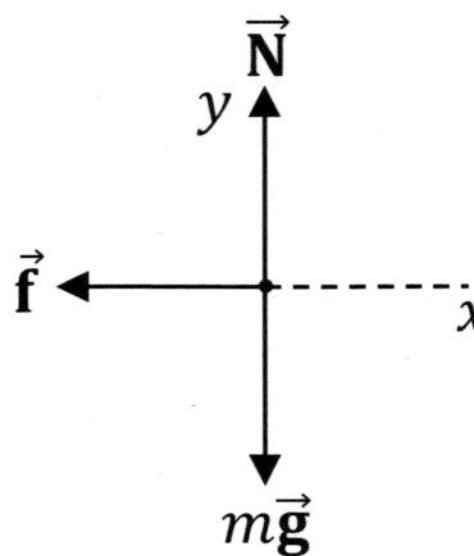

- Sum the y-components of the forces to solve for normal force: $\sum F_y = ma_y$.

$$N - mg = 0$$

- Note: $a_y = 0$ because the box doesn't accelerate up or down.
- Solve for normal force in the above equation: $N = mg$.
- Plug the expression for normal force into the equation prior to the FBD.
- Mass cancels. Solve for s.

$$s = \frac{h_0}{\mu}$$

- The distance between points A and B is $s = 100$ m.

---

136. Put i where the box starts and f just before the bottom. Put RH at the bottom.
- $PE_{g0} = mgh_0$ since i is **not** at the same height as RH.
- $PE_{s0} = 0$ since at i there isn't a spring compressed or stretched from equilibrium.
- $PE_0 = PE_{g0} + PE_{s0} = mgh_0$.
- $KE_0 = 0$ since the box is at **rest** at i.
- $W_{nc} = -\mu Ns$ because there is **friction** between i and f.
- $PE_g = 0$ since f is at the same height as RH.
- $PE_s = 0$ since at f there isn't a spring compressed or stretched from equilibrium.
- $PE = PE_g + PE_s = 0$.
- $KE = \frac{1}{2}mv^2$ since the box is moving at f.
- Write out conservation of energy: $PE_0 + KE_0 + W_{nc} = PE + KE$.
- Substitute the previous expressions into the conservation of energy equation.

$$mgh_0 - \mu Ns = \frac{1}{2}mv^2$$

- Draw a FBD to find **normal force**.

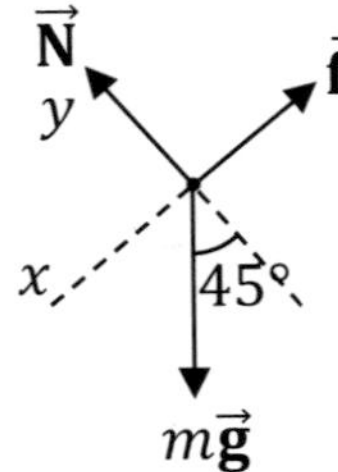

- Sum the y-components of the forces to solve for normal force: $\sum F_y = ma_y$.

$$N - mg\cos 45° = 0$$

- Note: $a_y = 0$ because the box doesn't accelerate perpendicular to the incline.
- Solve for normal force in the above equation.
- Plug the expression for normal force into the equation prior to the FBD.
- Mass cancels. Solve for the final speed.

$$v = \sqrt{2gh_0 - \frac{gs\cos 45°}{2}}$$

- The problem gives you the distance traveled: $s = 240\sqrt{2}$ m.
- Draw a right triangle and apply trig to solve for h_0.

$$\sin 45° = \frac{h_0}{s}$$

- Solve for the initial height.
- The initial height is $h_0 = 240$ m.
- Plug numbers into the above equation for v.
- The final speed is $v \approx 60$ m/s. (It's 59 m/s if you don't round gravity.)

---

137. (A) Put i where it is moving 50 m/s and f at point A. Put RH at the bottom.
   - $PE_{g0} = 0$ since i is at the same height as RH.
   - $PE_{s0} = 0$ since at i there isn't a spring compressed or stretched from equilibrium.
   - $PE_0 = PE_{g0} + PE_{s0} = 0$.
   - $KE_0 = \frac{1}{2}mv_0^2$ since the roller coaster is moving at i.
   - $W_{nc} = 0$ since there are no frictional forces.
   - $PE_g = mgh$ since f is **not** at the same height as RH.
   - $PE_s = 0$ since at f there isn't a spring compressed or stretched from equilibrium.
   - $PE = PE_g + PE_s = mgh$.
   - $KE = \frac{1}{2}mv^2$ since the roller coaster is moving at f.
   - Write out conservation of energy: $PE_0 + KE_0 + W_{nc} = PE + KE$.
   - Substitute the previous expressions into the conservation of energy equation.

$$\frac{1}{2}mv_0^2 = mgh + \frac{1}{2}mv^2$$

   - Mass cancels. Solve for the final speed.

$$v = \sqrt{v_0^2 - 2gh}$$

   - The speed at point A is $v \approx 30\sqrt{2}$ m/s. (It's 43 m/s if you don't round gravity.)

(B) Study the **example** on pages 214-216.

   - Draw a FBD at the top of the loop to find the **normal force** at point A.

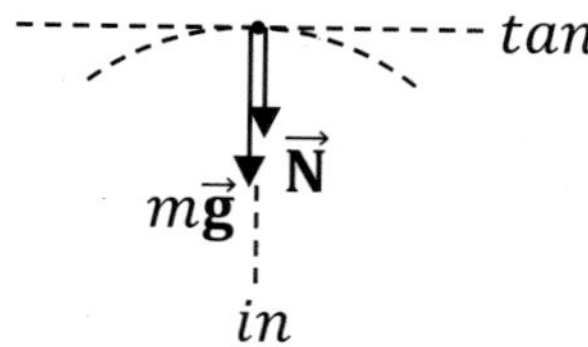

   - Since the roller coaster is traveling in a circle at point A, like we did in Chapter 17, sum the inward components of the forces to solve for normal force: $\sum F_{in} = ma_c$.

$$N + mg = ma_c$$

- Use the equation $a_c = \frac{v^2}{R}$. Plug this expression into the previous equation. Solve for normal force.

$$N = m\left(\frac{v^2}{R} - g\right)$$

- Plug numbers into the previous equation.
- The normal force at the top of the loop is $N = 19.5$ kN (or 19,500 N).

---

138. Put i at the fully compressed (FC) position and f at the top. Put RH at i.

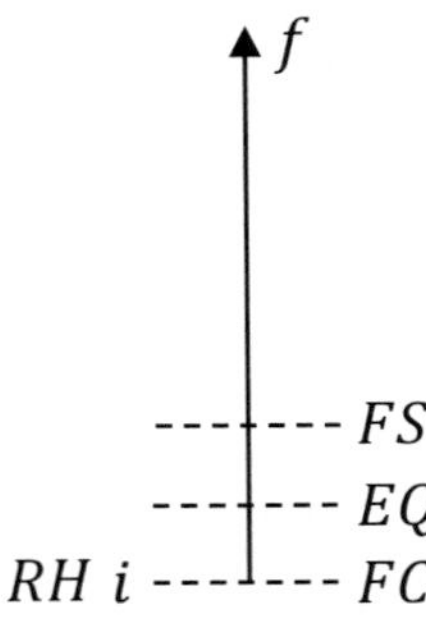

- $PE_{g0} = 0$ since i is at the same height as RH.
- $PE_{s0} = \frac{1}{2}kx_0^2$ since at i the **spring is compressed** from equilibrium.
- $PE_0 = PE_{g0} + PE_{s0} = \frac{1}{2}kx_0^2$.
- $KE_0 = 0$ since the pellet is at **rest** at i.
- $W_{nc} = 0$ since there are no frictional forces.
- $PE_g = mgh$ since f is **not** at the same height as RH.
- $PE_s = 0$ since at f the pellet is no longer attached to the spring.
- $PE = PE_g + PE_s = 0$.
- $KE = 0$ since the pellet is at **rest** at f (otherwise it would rise higher).
- Write out conservation of energy: $PE_0 + KE_0 + W_{nc} = PE + KE$.
- Substitute the previous expressions into the conservation of energy equation.

$$\frac{1}{2}kx_0^2 = mgh$$

- Solve for the spring constant.

$$k = \frac{2mgh}{x_0^2}$$

- The spring constant is $k \approx 288$ N/m. (It's 283 N/m if you don't round gravity.)

---

139. Put i at the fully compressed (FC) position, f_A at the equilibrium (EQ) position, and f_B between EQ and the fully stretched (FS) position. Put RH on the ground.

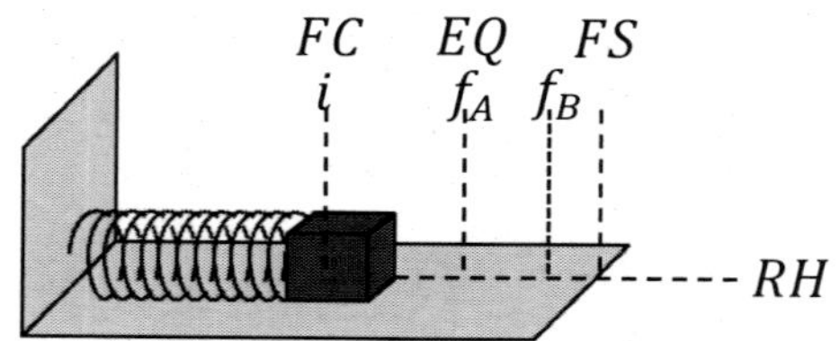

- $PE_{g0} = 0$ since i is at the same height as RH.
- $PE_{s0} = \frac{1}{2}kx_0^2$ since at i the **spring is compressed** from equilibrium.
- $PE_0 = PE_{g0} + PE_{s0} = \frac{1}{2}kx_0^2$.
- $KE_0 = 0$ since the box is at **rest** at i.
- $W_{nc} = 0$ since there are no frictional forces.

(A) In part (A), the final position (f_A) is at the equilibrium (EQ) position.

- $PE_g = 0$ since f_A is at the same height as RH.
- $PE_s = 0$ since at f_A the spring is at **equilibrium**.
- $PE = PE_g + PE_s = 0$.
- $KE = \frac{1}{2}mv_A^2$ since the box is moving at f_A (that's where it moves fastest).
- Write out conservation of energy: $PE_0 + KE_0 + W_{nc} = PE_A + KE_A$.
- Substitute the previous expressions into the conservation of energy equation.

$$\frac{1}{2}kx_0^2 = \frac{1}{2}mv_A^2$$

- Solve for the final speed.

$$v_A = x_0\sqrt{\frac{k}{m}}$$

- The speed at equilibrium is $v_A = 25$ m/s.

(B) In part (B), the final position (f_B) is **between** EQ and the fully stretched (FS) position.

- $PE_g = 0$ since f_B is at the same height as RH.
- $PE_s = \frac{1}{2}kx_B^2$ since at f_B the **spring is stretched** from equilibrium.
- $PE = PE_g + PE_s = 0$.
- $KE = \frac{1}{2}mv_B^2$ since the box is moving at f_B (though not as fast as at f_A).
- Write out conservation of energy: $PE_0 + KE_0 + W_{nc} = PE_B + KE_B$.
- Substitute the previous expressions into the conservation of energy equation.

$$\frac{1}{2}kx_0^2 = \frac{1}{2}kx_B^2 + \frac{1}{2}mv_B^2$$

- Solve for the final speed.

$$v_B = \sqrt{\frac{k}{m}(x_0^2 - x_B^2)}$$

- The speed at point B is $v_B = 15$ m/s.

140. Study the **example** on pages 217-218. This problem is nearly identical to that example.

- Put i at the surface of the planet and f at $R \to \infty$. Put RH at $R \to \infty$.
- (The reference height is where potential energy equals zero. PE_g approaches zero as R approaches infinity.)
- $PE_{g0} = -\frac{Gm_p m}{R_0}$ since i is not at the same height as RH.
- (Use the formula for an **astronomical** change in altitude. **Don't** use mgh_0.)
- $PE_{s0} = 0$ since at i there isn't a spring compressed or stretched from equilibrium.
- $PE_0 = PE_{g0} + PE_{s0} = -\frac{Gm_p m}{R_0}$.
- $KE_0 = \frac{1}{2}mv_0^2$ since the projectile must be moving at i in order to leave the planet.
- $W_{nc} = 0$ since we neglect air resistance unless stated otherwise.
- $PE_g = 0$ since f is at the same height as RH (and since $-\frac{Gm_p m}{R}$ approaches zero as R approaches infinity).
- $PE_s = 0$ since at f there isn't a spring compressed or stretched from equilibrium.
- $PE = PE_g + PE_s = 0$.
- $KE = 0$ since escape speed is the minimum speed needed to reach f.
- Write out conservation of energy: $PE_0 + KE_0 + W_{nc} = PE + KE$.
- Substitute the previous expressions into the conservation of energy equation.

$$-\frac{Gm_p m}{R_0} + \frac{1}{2}mv_0^2 = 0$$

- The mass of the projectile cancels. Solve for the initial speed.

$$v_0 = \sqrt{\frac{2Gm_p}{R_0}}$$

- The escape speed is $v = 100$ km/s. (It's the same as 10^5 m/s.)

---

141. Put i where $R_0 = 8.0 \times 10^7$ m and f where $R = 4.0 \times 10^7$ m. Put RH at $R \to \infty$.

- $PE_{g0} = -\frac{Gm_p m}{R_0}$ since i is not at the same height as RH.
- (Use the formula for an **astronomical** change in altitude. **Don't** use mgh_0.)
- $PE_{s0} = 0$ since at i there isn't a spring compressed or stretched from equilibrium.
- $PE_0 = PE_{g0} + PE_{s0} = -\frac{Gm_p m}{R_0}$.
- $KE_0 = \frac{1}{2}mv_0^2$ since the rocket is moving at i.
- $W_{nc} = 0$ since there are no resistive forces.
- $PE_g = -\frac{Gm_p m}{R}$ since f is not at the same height as RH.
- $PE_s = 0$ since at f there isn't a spring compressed or stretched from equilibrium.
- $PE = PE_g + PE_s = 0$.

- $KE = \frac{1}{2}mv^2$ since the rocket is moving at f.
- Write out conservation of energy: $PE_0 + KE_0 + W_{nc} = PE + KE$.
- Substitute the previous expressions into the conservation of energy equation.

$$-\frac{Gm_p m}{R_0} + \frac{1}{2}mv_0^2 = -\frac{Gm_p m}{R} + \frac{1}{2}mv^2$$

- The mass of the rocket cancels. Solve for the final speed. Plug in numbers.
- If you find yourself taking the squareroot of a negative number, you either made a mistake in your algebra, a mistake plugging in numbers, or a calculation mistake.
- The final speed is $v = 9.0$ km/s. (It's the same as 9000 m/s.)

Chapter 23: One-dimensional Collisions

142. You **don't** need to convert grams (g) to kilograms. Just be consistent.
   - Use the equation for a **perfectly inelastic** collision:

$$m_1\vec{v}_{10} + m_2\vec{v}_{20} = (m_1 + m_2)\vec{v}$$

   - The initial velocity of the banana is zero: $\vec{v}_{10} = 0$. Plug in numbers.
   - Solve for $\vec{v}$. The final speed of the banana and arrow is $v = 15$ m/s.

---

143. Choose **north** to be the positive direction. Then south will be negative.
   - Use the equation for an **inverse perfectly inelastic** collision:

$$(m_1 + m_2)\vec{v}_0 = m_1\vec{v}_1 + m_2\vec{v}_2$$

   - The initial velocity of the system is zero: $\vec{v}_0 = 0$.
   - Since the monkey walks south, $\vec{v}_1 = -3\hat{y}$ m/s is negative.
   - Solve for $\vec{v}_2$. The final velocity of the canoe is $\vec{v}_2 = \frac{3}{5}\hat{y}$ m/s (to the **north**).

---

144. Choose **east** to be the positive direction. Then west will be negative.
   - Since one box travels east while the other box travels west, one of the initial velocities will be negative: $\vec{v}_{10} = 5.0\hat{x}$ m/s and $\vec{v}_{20} = -4.0\hat{x}$ m/s.

(A) Use the equation for a **perfectly inelastic** collision:

$$m_1\vec{v}_{10} + m_2\vec{v}_{20} = (m_1 + m_2)\vec{v}$$

- Plug in numbers. Solve for $\vec{v}$.
- The final velocity of the boxes is $\vec{v} = -1.0\hat{x}$ m/s. The boxes travel to the **west**.

(B) Use the equations for an **elastic** collision. (To review what $\hat{x}$ means, see Chapter 10.)

$$m_1\vec{v}_{10} + m_2\vec{v}_{20} = m_1\vec{v}_1 + m_2\vec{v}_2$$

$$\vec{v}_{10} + \vec{v}_1 = \vec{v}_{20} + \vec{v}_2$$

- Plug the masses and initial velocities into the two equations above. Simplify.

$$-18\hat{x} = 6\vec{v}_1 + 12\vec{v}_2$$

$$5\hat{x} + \vec{v}_1 = -4\hat{x} + \vec{v}_2$$

- Isolate $\vec{v}_2$ in the bottom equation.

$$\vec{v}_2 = 9\hat{x} + \vec{v}_1$$

- Substitute this expression for $\vec{v}_2$ into the top equation. Solve for $\vec{v}_1$. (Algebraically, we're applying the method of substitution reviewed in Chapter 1.)
- The final velocity of the box of bananas is $\vec{v}_1 = -7.0\hat{x}$ m/s.
- Plug $\vec{v}_1$ into the equation above.
- The final velocity of the box of applies is $\vec{v}_2 = 2.0\hat{x}$ m/s.
- Since we chose east to be positive, the box of bananas travels **7.0 m/s to the west** after the collision while the box of apples travels **2.0 m/s to the east**.

---

145. Use the equations for an **elastic** collision:

$$m_1\vec{v}_{10} + m_2\vec{v}_{20} = m_1\vec{v}_1 + m_2\vec{v}_2$$
$$\vec{v}_{10} + \vec{v}_1 = \vec{v}_{20} + \vec{v}_2$$

- Choose **north** to be the positive direction. Then south will be negative.
- Since one box travels north while the other box travels south, one of the initial velocities will be negative: $\vec{v}_{10} = 16.0\hat{y}$ m/s and $\vec{v}_{20} = -6.0\hat{y}$ m/s.
- Plug the masses and initial velocities into the two equations above. Simplify.

$$-6\hat{y} = 3\vec{v}_1 + 9\vec{v}_2$$
$$16\hat{y} + \vec{v}_1 = -6\hat{y} + \vec{v}_2$$

- Isolate $\vec{v}_2$ in the bottom equation.

$$\vec{v}_2 = 22\hat{y} + \vec{v}_1$$

- Substitute this expression for $\vec{v}_2$ into the top equation. Solve for $\vec{v}_1$. (Algebraically, we're applying the method of substitution reviewed in Chapter 1.)
- The final velocity of the box of coconuts is $\vec{v}_1 = -17.0\hat{y}$ m/s.
- Plug $\vec{v}_1$ into the equation above.
- The final velocity of the box of grapefruit is $\vec{v}_2 = 5.0\hat{y}$ m/s.
- Since we chose north to be positive, the box of coconuts travels **17.0 m/s to the south** after the collision while the box of grapefruit travels **5.0 m/s to the north**.

(B) One of the impulse equations involves the average collision force.

$$\vec{J} = \Delta\vec{p} = \vec{p} - \vec{p}_0 = m(\vec{v} - \vec{v}_0) = \vec{F}_c\Delta t$$

- Apply the impulse equations to a single object (just one box).
- Solve for the average collision force exerted on object 1. The average collision force exerted on object 2 will be equal and opposite according to Newton's third law.

$$F_c = \frac{m_1|v_1 - v_{10}|}{\Delta t}$$

- (The "magnitude" of a vector is positive. That's why there are absolute values.)
- Plug in numbers. Convert the time to seconds: $\Delta t = 0.25 \text{ s} = \frac{1}{4}$ s.
- The average collision force is $F_c = 396$ N.

---

146. Solve this problem in stages. Stage 1: Apply the law of conservation of energy to solve for the speed of the box of bananas at point A just prior to the collision. Stage 2: Apply the law of conservation of momentum to the collision.

Stage 1: Conserve energy to find the speed of the box of bananas just prior to the collision.

- Put i at the top of the hill and f_{bc} just before point A. Put RH on the horizontal.
- Notation: The subscript "bc" stands for "**before collision**."
- $PE_{g0} = mgh_0$ since i is **not** at the same height as RH.
- $PE_{s0} = 0$ since at i there isn't a spring compressed or stretched from equilibrium.
- $PE_0 = PE_{g0} + PE_{s0} = mgh_0$.
- $KE_0 = 0$ since the box is at **rest** at i.
- $W_{nc} = 0$ since there are no frictional forces.
- $PE_g = 0$ since f_{bc} is at the same height as RH.
- $PE_s = 0$ since at f_{bc} there isn't a spring compressed or stretched from equilibrium.
- $PE = PE_g + PE_s = 0$.
- $KE = \frac{1}{2}mv_{bc}^2$ since the box is moving at f_{bc}.
- Write out conservation of energy: $PE_0 + KE_0 + W_{nc} = PE + KE$.
- Substitute the previous expressions into the conservation of energy equation.

$$mgh_0 = \frac{1}{2}mv_{bc}^2$$

- Mass cancels. Solve for the speed of the box just before the collision.

$$v_{bc} = \sqrt{2gh_0}$$

- The initial height equals the radius of the circle: $h_0 = R = 5.0$ m.
- The speed of the box just before the collision is $v_{bc} \approx 10$ m/s (it's really 9.9 m/s).

Stage 2: Conserve momentum for the collision.

- Use the equation for a **perfectly inelastic** collision:

$$m_1\vec{v}_{bc} + m_2\vec{v}_{20} = (m_1 + m_2)\vec{v}$$

- Recall that the subscript "bc" stands for "**before collision**."
- The initial velocity of the 30-kg box of pineapples is zero: $\vec{v}_{20} = 0$.
- The final speed of the boxes is $v = 4.0$ m/s.

Chapter 24: Two-dimensional Collisions

147. We choose $+x$ to point **east** and $+y$ to point **north**.

- The given values are: $v_{10} = 20$ m/s, $v_{20} = 40$ m/s, $\theta_{10} = 90°$, and $\theta_{20} = 0°$.
- $\theta_{10} = 90°$ since $\vec{v}_{10}$ is north (along $+y$). $\theta_{20} = 0°$ since $\vec{v}_{20}$ is east (along $+x$).
- Find the x- and y-components of the initial velocities.

$$v_{10x} = v_{10}\cos\theta_{10} \quad , \quad v_{20x} = v_{20}\cos\theta_{20}$$
$$v_{10y} = v_{10}\sin\theta_{10} \quad , \quad v_{20y} = v_{20}\sin\theta_{20}$$

- The initial components are $v_{10x} = 0$, $v_{10y} = 20$ m/s, $v_{20x} = 40$ m/s., and $v_{20y} = 0$.
- Use the equations for a two-dimensional **perfectly inelastic** collision.

$$m_1 v_{10x} + m_2 v_{20x} = (m_1 + m_2) v_x$$
$$m_1 v_{10y} + m_2 v_{20y} = (m_1 + m_2) v_y$$

- Solve for the components of the final velocity: $v_x = 12$ m/s, $v_y = 16$ m/s.
- Apply the Pythagorean theorem to find the final speed.

$$v = \sqrt{v_x^2 + v_y^2}$$

- The final speed of the bananamobiles is 20 m/s.

---

148. In this problem, we know the final velocity and we're looking for one of the **initial** velocities. This will make the solution a little different than the previous problem.

- We choose $+x$ to point **east** and $+y$ to point **north**.
- The given values are: $v_{10} = 20$ m/s, $v = 10\sqrt{2}$ m/s, $\theta_{10} = 270°$, and $\theta = 225°$.
- $\theta_{10} = 270°$ since $\vec{v}_{10}$ is south (along $-y$). $\theta = 225°$ since $\vec{v}$ is southwest.
- In this problem, we're looking for v_{20} and θ_{20}.
- Find the x- and y-components of the initial velocity of object 1.

$$v_{10x} = v_{10} \cos\theta_{10} \quad , \quad v_{10y} = v_{10} \sin\theta_{10}$$

- The components of object 1's initial velocity are $v_{10x} = 0$ and $v_{10y} = -20$ m/s.
- Find the x- and y-components of the final velocity.

$$v_x = v\cos\theta \quad , \quad v_y = v\sin\theta$$

- The components of the final velocity are $v_x = -10$ m/s and $v_y = -10$ m/s.
- Use the equations for a two-dimensional **perfectly inelastic** collision.

$$m_1 v_{10x} + m_2 v_{20x} = (m_1 + m_2) v_x$$
$$m_1 v_{10y} + m_2 v_{20y} = (m_1 + m_2) v_y$$

- Solve for the components of object 2's initial velocity: $v_{20x} = -20$ m/s, $v_{20y} = 0$.
- Apply the Pythagorean theorem to find the initial speed of object 2.

$$v_{20} = \sqrt{v_{20x}^2 + v_{20y}^2}$$

- The initial speed of object 2 is 20 m/s.
- Apply trig to determine the direction of the initial velocity of object 2.

$$\theta_{20} = \tan^{-1}\left(\frac{v_{20y}}{v_{20x}}\right)$$

- The direction of the initial velocity of object 2 is 180° (which means to the **west**).

---

149. Begin by making a list of the given symbols.

- Orient $+x$ along the path of the billiard ball that is moving initially.
- What's given? $v_{10} = 6\sqrt{3}$ m/s, $\theta_{10} = 0°$, $v_{20} = 0$, $v_1 = 3\sqrt{3}$ m/s, and $\theta_{10} = 60°$.

- Find the x- and y-components of the initial velocities.

$$v_{10x} = v_{10} \cos \theta_{10} \quad , \quad v_{20x} = v_{20} \cos \theta_{20}$$
$$v_{10y} = v_{10} \sin \theta_{10} \quad , \quad v_{20y} = v_{20} \sin \theta_{20}$$

- The initial components are $v_{10x} = 6\sqrt{3}$ m/s, $v_{10y} = 0$, $v_{20x} = 0$, and $v_{20y} = 0$.
- Find the x- and y-components of the final velocity of object 1.

$$v_{1x} = v_1 \cos \theta_1 \quad , \quad v_{1y} = v_1 \sin \theta_1$$

- The components of the final velocity of object 1 are $v_{1x} = \frac{3\sqrt{3}}{2}$ m/s and $v_{1y} = \frac{9}{2}$ m/s.
- Use the equations for a two-dimensional **elastic** collision with equal masses.

$$v_{10x} + v_{20x} = v_{1x} + v_{2x}$$
$$v_{10y} + v_{20y} = v_{1y} + v_{2y}$$
$$\theta_{1ref} + \theta_{2ref} = 90°$$

- Find the components of the final velocity of object 2: $v_{2x} = \frac{9\sqrt{3}}{2}$ m/s, $v_{2y} = -\frac{9}{2}$ m/s.
- Apply the Pythagorean theorem to find the final speed of object 2.

$$v_2 = \sqrt{v_{2x}^2 + v_{2y}^2}$$

- The final speed of object 2 is 9.0 m/s.
- Apply trig to determine the direction of the final velocity of object 2.

$$\theta_2 = \tan^{-1}\left(\frac{v_{2y}}{v_{2x}}\right)$$

- The direction of the final velocity of object 2 is 330°.
- (Since $v_{2x} > 0$ and $v_{2y} < 0$, the answer lies in Quadrant IV.)

Chapter 25: Rocket Propulsion

150. Begin by making a list of the given symbols along with their numerical values.

$m_p = 2000$ kg, $r_p = 25\% = \frac{1}{4}$, $v_{y0} = 200$ m/s, $u = 1500$ m/s, $v_y = 600$ m/s, $R_b = 50$ kg/s

(A) Your goal is to find m_f, the **final mass of fuel** remaining in the rocket.

- Use the equation $r_p = \frac{m_p}{m_0}$ to determine the initial mass of the rocket including fuel.
- You should find that $m_0 = 8000$ kg.
- Use the zero-gravity equation $v_y = v_{y0} + u \ln\left(\frac{m_0}{m}\right)$ to determine m, the **mass of the rocket plus the unburned fuel**.
- This intermediate answer is $m = 8000\, e^{-4/15} = 6127$ kg.
- Now we need to find the initial and final mass of the fuel.
- Use the equation $m_0 = m_p + m_{f0}$ to solve for m_{f0}, the **initial mass of the fuel**.
- You should find that $m_{f0} = 6000$ kg.

- Use the equation $m = m_p + m_f$ to solve for m_f, the **final mass of the fuel**.
- You should find that $m_f = 4127$ kg.
- The percentage of fuel spent is given by:

$$\frac{m_{f0} - m_f}{m_{f0}} 100\%$$

- The final answer is that 31% of the fuel is spent.

(B) Use the equation that relates the **burn time** to the burn rate.

- Use the equation $m - m_0 = -R_b t_b$.
- The burn time is $t_b = 37$ s.

---

151. Begin by making a list of the given symbols along with their numerical values.

$$m_p = 3000 \text{ kg}, m_{f0} = 6000 \text{ kg}, v_{y0} = 100 \text{ m/s}, u = 2000 \text{ m/s}$$
$$v_y = 500 \text{ m/s}, t_b = 12 \text{ s}, g = 9.81 \text{ m/s}^2 \approx 10 \text{ m/s}^2$$

(A) Your goal is to find m_f, the **final mass of fuel** remaining in the rocket.

- Use the equation $m_0 = m_p + m_{f0}$ to find the **initial mass of the rocket including fuel**.
- You should find that $m_0 = 9000$ kg.
- Use the uniform gravity equation $v_y = v_{y0} - g t_b + u \ln\left(\frac{m_0}{m}\right)$ to find m, the **mass of the rocket plus the unburned fuel**.
- This intermediate answer is $m \approx 9000\, e^{-13/50}$ (or $m = 6947$ kg using 9.81 m/s²).
- Use the equation $m = m_p + m_f$ to solve for m_f, the **final mass of the fuel**.
- You should find that $m_f = 3947$ kg.
- The percentage of fuel spent is given by:

$$\frac{m_{f0} - m_f}{m_{f0}} 100\%$$

- The answer is that 34% of the fuel is spent.

(B) Use the equation that relates the burn time to the **burn rate**.

- Use the equation $m - m_0 = -R_b t_b$.
- The burn rate is $R_b = 171$ kg/s.

(C) Use the equation for **thrust**.

- Use the equation $F_t = -u \frac{dm}{dt}$, where $\frac{dm}{dt} = -R_b$.
- $\frac{dm}{dt}$ is negative because the rocket's mass is a decreasing function of time. (Therefore the thrust is positive: Two minus signs make a plus sign.)
- The thrust is $F_t = 342$ kN. (It's the same as 342,000 N.)

(D) Apply the equation $F_{thrust} - m_0 g = m_0 a_{y0}$.

- Use the initial mass to find the initial acceleration: $F_{thrust} - m_0 g = m_0 a_{y0}$.
- The initial acceleration is $a_{y0} = 28 \text{ m/s}^2$.

(E) Apply the equation $F_{thrust} - mg = ma_y$.

- Use the final mass to find the final acceleration: $F_{thrust} - mg = ma_y$.
- The final acceleration is $a_y = 39 \text{ m/s}^2$.

Chapter 26: Techniques of Integration and Coordinate Systems

152. Make the substitution $u = \frac{x}{5} - 1$.

- Take an implicit derivative of both sides to get $du = \frac{dx}{5}$.
- Solve for dx to get $dx = 5du$.
- Plug the limits of x into the equation $u = \frac{x}{5} - 1$.
- The new limits are from $u = -1$ to $u = 2$.
- After making these substitutions, the integral becomes $\int_{u=-1}^{2} 5u^4 \, du$.
- The anti-derivative is u^5.
- Evaluate the anti-derivative over the limits.
- $[u^5]_{u=-1}^{2} = (2)^5 - (-1)^5$. Note that $(-1)^5 = -1$ and $-(-1) = +1$.
- The final answer is 33.

---

153. Make the substitution $u = \frac{\pi}{2} - \frac{x}{6}$.

- Take an implicit derivative of both sides to get $du = -\frac{dx}{6}$.
- Solve for dx to get $dx = -6du$.
- Plug the limits of x into the equation $u = \frac{\pi}{2} - \frac{x}{6}$.
- The new limits are from $u = \frac{\pi}{2}$ to $u = \frac{\pi}{6}$.
- After making these substitutions, the integral becomes $\int_{u=\pi/2}^{\pi/6} (-6\cos u) \, du$.
- The anti-derivative is $-6 \sin u$.
- Evaluate the anti-derivative over the limits.
- $[-6 \sin u]_{u=-\pi/2}^{\pi/6} = -6 \sin\left(\frac{\pi}{6}\right) + 6 \sin\left(\frac{\pi}{2}\right)$. Note that the angles are in **radians**.
- The final answer is 3.

---

154. Make the substitution $x = \tan u$.

- Take an implicit derivative of both sides to get $dx = \sec^2 u \, du$.
- Plug the limits of x into the equation $u = \tan^{-1} x$.
- The new limits are from $u = 0°$ to $u = 60°$.
- After making these substitutions, the integral becomes $\int_{u=0°}^{60°} \frac{\sec^2 u}{\sqrt{1+\tan^2 u}} du$.
- Apply the trig identity $1 + \tan^2 u = \sec^2 u$. Simplify.

- You should get $\int_{u=0°}^{60°} \sec u \, du$.
- Find the anti-derivative of secant at the end of Chapter 9.
- The anti-derivative is $\ln|\sec u + \tan u|$.
- Evaluate the anti-derivative over the limits.
- $[\ln|\sec u + \tan u|]_{u=0°}^{60°} = \ln|\sec 60° + \tan 60°| - \ln|\sec 0° + \tan 0°|$.
- The last term vanishes because $\ln|1| = 0$.
- The final answer is $\ln|2 + \sqrt{3}|$.

---

155. Make the substitution $x = 2 \sin u$.
- Take an implicit derivative of both sides to get $dx = 2 \cos u \, du$.
- Plug the limits of x into the equation $u = \sin^{-1}\left(\frac{x}{2}\right)$.
- The new limits are from $u = 0°$ to $u = 90°$.
- After making these substitutions, the integral becomes $\int_{u=0°}^{90°} \sqrt{4 - 4\sin^2 u} \; 2\cos u \, du$.
- Apply the trig identity $1 - \sin^2 u = \cos^2 u$. Simplify.
- You should get $4 \int_{u=0°}^{90°} \cos^2 u \, du$.
- Apply the trig identity $\cos^2 u = \frac{1+\cos 2u}{2}$.
- After this substitution, separate the integral into two integrals, one for each term.
- You should get $4 \int_{u=0°}^{90°} \frac{1}{2} du + 4 \int_{u=0°}^{90°} \frac{\cos 2u}{2} du = 2 \int_{u=0°}^{90°} du + 2 \int_{u=0°}^{90°} \cos 2u \, du$.
- The anti-derivatives are $2u$ and $\sin 2u$ (where the coefficients have been included).
- Convert from degrees to **radians**. The new limits are from $u = 0$ to $u = \frac{\pi}{2}$.
- Evaluate the anti-derivatives over the limits.
- The final answer is π.

---

156. First integrate over y (since the upper limit of y is x^3).
- Factor the x out of the y-integration: $\int_{x=0}^{2} x \left(\int_{y=0}^{x^3} y \, dy\right) dx$.
- The anti-derivative for the y-integration is $\frac{y^2}{2}$. Evaluate $\frac{y^2}{2}$ over the limits.
- The answer to the definite integral over y is $\frac{x^6}{2}$.
- Now you should have $\int_{x=0}^{2} \frac{x^7}{2} dx$.
- The anti-derivative for the x-integration is $\frac{x^8}{16}$. Evaluate $\frac{x^8}{16}$ over the limits.
- The final answer to the double integral is 16.

---

157. You can do these integrals in any order. We will integrate over y first.
- Factor the x^3 out of the y-integration: $\int_{x=0}^{5} x^3 \left(\int_{y=0}^{4} y^4 \, dy\right) dx$.

- The anti-derivative for the y-integration is $\frac{y^5}{5}$. Evaluate $\frac{y^5}{5}$ over the limits.
- The answer to the definite integral over y is $\frac{4^5}{5}$.
- Now you should have $\int_{x=0}^{5} \frac{4^5 x^3}{5} dx$. The $\frac{4^5}{5}$ part is just a constant coefficient.
- The anti-derivative for the x-integration is $\frac{4^5}{5}\frac{x^4}{4} = \frac{4^4 x^4}{5}$. Evaluate it over the limits.
- The final answer to the double integral is 32,000. It's the same as $5^3 4^4$.

---

158. You can do these integrals in any order. We will integrate over y first.
  - Apply the rule $\sqrt{xy} = x^{1/2} y^{1/2}$.
  - Factor the $x^{1/2}$ out of the y-integration: $\int_{x=0}^{4} x^{1/2} \left(\int_{y=0}^{9} y^{1/2}\, dy\right) dx$.
  - The anti-derivative for the y-integration is $\frac{2y^{3/2}}{3}$. Evaluate $\frac{2y^{3/2}}{3}$ over the limits.
  - The answer to the definite integral over y is 18. Note that $9^{3/2} = \left(\sqrt{9}\right)^3 = 27$.
  - Now you should have $\int_{x=0}^{4} 18x^{1/2}\, dx$.
  - The anti-derivative for the x-integration is $12x^{3/2}$. Evaluate $12x^{3/2}$ over the limits.
  - The final answer to the double integral is 96. Note that $4^{3/2} = \left(\sqrt{4}\right)^3 = 8$.

---

159. First integrate over y (since the upper limit of y is x).
  - Split the double integral into two separate integrals using $6(x - y) = 6x - 6y$.

$$\int_{x=0}^{3} \int_{y=0}^{x} 6(x-y)\, dxdy = \int_{x=0}^{3} \int_{y=0}^{x} 6x\, dxdy - \int_{x=0}^{3} \int_{y=0}^{x} 6y\, dxdy$$

  - Factor the x out of the y-integration: $\int_{x=0}^{3} 6x \left(\int_{y=0}^{x} dy\right) dx - \int_{x=0}^{3} 6 \left(\int_{y=0}^{x} y\, dy\right) dx$.
  - The anti-derivatives for the y-integration are y and $\frac{y^2}{2}$. Evaluate over the limits.
  - The answers to the definite integrals over y are x and $\frac{x^2}{2}$.
  - Now you should have $\int_{x=0}^{3} 6x^2\, dx - \int_{x=0}^{3} 3x^2\, dx$.
  - The anti-derivatives for the x-integration are $2x^3$ and x^3. Evaluate over the limits.
  - The final answer to the double integral is 27.

---

160. First integrate over z: $\int_{x=0}^{2} \int_{y=0}^{x^2} 27\left(\int_{z=0}^{y} z^2\, dz\right) dy\, dx$.
  - The anti-derivative for the z-integration is $\frac{z^3}{3}$. Evaluate $\frac{z^3}{3}$ over the limits.
  - The answer to the definite integral over z is $\frac{y^3}{3}$.
  - Now you should have $\int_{x=0}^{2} \left(\int_{y=0}^{x^2} 9y^3\, dy\right) dx$. Integrate over y.

- The anti-derivative for the y-integration is $\frac{9y^4}{4}$. Evaluate $\frac{9y^4}{4}$ over the limits.
- The answer to the definite integral over y is $\frac{9x^8}{4}$.
- Now you should have $\int_{x=0}^{2} \frac{9x^8}{4} dx$. Integrate over x.
- The anti-derivative for the x-integration is $\frac{x^9}{4}$. Evaluate $\frac{x^9}{4}$ over the limits.
- The final answer to the triple integral is 128.

---

161. First integrate over y: $\int_{z=0}^{9} \int_{x=0}^{z} 14x \left(\int_{y=0}^{\sqrt{x}} y^2 \, dy \right) dx \, dz$.
   - The anti-derivative for the y-integration is $\frac{y^3}{3}$. Evaluate $\frac{y^3}{3}$ over the limits.
   - The answer to the definite integral over y is $\frac{x^{3/2}}{3}$.
   - Now you should have $\int_{z=0}^{9} \left(\int_{x=0}^{z} \frac{14x^{5/2}}{3} dx \right) dz$. Note that $xx^{3/2} = x^{5/2}$.
   - Next integrate over x.
   - The anti-derivative for the x-integration is $\frac{4x^{7/2}}{3}$. Evaluate $\frac{4x^{7/2}}{3}$ over the limits.
   - The answer to the definite integral over x is $\frac{4z^{7/2}}{3}$.
   - Now you should have $\int_{z=0}^{9} \frac{4z^{7/2}}{3} dz$. Integrate over z.
   - The anti-derivative for the z-integration is $\frac{8z^{9/2}}{27}$. Evaluate $\frac{8z^{9/2}}{27}$ over the limits.
   - Note that $9^{9/2} = \left(\sqrt{9}\right)^9 = 3^9 = 19{,}683$.
   - The final answer to the triple integral is 5832.

---

162. Integrate over area: $A = \int dA$. Study the **example** on page 275.
   - Express the differential area element using Cartesian coordinates: $dA = dxdy$.
   - Find the equation of the line for the hypotenuse.
   - The line for the hypotenuse has a slope of $-\frac{1}{2}$ and y-intercept of 3.
   - The equation for the line is therefore $y = -\frac{x}{2} + 3$.
   - Let x vary from 0 to 6. Then y will vary from 0 to $-\frac{x}{2} + 3$.
   - The double integral is:

$$A = \int_{x=0}^{6} \int_{y=0}^{-\frac{x}{2}+3} dydx$$

   - First integrate over y (since the upper limit of y is $-\frac{x}{2} + 3$).
   - The anti-derivative for the y-integration is y. Evaluate y over the limits.
   - The answer to the definite integral over y is $-\frac{x}{2} + 3$.

- Now you should have $\int_{x=0}^{6}\left(-\frac{x}{2}+3\right)dx$. Find the anti-derivative of each term.
- The anti-derivative for the x-integration is $-\frac{x^2}{4}+3x$. Evaluate this over the limits.
- The final answer to the double integral is $A = 9$.

---

163. Integrate over area: $A = \int dA$. Study the **example** on page 276.
- Express the differential area element using 2D polar coordinates: $dA = rdrd\theta$.
- The double integral is:

$$A = \int_{r=R_1}^{R_2}\int_{\theta=0}^{2\pi} rdrd\theta$$

- You can do these integrals in any order. We will integrate over θ first.
- Factor the r out of the θ-integation: $\int_{r=R_1}^{R_2} r\left(\int_{\theta=0}^{2\pi} d\theta\right)dr$.
- The anti-derivative for the θ-integration is θ. Evaluate θ over the limits.
- The answer to the definite integral over θ is 2π.
- Now you should have $\int_{r=R_1}^{R_2} 2\pi r\, dr$.
- The anti-derivative for the r-integration is πr^2. Evaluate πr^2 over the limits.
- The final answer to the double integral is $A = \pi(R_2^2 - R_1^2)$.

Chapter 27: Center of Mass

164. We put the origin on the 150-g banana with $+x$ pointing toward the bunch of bananas.
- The knowns are $m_1 = 150$ g, $x_1 = 0$, $m_2 = 350$ g, and $x_2 = 250$ cm.
- Use the equation for center of mass with $N = 2$ (for two objects).

$$x_{cm} = \frac{m_1x_1 + m_2x_2}{m_1 + m_2}$$

- The x-coordinate of the center of mass is $x_{cm} = 175$ cm.
- This means that the center of mass lies 175 cm from the 150-g banana.

---

165. Make a list of the given symbols.
- The masses are $m_1 = 200$ g, $m_2 = 500$ g, and $m_3 = 800$ g.
- The x-coordinates are $x_1 = 7.0$ m, $x_2 = 6.0$ m, and $x_3 = 2.0$ m.
- The y-coordinates are $y_1 = 1.0$ m, $y_2 = 3.0$ m, and $y_3 = -4.0$ m.
- Use the equations for center of mass with $N = 3$ (for 3 objects).

$$x_{cm} = \frac{m_1x_1 + m_2x_2 + m_3x_3}{m_1 + m_2 + m_3}$$

$$y_{cm} = \frac{m_1 y_1 + m_2 y_2 + m_3 y_3}{m_1 + m_2 + m_3}$$

- The coordinates of the center of mass are $x_{cm} = 4.0$ m and $y_{cm} = -1.0$ m.

---

166. We put the origin at the left end with $+x$ pointing to the right.

- Divide the T-shaped object into two pieces.
- One piece is the 20.0-cm long handle. The other piece is the 4.0-cm wide end.
- The given masses are $m_1 = 6.0$ kg and $m_2 = 18.0$ kg.
- Where is the center of each piece?
- The x-coordinates of their centers are $x_1 = 10.0$ cm and $x_2 = 22.0$ cm.
- (Notes: $x_1 = 10.0$ cm because the handle's center is half its length from the left end: $x_1 = \frac{L}{2} = \frac{20}{2} = 10.0$ cm. $x_2 = 22.0$ cm because the center of the second piece lies 2.0 cm past the end of the handle: $x_2 = L + \frac{W}{2} = 20 + \frac{4}{2} = 20 + 2 = 22.0$ cm.)
- Use the equation for center of mass with $N = 2$ (for two objects).

$$x_{cm} = \frac{m_1 x_1 + m_2 x_2}{m_1 + m_2}$$

- The x-coordinate of the center of mass is $x_{cm} = 19.0$ cm.
- This means that the center of mass lies 1.0 cm to the left of the point where the handle meets the ends (since $20 - 19 = 1$).

---

167. Divide the object into 5 squares. Make a list of the given symbols.

- All of the masses are identical: Call each mass m_s.
- The x-coordinates are $x_1 = 10$ m, $x_2 = 30$ m, $x_3 = 30$ m, $x_4 = 30$ m, and $x_5 = 10$ m.
- The y-coordinates are $y_1 = 10$ m, $y_2 = 10$ m, $y_3 = 30$ m, $y_4 = 50$ m, and $y_5 = 50$ m.
- Use the equations for center of mass with $N = 5$ (for 5 squares).

$$x_{cm} = \frac{m_s x_1 + m_s x_2 + m_s x_3 + m_s x_4 + m_s x_5}{m_s + m_s + m_s + m_s + m_s}$$

$$y_{cm} = \frac{m_s y_1 + m_s y_2 + m_s y_3 + m_s y_4 + m_s y_5}{m_s + m_s + m_s + m_s + m_s}$$

- Notes: The denominators both equal $5m_s$. The m_s will cancel out.
- The coordinates of the center of mass are $x_{cm} = 22$ m and $y_{cm} = 30$ m.

---

168. Study the **example** on pages 281-282. This problem is very similar to that example.

- Visualize the complete circle as the sum of the missing piece plus the shape with the hole cut out of it. See the diagram on the following page.
- Write an equation for the center of mass of the large circle.

$$y_{cm} = \frac{m_1 y_1 + m_2 y_2}{m_1 + m_2}$$

- $y_{cm} = 0$ because the large circle is centered about the origin.

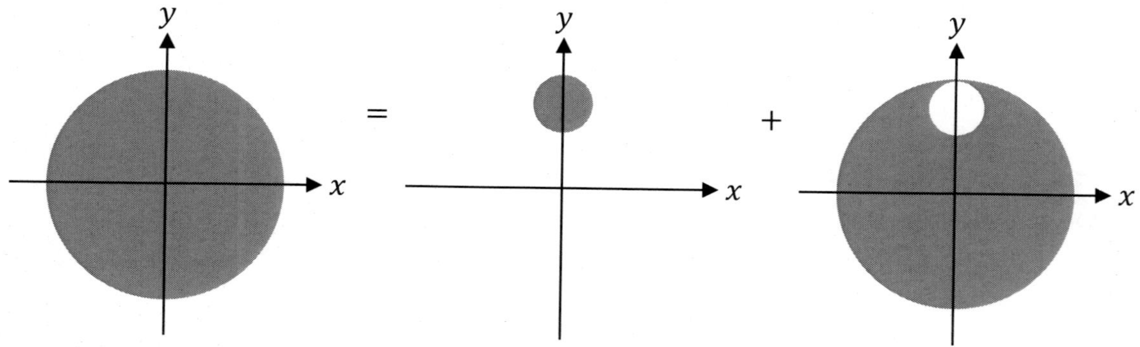

- Multiply both sides by $m_1 + m_2$. Note that $0(m_1 + m_2) = 0$.
$$0 = m_1 y_1 + m_2 y_2$$
- $y_1 = 6.0$ m because the center of the small circle lies 2.0 m from the edge of the large circle: $y_1 = R_L - R_S = 8 - 2 = 6.0$ m. (Notation: L is "large," S is "small.")
- The mass of each object is proportional to its area:
$$m_1 = \sigma \pi R_S^2$$
$$m_L = \sigma \pi R_L^2$$
$$m_2 = m_L - m_1$$
- The masses are $m_1 = 4\sigma\pi$, $m_L = 64\sigma\pi$, and $m_2 = 60\sigma\pi$.
- Plug all of these values into the equation above involving y_1 and y_2.
- σ and π will cancel out.
- The y-coordinate of the center of mass of the given shape is $y_2 = -\frac{2}{5}$ m.

---

169. Perform the center of mass integral.

- Integrate $y_{cm} = \frac{1}{m} \int y \, dm$. Note that $x_{cm} = 0$ since the entire rod lies on the y-axis.
- For a rod, write $dm = \lambda ds$. Don't pull λ out of the integral. Substitute $\lambda = \beta y^{1/2}$.
- Note that $\sqrt{y} = y^{1/2}$ and $yy^{1/2} = y^{3/2}$.
- For a line, work with Cartesian coordinates. The differential arc length is $ds = dy$.
- The limits of integration are from $y = 0$ to $y = 5$ m.
- After these substitutions, you should have:
$$y_{cm} = \frac{\beta}{m} \int_{y=0}^{5} y^{3/2} \, dx$$
- The anti-derivative is $\frac{2y^{5/2}}{5}$. Evaluate this over the limits.
- Check your intermediate answer: $y_{cm} = \frac{2\beta 5^{3/2}}{m}$. Note that $\frac{5^{5/2}}{5} = 5^{3/2}$.
- Integrate $m = \int dm$ using the same substitutions as before, starting with $dm = \lambda ds$.
- You should get $m = \beta \int_{y=0}^{5} y^{1/2} \, dy$.
- The anti-derivative is $\frac{2y^{3/2}}{3}$. Evaluate this over the limits.

- Check your intermediate answer: $m = \frac{2\beta 5^{3/2}}{3}$.
- Solve for β. You should get $\beta = \frac{3m}{2(5)^{3/2}}$.
- Substitute this expression for β into the previous expression for y_{cm}.
- The final answer is $y_{cm} = 3.0$ m. (This is 3 meters, **not** m for mass.)

---

170. Perform the center of mass integrals: $x_{cm} = \frac{1}{m}\int x\,dm$ and $y_{cm} = \frac{1}{m}\int y\,dm$.

- For a triangle, write $dm = \sigma dA$ and work with Cartesian coordinates. Write the differential area element as $dA = dxdy$.
- Since the triangle is uniform, σ is a constant and may come out of the integral.
- Find the equation of the line for the hypotenuse.
- The line for the hypotenuse has a slope of $-\frac{1}{2}$ and y-intercept of 3.
- The equation for the line is therefore $y = -\frac{x}{2} + 3$.
- Let x vary from 0 to 6. Then y will vary from 0 to $-\frac{x}{2} + 3$.
- After these substitutions, you should have:

$$x_{cm} = \frac{\sigma}{m}\int_{x=0}^{6}\int_{y=0}^{-\frac{x}{2}+3} x\,dy\,dx \quad , \quad y_{cm} = \frac{\sigma}{m}\int_{x=0}^{6}\int_{y=0}^{-\frac{x}{2}+3} y\,dy\,dx$$

- First integrate over y (since the upper limit of y is $-\frac{x}{2} + 3$).
- Factor x out of the y-integration (but not out of the x-integration).

$$x_{cm} = \frac{\sigma}{m}\int_{x=0}^{6} x\left(\int_{y=0}^{-\frac{x}{2}+3} dy\right)dx \quad , \quad y_{cm} = \frac{\sigma}{m}\int_{x=0}^{6}\left(\int_{y=0}^{-\frac{x}{2}+3} y\,dy\right)dx$$

- The anti-derivatives for the y-integration are y and $\frac{y^2}{2}$. Evaluate over the limits.
- Note that $\left[\frac{y^2}{2}\right]_{y=0}^{-\frac{x}{2}+3} = \frac{\left(-\frac{x}{2}+3\right)^2}{2} = \frac{1}{2}\left(\frac{x^2}{4} - 3x + 9\right)$.
- The answers to the definite integrals over y are $-\frac{x}{2} + 3$ and $\frac{1}{2}\left(\frac{x^2}{4} - 3x + 9\right)$.
- Note that $x\left(-\frac{x}{2} + 3\right) = -\frac{x^2}{2} + 3x$.
- You should get $x_{cm} = \frac{\sigma}{m}\int_{x=0}^{6}\left(-\frac{x^2}{2} + 3x\right)dx$ and $y_{cm} = \frac{\sigma}{2m}\int_{x=0}^{6}\left(\frac{x^2}{4} - 3x + 9\right)dx$.
- The anti-derivatives for the x-integration are $-\frac{x^3}{6} + \frac{3x^2}{2}$ and $\frac{x^3}{12} - \frac{3x^2}{2} + 9x$. Evaluate these expressions over the limits.
- Check your intermediate answers: $x_{cm} = \frac{18\sigma}{m}$ and $y_{cm} = \frac{9\sigma}{m}$.
- Integrate $m = \int dm$ using the same substitutions as before, starting with $dm = \sigma dA$.

- You should get:

$$m = \sigma \int_{x=0}^{6} \int_{y=0}^{-\frac{x}{2}+3} dy\, dx$$

- We did the same integral in problem 162. Review the hints to that problem if you need help with this double integral.
- Check your intermediate answer: $m = 9\sigma$.
- Solve for σ. You should get $\sigma = \frac{m}{9}$.
- Substitute this expression for σ into the previous expressions for x_{cm} and y_{cm}.
- The final answers are $x_{cm} = 2.0$ m and $y_{cm} = 1.0$ m. (These are meters, **not** mass.)

---

171. Perform the center of mass integral.

- Integrate $y_{cm} = \frac{1}{m}\int y\, dm$. Note that $x_{cm} = 0$ due to symmetry about the y-axis.
- For a thin arc, write $dm = \lambda ds$ and work with 2D polar coordinates. The differential arc length is $ds = Rd\theta$. Since the arc is uniform, λ is a constant.
- You may pull R out of the integral since it is a very thin ring (not a solid semicircle).
- The limits of integration are from $\theta = 0$ to $\theta = \pi$ **radians**. (Don't use degrees.)
- After these substitutions, you should have:

$$y_{cm} = \frac{\lambda R^2}{m} \int_{\theta=0}^{\pi} \sin\theta\, d\theta$$

- The anti-derivative is $-\cos\theta$. Evaluate this over the limits.
- Check your intermediate answer: $y_{cm} = \frac{2\lambda R^2}{m}$.
- Integrate $m = \int dm$ using the same substitutions as before, starting with $dm = \lambda ds$.
- You should get $m = \lambda R \int_{\theta=0}^{\pi} d\theta$.
- The anti-derivative is θ. Evaluate this over the limits.
- Check your intermediate answer: $m = \pi\lambda R$.
- Solve for λ. You should get $\lambda = \frac{m}{\pi R}$.
- Substitute this expression for λ into the previous expression for y_{cm}.
- The final answer is $y_{cm} = \frac{12}{\pi}$ cm.

---

172. Perform the center of mass integral.

- Integrate $y_{cm} = \frac{1}{m}\int y\, dm$. Note that $x_{cm} = 0$ due to symmetry about the y-axis.
- For a region bounded by a straight line and parabola, write $dm = \sigma dA$ and work with Cartesian coordinates. Write the differential area element as $dA = dxdy$.
- Since the region is uniform, σ is a constant and may come out of the integral.

- Set x^2 equal to 4 to find where the straight line intersects the parabola.
- The answers to $x^2 = 4$ are $x = \pm 2$.
- Let x vary from -2 to 2. Then y will vary from x^2 to 4. These choices map out every point between the parabola and the horizontal line.
- After these substitutions, you should have:

$$y_{cm} = \frac{\sigma}{m} \int_{x=-2}^{2} \int_{y=x^2}^{4} y\, dy\, dx$$

- First integrate over y (since the lower limit of y is x^2).

$$y_{cm} = \frac{\sigma}{m} \int_{x=-2}^{2} \left(\int_{y=x^2}^{4} y\, dy \right) dx$$

- The anti-derivative for the y-integration is $\frac{y^2}{2}$. Evaluate this over the limits.
- The answer to the definite integral over y is $\left[\frac{y^2}{2}\right]_{y=x^2}^{4} = \frac{4^2}{2} - \frac{x^4}{2} = 8 - \frac{x^4}{2}$.
- You should get $y_{cm} = \frac{\sigma}{m} \int_{x=-2}^{2} \left(8 - \frac{x^4}{2}\right) dx$.
- The anti-derivative for the x-integration is $8x - \frac{x^5}{10}$. Evaluate this over the limits.
- Check your intermediate answer: $y_{cm} = \frac{128\sigma}{5m}$. (If you plug in $\sigma = 2$, it's $\frac{256}{5m}$.)
- Integrate $m = \int dm$ using the same substitutions as before, starting with $dm = \sigma dA$.
- You should get:

$$m = \sigma \int_{x=-2}^{2} \int_{y=x^2}^{4} dy\, dx$$

- Check your intermediate answer: $m = \frac{32\sigma}{3}$. (If you plug in $\sigma = 2$, it's $\frac{64}{3}$.)
- Substitute the expression for mass into the previous expression for y_{cm}.
- The final answer is $y_{cm} = \frac{12}{5}$ m. (This is a meter, **not** m for mass.)

---

173. Perform the center of mass integral.

- Integrate $x_{cm} = \frac{1}{m} \int x\, dm$. Note that $y_{cm} = 0$ due to symmetry about the x-axis.
- For a thick circular ring, write $dm = \sigma dA$ and work with 2D polar coordinates. The differential area element is $dA = r dr d\theta$.
- Don't pull σ out of the integral. Substitute $\sigma = \beta r$.
- The limits of integration are from $r = 7$ cm to $r = 14$ cm and $\theta = -\frac{\pi}{2}$ to $\theta = \frac{\pi}{2}$ **radians**. (Don't use degrees.)
- After these substitutions, you should have:

$$x_{cm} = \frac{\beta}{m} \int_{r=7}^{14} \int_{\theta=-\frac{\pi}{2}}^{\frac{\pi}{2}} x\, r^2 dr\, d\theta$$

- (Note that one r came from $dA = r dr d\theta$ and the other r came from $\sigma = \beta r$.)
- Use the equation $x = r\cos\theta$ for 2D polar coordinates (from Chapter 26).

$$x_{cm} = \frac{\beta}{m} \int_{r=7}^{14} \int_{\theta=-\frac{\pi}{2}}^{\frac{\pi}{2}} r^3 \cos\theta\, dr\, d\theta$$

- You may do these integrals in any order since the limits are constant. We will integrate over θ first. Factor r out of the θ-integration (but not the r-integration).

$$x_{cm} = \frac{\beta}{m} \int_{r=7}^{14} r^3 \left(\int_{\theta=-\frac{\pi}{2}}^{\frac{\pi}{2}} \cos\theta\, d\theta \right) dr$$

- The anti-derivative for the θ-integration is $\sin\theta$. Evaluate this over the limits.
- The answer to the definite integral over θ is 2.
- You should get $x_{cm} = \frac{\beta}{m} \int_{r=7}^{14} 2r^3\, dr$.
- The anti-derivative for the r-integration is $\frac{r^4}{2}$. Evaluate this over the limits.
- Check your intermediate answer: $x_{cm} = \frac{36015\beta}{2m}$. (The number in the top is 36,015.)
- Integrate $m = \int dm$ using the same substitutions as before, starting with $dm = \sigma dA$.
- You should get:

$$m = \beta \int_{r=7}^{14} \int_{\theta=-\frac{\pi}{2}}^{\frac{\pi}{2}} r^2\, dr\, d\theta$$

- (Note that one r came from $dA = r dr d\theta$ and the other r came from $\sigma = \beta r$.)
- Check your intermediate answer: $m = \frac{2401\pi\beta}{3}$.
- Solve for β. You should get $\beta = \frac{3m}{2401\pi}$.
- Substitute this expression for β into the previous expression for y_{cm}.
- The final answer is $y_{cm} = \frac{45}{2\pi}$ cm.

Chapter 28: Uniform Angular Acceleration

174. All three knowns are given as numbers. The units should help identify the knowns.

- The three knowns are $\omega_0 = 8.0$ rev/s, $\alpha = 16.0$ rev/s^2, and $\Delta\theta = 6$ rev.

(A) Use the equation $\Delta\theta = \omega_0 t + \frac{1}{2}\alpha t^2$.

- After plugging in numbers and simplifying, you should get $8t^2 + 8t - 6 = 0$.
- The constants are $a = 8$, $b = 8$, and $c = -6$. (It would also be okay to use $a = -8$, $b = -8$, and $c = 6$, but don't mix and match. You can also divide these by 2.)
- The two solutions to the **quadratic** are $\frac{1}{2}$ s and $-\frac{3}{2}$ s. The correct time is $t = \frac{1}{2}$ s.

(B) Use the equation $\omega = \omega_0 + \alpha t$.

- The final angular speed is 16 rev/s.

---

175. All three knowns are given as numbers. The units should help identify the knowns.

- The three knowns are $\omega_0 = \frac{1}{5}$ rev/s, $\alpha = \frac{1}{20}$ rev/s^2, and $t = 1$ min.
- Convert the time to seconds: $t = 60$ s.

(A) To find the number of **revolutions** completed, solve for $\Delta\theta$.

- Use the equation $\Delta\theta = \omega_0 t + \frac{1}{2}\alpha t^2$.
- The number of revolutions completed is $\Delta\theta = 102$ rev.

(B) First find the final angular speed. Use the equation $\omega = \omega_0 + \alpha t$.

- The final angular speed is $\frac{16}{5}$ rev/s (or 3.2 rev/s). Convert to $\frac{32\pi}{5}$ rad/s.
- Use the equation $v = R\omega$. (Be sure to use **rad**/s, and not rev/s.)
- The final speed is $v = 96\pi$ m/s (or 302 m/s).

Chapter 29: The Vector Product

176. Identify the components of the given vectors.

- The components are: $A_x = 5, A_y = 2, A_z = 4, B_x = 3, B_y = 4$, and $B_z = 6$.
- Use the determinant form of the vector product.

$$\vec{\mathbf{A}} \times \vec{\mathbf{B}} = \begin{vmatrix} \hat{\mathbf{x}} & \hat{\mathbf{y}} & \hat{\mathbf{z}} \\ A_x & A_y & A_z \\ B_x & B_y & B_z \end{vmatrix} = \hat{\mathbf{x}} \begin{vmatrix} A_y & A_z \\ B_y & B_z \end{vmatrix} - \hat{\mathbf{y}} \begin{vmatrix} A_x & A_z \\ B_x & B_z \end{vmatrix} + \hat{\mathbf{z}} \begin{vmatrix} A_x & A_y \\ B_x & B_y \end{vmatrix}$$

- The 2×2 determinants work out to $12 - 16 = -4$, $30 - 12 = 18$, and $20 - 6 = 14$.
- Note the minus sign in the middle term. You will get $-18\,\hat{\mathbf{y}}$.
- The vector product is $\vec{\mathbf{A}} \times \vec{\mathbf{B}} = -4\,\hat{\mathbf{x}} - 18\,\hat{\mathbf{y}} + 14\,\hat{\mathbf{z}}$.

---

177. Identify the components of the given vectors.
- The components are: $A_x = 3, A_y = 0, A_z = -2, B_x = 8, B_y = -1$, and $B_z = -4$.
- (Observe that $\vec{\mathbf{A}}$ doesn't have a $\hat{\mathbf{y}}$: That's why $A_y = 0$.)
- Use the determinant form of the vector product.

$$\vec{\mathbf{A}} \times \vec{\mathbf{B}} = \begin{vmatrix} \hat{\mathbf{x}} & \hat{\mathbf{y}} & \hat{\mathbf{z}} \\ A_x & A_y & A_z \\ B_x & B_y & B_z \end{vmatrix} = \hat{\mathbf{x}} \begin{vmatrix} A_y & A_z \\ B_y & B_z \end{vmatrix} - \hat{\mathbf{y}} \begin{vmatrix} A_x & A_z \\ B_x & B_z \end{vmatrix} + \hat{\mathbf{z}} \begin{vmatrix} A_x & A_y \\ B_x & B_y \end{vmatrix}$$

- The 2×2 determinants work out to $0 - 2 = -2, -12 + 16 = 4$, and $-3 - 0 = -3$.
- Note the minus sign in the middle term. You will get $-4\,\hat{\mathbf{y}}$.
- The vector product is $\vec{\mathbf{A}} \times \vec{\mathbf{B}} = -2\,\hat{\mathbf{x}} - 4\,\hat{\mathbf{y}} - 3\,\hat{\mathbf{z}}$.

---

178. Identify the components of the given vectors.
- The components are: $A_x = 0, A_y = -1, A_z = 1, B_x = 1, B_y = 0$, and $B_z = -1$.
- (Observe that $\vec{\mathbf{A}}$ doesn't have an $\hat{\mathbf{x}}$ and $\vec{\mathbf{B}}$ doesn't have a $\hat{\mathbf{y}}$.)
- Use the determinant form of the vector product.

$$\vec{\mathbf{A}} \times \vec{\mathbf{B}} = \begin{vmatrix} \hat{\mathbf{x}} & \hat{\mathbf{y}} & \hat{\mathbf{z}} \\ A_x & A_y & A_z \\ B_x & B_y & B_z \end{vmatrix} = \hat{\mathbf{x}} \begin{vmatrix} A_y & A_z \\ B_y & B_z \end{vmatrix} - \hat{\mathbf{y}} \begin{vmatrix} A_x & A_z \\ B_x & B_z \end{vmatrix} + \hat{\mathbf{z}} \begin{vmatrix} A_x & A_y \\ B_x & B_y \end{vmatrix}$$

- The 2×2 determinants work out to $1 - 0 = 1, 0 - 1 = -1$, and $0 + 1 = 1$.
- Note the minus sign in the middle term. You will get $+\,\hat{\mathbf{y}}$.
- The vector product is $\vec{\mathbf{A}} \times \vec{\mathbf{B}} = \hat{\mathbf{x}} + \hat{\mathbf{y}} + \hat{\mathbf{z}}$.

---

179. Use the equation $C = \|\vec{\mathbf{A}} \times \vec{\mathbf{B}}\| = AB \sin\theta$.
- Note that A, B, and θ are all given directly in the problem.
- The answer is $C = 6$.

Chapter 30: Torque

180. Identify the known quantities.

(A) $\vec{\mathbf{r}}_1$ extends from the fulcrum to the white box.
- $r_1 = L - 4 = 20 - 4 = 16.0$ m and $m_1 = 30$ kg.
- The force is $F_1 = m_1 g$.
- The force comes out to $F_1 \approx 300$ N. (It's 294 N if you don't round gravity.)
- $\theta_1 = 90°$ since $\vec{\mathbf{r}}_1$ is horizontal and $\vec{\mathbf{F}}_1$ is vertical.
- Use the torque equation: $\tau_1 = r_1 F_1 \sin\theta_1$.
- The torque is $\tau_1 \approx 4800$ Nm. (It's 4.71×10^3 Nm if you don't round gravity.)

(B) $\vec{\mathbf{r}}_2$ extends from the fulcrum to the center of the plank (where gravity acts on **average**).

- $r_2 = \frac{L}{2} - 4 = \frac{20}{2} - 4 = 10 - 4 = 6.0$ m and $m_2 = 60$ kg.
- The force is $F_2 = m_2 g$.
- The force comes out to $F_2 \approx 600$ N. (It's 589 N if you don't round gravity.)
- $\theta_2 = 90°$ since $\vec{\mathbf{r}}_2$ is horizontal and $\vec{\mathbf{F}}_2$ is vertical.
- Use the torque equation: $\tau_2 = r_2 F_2 \sin\theta_2$.
- The torque is $\tau_2 \approx 3600$ Nm. (It's 3.53×10^3 Nm if you don't round gravity.)

---

181. $\vec{\mathbf{r}}$ extends from the hinge to the center of the rod (where gravity acts on **average**).

- $r = \frac{L}{2} = \frac{6}{2} = 3.0$ m and $m = 7.0$ kg.
- The force is $F = mg$.
- The force comes out to $F \approx 70$ N. (It's 69 N if you don't round gravity.)
- $\theta = 30°$. θ is the angle between $\vec{\mathbf{r}}$ (along the rod) and $\vec{\mathbf{F}}$ (straight down).
- Use the torque equation: $\tau = rF\sin\theta$.
- The torque is $\tau \approx 105$ Nm. (It's 103 Nm if you don't round gravity.)

---

182. Identify the known quantities.

(A) $\vec{\mathbf{r}}_1$ extends from the fulcrum to the hanging monkey.

- $r_1 = \frac{L}{2} = \frac{12}{2} = 6.0$ m and $m_1 = 60$ kg.
- The force is $F_1 = m_1 g$.
- The force comes out to $F_1 \approx 600$ N. (It's 589 N if you don't round gravity.)
- $\theta_1 = 120°$ since $\vec{\mathbf{r}}_1$ is along the plank (up to the right) and $\vec{\mathbf{F}}_1$ is straight down.
- Use the torque equation: $\tau_1 = r_1 F_1 \sin\theta_1$.
- The torque is $\tau_1 \approx 1800\sqrt{3}$ Nm. (It's 3.06×10^3 Nm if you don't round gravity.)

(B) $\vec{\mathbf{r}}_2$ extends from the fulcrum to the standing monkey.

- $r_2 = \frac{L}{2} - 4 = \frac{12}{2} - 4 = 6 - 4 = 2.0$ m and $m_2 = 80$ kg.
- The force is $F_2 = m_2 g$.
- The force comes out to $F_2 \approx 800$ N. (It's 785 N if you don't round gravity.)
- $\theta_2 = 60°$ since $\vec{\mathbf{r}}_2$ is along the plank (down to the left) and $\vec{\mathbf{F}}_2$ is straight down.
- Use the torque equation: $\tau_2 = -r_2 F_2 \sin\theta_2$.
- The torque is $\tau_2 \approx -800\sqrt{3}$ Nm. (It's -1.36×10^3 Nm if you don't round gravity.)
- The minus **sign** represents that τ_2 is **counterclockwise**, whereas τ_1 is clockwise.

---

183. Identify the known quantities.

(A) $\vec{\mathbf{r}}_1$ extends from the hinged edge of the door to the doorknob.

- $r_1 = 45$ cm $= 0.45$ m $= \frac{45}{100}$ m $= \frac{9}{20}$ m and $F_1 = 90$ N.

- $\theta_1 = 0°$ since $\vec{\mathbf{r}}_1$ and $\vec{\mathbf{F}}_1$ are both directed away from the hinges.
- Use the torque equation: $\tau_1 = r_1 F_1 \sin\theta_1$.
- The torque is $\tau_1 = 0$ (since $\sin 0° = 0$).

(B) $\vec{\mathbf{r}}_2$ extends from the hinged edge of the door to the doorknob.

- $r_2 = 45\text{ cm} = 0.45\text{ m} = \frac{45}{100}\text{ m} = \frac{9}{20}\text{ m}$ and $F_2 = 60$ N.
- $\theta_2 = 90°$ since $\vec{\mathbf{r}}_2$ is along the width of the door and $\vec{\mathbf{F}}_2$ is perpendicular to the door.
- Use the torque equation: $\tau_2 = r_2 F_2 \sin\theta_2$.
- The torque is $\tau_2 = 27$ Nm.

(C) $\vec{\mathbf{r}}_3$ extends from the hinged edge of the door to the geometric **center** of the door.

- $r_3 = \frac{W}{2} = \frac{50}{2}\text{ cm} = 25\text{ cm} = \frac{1}{4}\text{ m}$ and $F_3 = 80$ N.
- $\theta_3 = 60°$ or $120°$. Since $\vec{\mathbf{F}}_3$ makes an angle of 30° with the normal, it makes either an angle of 60° or 120° with the plane of the door (and $\vec{\mathbf{r}}_3$ lies in the plane of the door). It won't matter whether it's 60° or 120° since $\sin 60° = \sin 120°$.
- Use the torque equation: $\tau_3 = r_3 F_3 \sin\theta_3$.
- The torque is $\tau_3 = 10\sqrt{3}$ Nm.

---

184. Find the vector product between $\vec{\mathbf{r}}$ and $\vec{\mathbf{F}}$.

- Identify the components of the given vectors.
- The components are: $x = 6, y = 1, z = -5, F_x = 3, F_y = 2$, and $F_z = 4$.
- Use the determinant form of the vector product.

$$\vec{\mathbf{r}} \times \vec{\mathbf{F}} = \begin{vmatrix} \hat{\mathbf{x}} & \hat{\mathbf{y}} & \hat{\mathbf{z}} \\ x & y & z \\ F_x & F_y & F_z \end{vmatrix} = \hat{\mathbf{x}} \begin{vmatrix} y & z \\ F_y & F_z \end{vmatrix} - \hat{\mathbf{y}} \begin{vmatrix} x & z \\ F_x & F_z \end{vmatrix} + \hat{\mathbf{z}} \begin{vmatrix} x & y \\ F_x & F_y \end{vmatrix}$$

- The 2×2 determinants work out to $4 + 10 = 14$, $24 + 15 = 39$, and $12 - 3 = 9$.
- Note the minus sign in the middle term. You will get $-39\,\hat{\mathbf{y}}$.
- The vector product is $\vec{\mathbf{r}} \times \vec{\mathbf{F}} = 14\,\hat{\mathbf{x}} - 39\,\hat{\mathbf{y}} + 9\,\hat{\mathbf{z}}$.

Chapter 31: Static Equilibrium

185. Draw an extended FBD showing where each force acts on the plank.

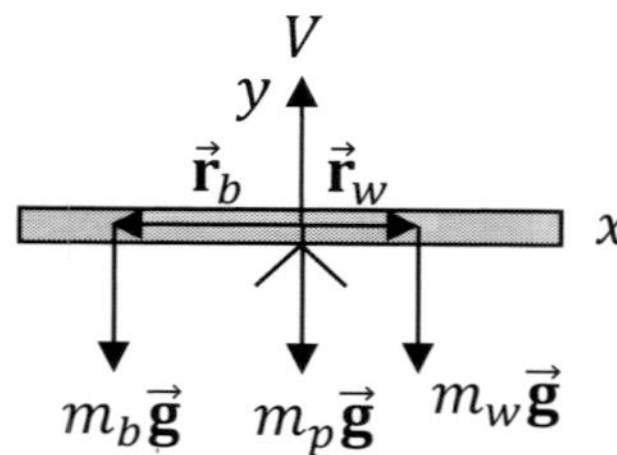

- The distance is $r_b = 5.0$ m. The angles are $\theta_b = 90°$ and $\theta_w = 90°$.
- The torques are $\tau_b = -r_b m_b g \sin\theta_b$ and $\tau_w = r_w m_w g \sin\theta_w$.
- Set the net torque equal to zero: $\sum \tau = 0$.
- Simplify. You should get $r_b m_b = r_w m_w$. Solve for r_w.
- The answer is $r_w = 4.0$ m.

---

186. Draw an extended FBD showing where each force acts on the plank.

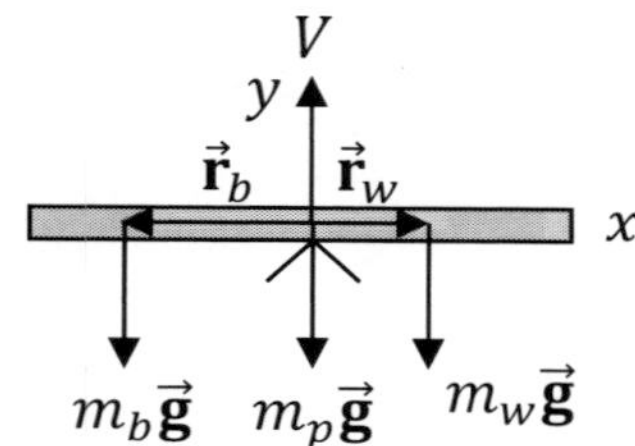

- The distances are $r_b = 5.0$ m and $r_w = 2.0$ m. The angles are $\theta_b = 90°$ and $\theta_w = 90°$.
- The torques are $\tau_b = -r_b m_b g \sin\theta_b$ and $\tau_w = r_w m_w g \sin\theta_w$.
- Set the net torque equal to zero: $\sum \tau = 0$.
- Simplify. You should get $r_b m_b = r_w m_w$. Solve for m_b.
- The answer is $m_b = 8.0$ kg.

---

187. Draw an extended FBD showing where each force acts on the plank.

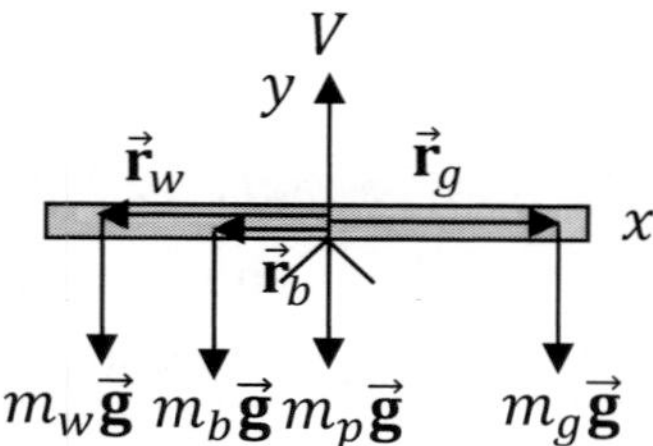

- The distances are $r_b = 4.0$ m and $r_w = 16.0$ m. The angles are all 90°.
- The torques are $\tau_b = -r_b m_b g \sin\theta_b$, $\tau_w = -r_w m_w g \sin\theta_w$, and $\tau_g = r_g m_g g \sin\theta_g$.
- Set the net torque equal to zero: $\sum \tau = 0$.

- Simplify. You should get $r_b m_b + r_w m_w = r_g m_g$. Solve for r_g.
- The answer is $r_g = 20.0$ m.

188. Draw an extended FBD showing where each force acts on the plank.

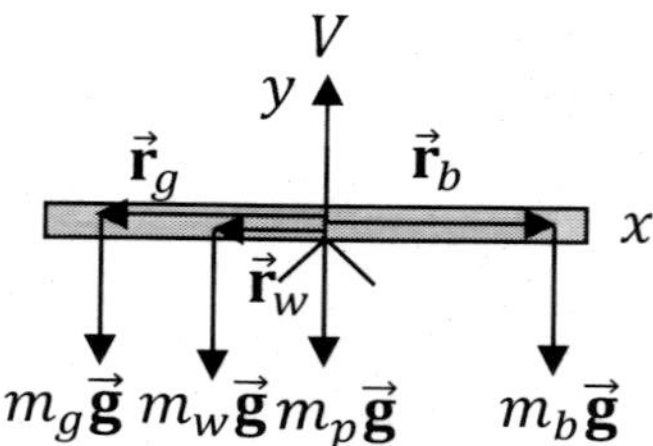

- The distances are $r_b = 7.5$ m and $r_w = 2.5$ m. The angles are all 90°.
- The torques are $\tau_b = r_b m_b g \sin\theta_b$, $\tau_w = -r_w m_w g \sin\theta_w$, and $\tau_g = -r_g m_g g \sin\theta_g$.
- Set the net torque equal to zero: $\sum \tau = 0$.
- Simplify. You should get $r_w m_w + r_g m_g = r_b m_b$. Solve for r_g.
- The answer is $r_g = 9.0$ m.

189. Draw a FBD for the **knot** (see the similar **example** on page 324).

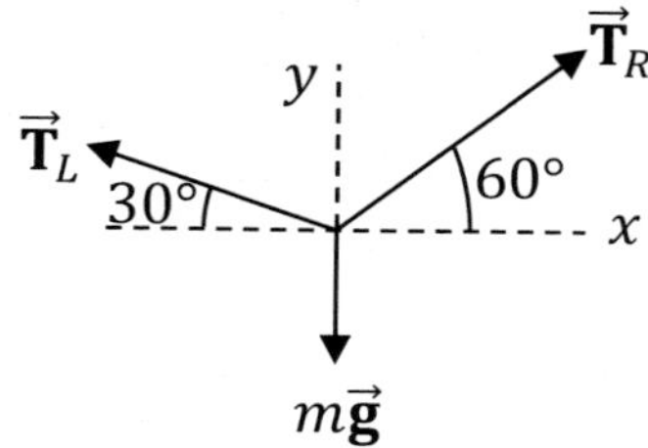

- Unlike the example on page 324, the tensions are **different** because the angles differ.
- Set the sums of the components of the forces equal to zero: $\sum F_x = 0$ and $\sum F_y = 0$.
- You should get $T_R \cos 60° - T_L \cos 30° = 0$ and $T_R \sin 60° + T_L \sin 30° - mg = 0$.
- Solve for T_R in terms of T_L in the x-sum.
- You should get $T_R = T_L\sqrt{3}$.
- Plug this expression in for T_R in the y-sum. Note that $\sqrt{3}\sqrt{3} = 3$.
- Solve for T_L. The answer is $T_L \approx 200$ N. (It's 196 N if you don't round gravity.)
- Use this answer to find T_R. The answer is $T_R \approx 200\sqrt{3}$ N (or 340 N).
- The bottom tension equals the weight of the monkey: $T_B = mg$.
- The bottom tension is $T_B \approx 400$ N. (It's 392 N if you don't round gravity.)

190. First study the **example** on pages 325-326. Draw an extended FBD for the boom.
- There is tension ($\vec{\mathbf{T}}$) in the tie rope. The weight ($m_L\vec{\mathbf{g}}$) of the load acts at the end of the boom. The weight of the boom ($m_B\vec{\mathbf{g}}$) acts on **average** at the **center** of the boom. The hingepin force has horizontal (H) and vertical (V) components.

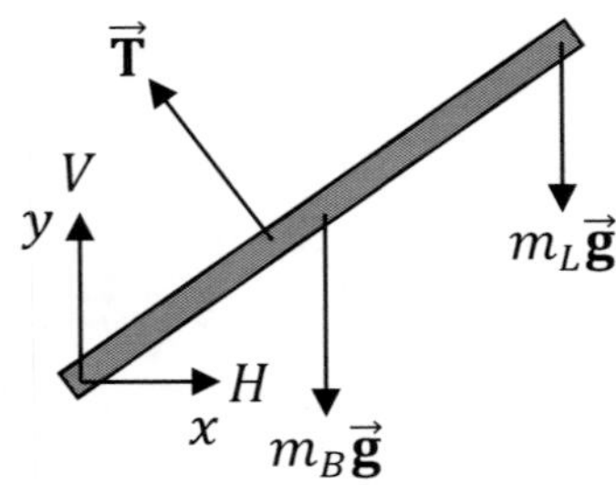

- Note that the tension is 1.0 m closer to the hinge than the center of the boom is. That is, r_T is less than r_B.
- The needed distances are $r_T = 4.0$ m, $r_B = \frac{L}{2} = \frac{10}{2} = 5.0$ m, and $r_L = L = 10.0$ m.
- The angles are $\theta_T = 90°$, $\theta_B = 150°$, and $\theta_L = 150°$.
- Note: $\theta_B = 60° + 90° = 150°$, for example, since the weight of the boom is straight down and $\vec{\mathbf{r}}_B$ is along the boom (up to the right, 60° above the horizontal).
- Sum the components of the forces and the torques: $\sum F_x = 0$, $\sum F_y = 0$, and $\sum \tau = 0$.

$$H - T\cos 30° = 0$$
$$V + T\sin 30° - m_B g - m_L g = 0$$
$$r_B m_B g \sin 150° + r_L m_L g \sin 150° - r_T T \sin 90° = 0$$

- Note: $\vec{\mathbf{T}}$ is 30° above the x-axis, such that $T\cos 30°$ and $T\sin 30°$ appear in the sums of the components of the forces. In the sum of the torques, $\theta_T = 90°$ because the tie rope is perpendicular to the boom.
- Solve for m_L in the equation above (from the torque sum). Use the maximum tension to find the maximum load.
- The maximum load is $m_L = 23$ kg.
- Solve for H and V in the equations above.
- The horizontal and vertical components of the hingepin force are $H = 300\sqrt{3}$ N (or 520 N) and $V \approx 430$ N (or 425 N if you don't round gravity).

Chapter 32: Moment of Inertia

191. What are the R's?

(A) Each R is the distance from the mass to the x-axis.

- The distances are $R_1 = 0$, $R_2 = 2\sqrt{3}$ m, and $R_3 = 0$.
- To see that $R_2 = 2\sqrt{3}$ m, divide the given triangle into two right triangles. The given triangle is an equilateral triangle, so each right triangle has a base of 2.0 m and a hypotenuse of 4.0 m. Use the Pythagorean theorem to solve for the height (R_2).
- Use the formula $I = m_1 R_1^2 + m_2 R_2^2 + m_3 R_3^2$.
- The moment of inertia about the x-axis is $I = 72$ kg·m^2.

(B) Each R is the distance from the mass to the y-axis.

- The distances are $R_1 = 0$, $R_2 = \frac{L}{2} = \frac{4}{2} = 2.0$ m, and $R_3 = L = 4.0$ m.
- Note that mass 2 is half an edge from the y-axis.
- Use the formula $I = m_1 R_1^2 + m_2 R_2^2 + m_3 R_3^2$.
- The moment of inertia about the y-axis is $I = 120$ kg·m^2.

(C) Each R is the distance from the mass to the origin.

- The distances are $R_1 = 0$, $R_2 = L = 4.0$ m, and $R_3 = L = 4.0$ m.
- Note that masses 2 and 3 are both exactly one edge from the origin.
- Use the formula $I = m_1 R_1^2 + m_2 R_2^2 + m_3 R_3^2$.
- The moment of inertia about the z-axis is $I = 192$ kg·m^2.

---

192. What are the masses and what are the R's?

- Convert the masses from grams to kilograms since the answers are given in kg·m^2.
- The masses are $m_1 = 0.4 \text{ kg} = \frac{2}{5}$ kg, $m_2 = 0.3 \text{ kg} = \frac{3}{10}$ kg, and $m_3 = 0.5 \text{ kg} = \frac{1}{2}$ kg.

(A) Each R is the distance from the mass to the x-axis.

- The distances are $R_1 = 0$, $R_2 = 2.0$ m, and $R_3 = 1.0$ m. These are the y-coordinates. Note that the y-coordinate tells you the distance to the x-axis.
- Use the formula $I = m_1 R_1^2 + m_2 R_2^2 + m_3 R_3^2$.
- The moment of inertia about the x-axis is $I = \frac{17}{10}$ kg·m^2.

(B) Each R is the distance from the mass to the y-axis.

- The distances are $R_1 = 3.0$ m, $R_2 = 0$, and $R_3 = 4.0$ m. These are the x-coordinates in absolute values. Note that the x-coordinate tells you the distance to the y-axis.
- Use the formula $I = m_1 R_1^2 + m_2 R_2^2 + m_3 R_3^2$.
- The moment of inertia about the y-axis is $I = \frac{58}{5}$ kg·m^2.

(C) Each R is the distance from the mass to the origin.

- Use the Pythagorean theorem to find the distance from each mass to the origin: $R = \sqrt{x^2 + y^2}$, where x and y are the given coordinates (x, y) for each point.
- The distances are $R_1 = 3.0$ m, $R_2 = 2.0$ m, and $R_3 = \sqrt{17}$ m.
- Use the formula $I = m_1 R_1^2 + m_2 R_2^2 + m_3 R_3^2$.
- The moment of inertia about the z-axis is $I = \frac{133}{10}$ kg·m^2.

---

193. Study the **example** at the bottom of page 339.

- First find the moment of inertia of each piece about an axis that passes through its center, and then apply the parallel-axis theorem.
- Look up the formulas for the moment of inertia of a rod and of a hollow sphere rotating about an axis passing through its center.

- These formulas are $I_{rod}^{cm} = \frac{1}{12} m_r L^2$ and $I_{hs}^{cm} = \frac{2}{3} m_{hs} R^2$.
- Plug the given numbers into these formulas.
- You should get $I_{middle}^{cm} = 32$ kg·m², $I_{left}^{cm} = 24$ kg·m², and $I_{right}^{cm} = 2$ kg·m².
- Since the hollow spheres aren't really rotating about axes through their centers, we must apply the parallel-axis theorem.

$$I_{left} = I_{left}^{cm} + m_L R_L^2 \quad , \quad I_{right} = I_{right}^{cm} + m_R R_R^2$$

- These answers are $I_{left} = 348$ kg·m² and $I_{right} = 77$ kg·m².
- Add the three moments of inertia together:

$$I = I_{middle}^{cm} + I_{left} + I_{right}$$

- The moment of inertia of the object is $I = 457$ kg·m².

---

194. Perform the moment of inertia integral: $I = \int r_\perp^2 \, dm$.

- For a rod, write $dm = \lambda ds$. Don't pull λ out of the integral. Substitute $\lambda = \beta y^{1/2}$.
- For a line, work with Cartesian coordinates. The differential arc length is $ds = dy$.
- The limits of integration are from $y = 0$ to $y = L$.
- Draw a representative dm. Label $r_\perp$ for that dm (where $r_\perp$ is the shortest distance from dm to the axis of rotation). For this problem, $r_\perp = y$.
- Note that $\sqrt{y} = y^{1/2}$ and $y^2 y^{1/2} = y^{5/2}$.
- After these substitutions, you should have:

$$I = \beta \int_{y=0}^{L} y^{5/2} \, dy$$

- Check your intermediate answer: $I = \frac{2\beta L^{7/2}}{7}$.
- Integrate $m = \int dm$ using the same substitutions as before, starting with $dm = \lambda ds$.
- You should get $m = \beta \int_{y=0}^{L} y^{1/2} \, dy$.
- Check your intermediate answer: $m = \frac{2\beta L^{3/2}}{3}$.
- Solve for β. You should get $\beta = \frac{3m}{2L^{3/2}}$.
- Substitute this expression for β into the previous expression for I.
- The final answer is $I = \frac{3}{7} m L^2$. (Note that $\frac{L^{7/2}}{L^{3/2}} = L^2$.)

---

195. Perform the moment of inertia integral: $I = \int r_\perp^2 \, dm$.

- For a triangle, write $dm = \sigma dA$ and work with Cartesian coordinates. Write the differential area element as $dA = dxdy$.
- Since the triangle is uniform, σ is a constant and may come out of the integral.
- Find the equation of the line for the hypotenuse.

- The line for the hypotenuse has a slope of $\frac{h}{b}$ and y-intercept of 0.
- The equation for the line is therefore $y = \frac{hx}{b}$.
- Let x vary from 0 to b. Then y will vary from 0 to $\frac{hx}{b}$.
- Draw a representative dm. Label $r_\perp$ for that dm (where $r_\perp$ is the shortest distance from dm to the axis of rotation). For this problem, $r_\perp = y$.
- After these substitutions, you should have:

$$I = \sigma \int_{x=0}^{b} \int_{y=0}^{\frac{hx}{b}} y^2 \, dxdy$$

- First integrate over y (since the upper limit of y is $\frac{hx}{b}$).
- The y-integration gives $\frac{h^3x^3}{3b^3}$. Next integrate over x.
- Check your intermediate answer: $I = \frac{\sigma bh^3}{12}$.
- Integrate $m = \int dm$ using the same substitutions as before, starting with $dm = \sigma dA$.
- You should get:

$$m = \sigma \int_{x=0}^{b} \int_{y=0}^{\frac{hx}{b}} dy \, dx$$

- Check your intermediate answer: $m = \frac{\sigma bh}{2}$. (If you need help with this integral, see the example on page 275, which features a very similar integral.)
- Solve for σ. You should get $\sigma = \frac{2m}{bh}$.
- Substitute this expression for σ into the previous expression for I.
- The final answer is $I = \frac{mh^2}{6}$.

---

196. Perform the moment of inertia integral: $I = \int r_\perp^2 \, dm$.

- For a solid disc (unlike a thin ring), write $dm = \sigma dA$ and work with 2D polar coordinates. The differential area element is $dA = rdrd\theta$.
- Since the disc is uniform, σ is a constant and may come out of the integral.
- The limits of integration are from $r = 0$ to $r = R$ and $\theta = 0$ to $\theta = 2\pi$ **radians**. (Don't use degrees.) Note that R is a constant, whereas r is a variable.
- Draw a representative dm. Label $r_\perp$ for that dm (where $r_\perp$ is the shortest distance from dm to the axis of rotation). For this problem, $r_\perp = x$.
- Use the equation $x = r\cos\theta$ for 2D polar coordinates (from Chapter 26).
- Then $r_\perp^2 = x^2 = r^2\cos^2\theta$.
- After these substitutions, you should have:

$$I = \sigma \int_{r=0}^{R} \int_{\theta=0}^{2\pi} r^3 \cos^2\theta \, dr \, d\theta$$

- (Note: One r came from $dA = rdrd\theta$ and the other r's came from $r_\perp^2 = r^2\cos^2\theta$.)
- You may do these integrals in any order since the limits are constant. We will integrate over θ first. Factor r out of the θ-integration (but not the r-integration).

$$I = \sigma \int_{r=0}^{R} r^3 \left(\int_{\theta=0}^{2\pi} \cos^2\theta \, d\theta \right) dr$$

- Apply the trig identity $\cos^2\theta = \frac{1+\cos 2\theta}{2}$.
- After this substitution, separate the integral into two integrals, one for each term.
- You should get $\int_{\theta=0}^{2\pi} \frac{1}{2} d\theta + \int_{\theta=0}^{2\pi} \frac{\cos 2\theta}{2} d\theta$.
- The anti-derivatives are $\frac{\theta}{2}$ and $\frac{\sin 2\theta}{4}$. Evaluate these over the limits.
- The answer to the definite integral over θ is π.
- You should get $I = \frac{\sigma}{m} \int_{r=0}^{R} \pi r^3 \, dr$.
- The anti-derivative for the r-integration is $\frac{r^4}{2}$. Evaluate this over the limits.
- Check your intermediate answer: $I = \frac{\pi\sigma R^4}{4}$.
- Integrate $m = \int dm$ using the same substitutions as before, starting with $dm = \sigma dA$.
- You should get:

$$m = \sigma \int_{r=0}^{R} \int_{\theta=0}^{2\pi} r \, dr \, d\theta$$

- Check your intermediate answer: $m = \pi\sigma R^2$.
- Solve for σ. You should get $\sigma = \frac{m}{\pi R^2}$.
- Substitute this expression for σ into the previous expression for I.
- The final answer is $I = \frac{mR^2}{4}$. (Note that this is flipping mode, **not** rolling mode.)

---

197. Perform the moment of inertia integral: $I = \int r_\perp^2 \, dm$.
- Study the **example** on pages 348-349.
- For a solid sphere, write $dm = \rho dV$ and work with spherical coordinates. The differential volume element is $dV = r^2 \sin\theta \, drd\theta d\varphi$. (Review Chapter 26.)
- Since the sphere is non-uniform, don't pull ρ out of the integral. Substitute $\rho = \beta r$.
- The limits of integration are from $r = 0$ to R, $\theta = 0$ to π, and $\varphi = 0$ to 2π (as explained in the example on pages 348-349).
- We choose the axis of rotation to be the z-axis, as in the example on pages 348-349.
- Draw a representative dm. Label $r_\perp$ for that dm (where $r_\perp$ is the shortest distance

from dm to the axis of rotation). For this problem, $r_\perp = r\sin\theta$ (the distance from any dm to the z-axis). See the picture on page 348.

- After these substitutions, you should have:

$$I = \beta \int_{r=0}^{R} \int_{\varphi=0}^{2\pi} \int_{\theta=0}^{\pi} r^5 \sin^3\theta \, dr d\theta d\varphi$$

- (Note: Two r's came from $dV = r^2 \sin\theta \, dr d\theta d\varphi$, two r's came from $r_\perp^2 = r^2 \sin^2\theta$, and another r came from $\rho = \beta r$. Similarly, two $\sin\theta$'s came from $r_\perp^2 = r^2 \sin^2\theta$ and another $\sin\theta$ came from $dV = r^2 \sin\theta \, dr d\theta d\varphi$.)
- You may do these integrals in any order since the limits are constant. We will integrate over φ, then θ, and finally r. You may pull r out of the angular integrals (but not the r-integral), and you may pull $\sin\theta$ out of the r and φ integrals (but not the θ-integral).
- You should get:

$$I = \beta \int_{r=0}^{R} r^5 dr \int_{\theta=0}^{\pi} \sin^3\theta \, d\theta \int_{\varphi=0}^{2\pi} d\varphi$$

- The answer to the definite integral over φ is 2π.
- You should get:

$$I = 2\pi\beta \int_{r=0}^{R} r^5 dr \int_{\theta=0}^{\pi} \sin^3\theta \, d\theta$$

- Apply the trig identity $\sin^3\theta = \sin\theta \sin^2\theta = \sin\theta\,(1 - \cos^2\theta)$ like we did in the example on pages 348-349. Integrate both terms. For the second term, make the substitutions $u = \cos\theta$ and $du = -\sin\theta \, d\theta$ (we also did this in the example).
- The answer to the definite integral over θ is $\frac{4}{3}$.
- You should get $I = \frac{8\pi\beta}{3} \int_{r=0}^{R} r^5 dr$.
- The anti-derivative for the r-integration is $\frac{r^6}{6}$. Evaluate this over the limits.
- Check your intermediate answer: $I = \frac{4\pi\beta R^6}{9}$.
- Integrate $m = \int dm$ using the same substitutions as before, starting with $dm = \rho dV$.
- You should get:

$$m = \beta \int_{r=0}^{R} \int_{\varphi=0}^{2\pi} \int_{\theta=0}^{\pi} r^3 \sin\theta \, dr d\varphi d\theta$$

- (Note: Two r's came from $dV = r^2 \sin\theta \, dr d\theta d\varphi$ and another r came from $\rho = \beta r$.)
- Check your intermediate answer: $m = \pi\beta R^4$. This integral is very similar to the mass integral on page 349 (only the power of r is different).
- Solve for β. You should get $\beta = \frac{m}{\pi R^4}$.

- Substitute this expression for β into the previous expression for I.
- The final answer is $I = \frac{4mR^2}{9}$.

---

198. Perform the moment of inertia integral: $I = \int r_\perp^2 \, dm$.

- Study the **example** on pages 348-349.
- For a very thin hollow sphere (unlike a solid sphere or a thick spherical shell), write $dm = \sigma dA$ and work with spherical coordinates. The differential area element is $dA = R^2 \sin\theta \, d\theta d\varphi$. (Review Chapter 26.)
- Unlike the previous problem, here we work with R which is a **constant**, and don't work with r (both of which are different from $r_\perp$), since for a very thin hollow sphere, every point on the sphere is the same distance (R) from the origin.
- Since the sphere is uniform, σ is a constant and may come out of the integral.
- The limits of integration are from $\theta = 0$ to π and $\varphi = 0$ to 2π (as explained in the example on pages 348-349).
- We choose the axis of rotation to be the z-axis, as in the examples on pages 348-349.
- Draw a representative dm. Label $r_\perp$ for that dm (where $r_\perp$ is the shortest distance from dm to the axis of rotation). For this problem, $r_\perp = R\sin\theta$ (the distance from any dm on the surface of the sphere to the z-axis). See the picture on page 348.
- After these substitutions, you should have:

$$I = \sigma R^4 \int_{\theta=0}^{\pi} \int_{\varphi=0}^{2\pi} \sin^3\theta \, d\theta d\varphi$$

- (Note: One $\sin\theta$ came from $dV = R^2 \sin\theta \, dr d\theta d\varphi$ and two more $\sin\theta$'s came from $r_\perp^2 = R^2\sin^2\theta$. Similarly, two R's came from $dV = R^2 \sin\theta \, dr d\theta d\varphi$ and two more R's came from $r_\perp^2 = R^2\sin^2\theta$.)
- You may do these integrals in any order since the limits are constant. We will integrate over φ first. You may pull $\sin\theta$ out of the φ integral.
- You should get:

$$I = \sigma R^4 \int_{\theta=0}^{\pi} \sin^3\theta \, d\theta \int_{\varphi=0}^{2\pi} d\varphi$$

- The answer to the definite integral over φ is 2π.
- You should get:

$$I = 2\pi\sigma R^4 \int_{\theta=0}^{\pi} \sin^3\theta \, d\theta$$

- Apply the trig identity $\sin^3\theta = \sin\theta\sin^2\theta = \sin\theta\,(1-\cos^2\theta)$ like we did in the example on pages 348-349. Integrate both terms. For the second term, make the substitutions $u = \cos\theta$ and $du = -\sin\theta \, d\theta$ (we also did this in the example).

- The answer to the definite integral over θ is $\frac{4}{3}$.
- Check your intermediate answer: $I = \frac{8\pi\sigma R^4}{3}$.
- Integrate $m = \int dm$ using the same substitutions as before, starting with $dm = \sigma dA$.
- You should get:

$$m = \sigma R^2 \int_{\theta=0}^{\pi} \int_{\varphi=0}^{2\pi} \sin\theta \, d\theta d\varphi$$

- Check your intermediate answer: $m = 4\pi\sigma R^2$. Solve for σ. You should get $\sigma = \frac{m}{4\pi R^2}$.
- Substitute this expression for σ into the previous expression for I.
- The final answer is $I = \frac{2mR^2}{3}$.

Chapter 33: A Pulley Rotating without Slipping

199. Check your FBD's. You need two FBD's and one extended FBD for the pulley.

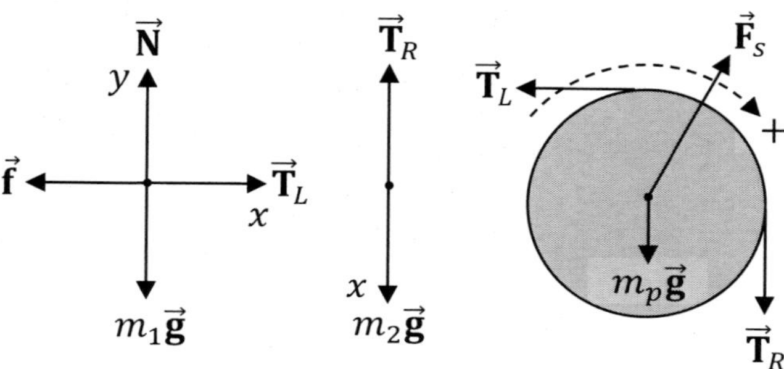

- Study the **example** on pages 357-358. This problem is similar to that example.
- Each object has weight ($m_1\vec{\mathbf{g}}$, $m_2\vec{\mathbf{g}}$, and $m_p\vec{\mathbf{g}}$) pulling straight down.
- A normal force ($\vec{\mathbf{N}}$) supports the box (perpendicular to the surface).
- Friction ($\vec{\mathbf{f}}$) acts opposite to the velocity of the box of bananas.
- There are two pairs of tension ($\vec{\mathbf{T}}_L$ and $\vec{\mathbf{T}}_R$) forces along the cord. As explained in Chapter 33, the tensions must be different in order for the cord to **rotate with the pulley without slipping** (so that the pulley's acceleration will match the cord's).
- Label $+x$ in the direction that each object accelerates. (Think of the pulley as bending the x-axis.) Label $+y$ perpendicular to x.
- Make the positive sense of rotation match the motion of the system: clockwise.
- Write out the sums: $\sum F_{1x} = m_1 a_x$, $\sum F_{1y} = m_1 a_y$, and $\sum F_{2x} = m_2 a_x$.

$$T_L - f = m_1 a_x \quad , \quad N - m_1 g = 0 \quad , \quad m_2 g - T_R = m_2 a_x$$

- Also sum the torques acting on the pulley: $\sum \tau = I\alpha$.

$$R_p T_R \sin 90° - R_p T_L \sin 90° = \frac{1}{2} m_p R_p^2 \alpha$$

- The angles are 90° because tension is tangential to the pulley (perpendicular to the radius). Find the moment of inertia of a solid disc in Chapter 32: $I_p = \frac{1}{2} m_p R_p^2$.
- Make the substitution $R_p \alpha = a_x$ (since the angular acceleration of the pulley must match the acceleration of the masses and cord in order for the cord to rotate with the pulley without slipping). Divide both sides of the torque equation by R_p and simplify. You should get:

$$T_R - T_L = \frac{m_p a_x}{2}$$

- Solve for normal force in the y-sum. Plug in numbers.
- Normal force is $N \approx 400$ N (or 392 N if you don't round g to 10 m/s^2).
- What is the equation for friction?
- Use the equation $f = \mu N$. Plug in numbers.
- Friction force is $f \approx 200$ N (or 196 N if you don't round g to 10 m/s^2.)
- Add the x-equations together in order to cancel the tension forces.

$$m_2 g - f = m_1 a_x + m_2 a_x + \frac{m_p a_x}{2}$$

- Plug in numbers and solve for acceleration.
- The answer is $a_x \approx 3.0$ m/s^2. (If you don't round gravity, $a_x = 2.94$ m/s^2.)

---

200. Check your FBD's. You need two FBD's and one extended FBD for the pulley.

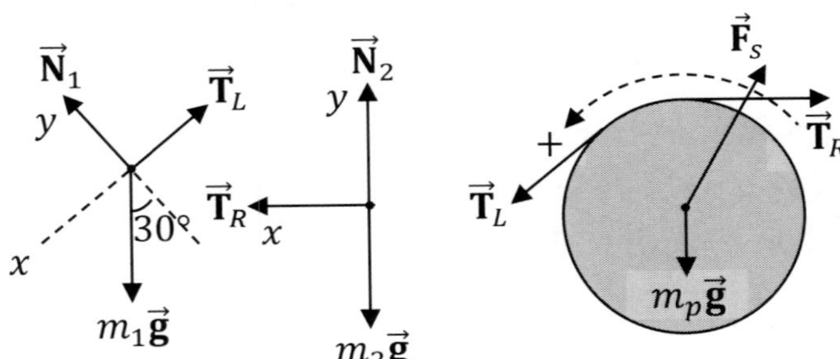

- Study the **example** on pages 357-358. This problem is similar to that example.
- Each object has weight ($m_1\vec{\mathbf{g}}$, $m_2\vec{\mathbf{g}}$, and $m_p\vec{\mathbf{g}}$) pulling straight down.
- Normal forces ($\vec{\mathbf{N}}_1$ and $\vec{\mathbf{N}}_2$) push perpendicular to the surfaces.
- There are two pairs of tension ($\vec{\mathbf{T}}_L$ and $\vec{\mathbf{T}}_R$) forces along the cord. As explained in Chapter 33, the tensions must be different in order for the cord to **rotate with the pulley without slipping** (so that the pulley's acceleration will match the cord's).
- Label $+x$ in the direction that the each monkey accelerates: **down the incline** for one and **to the left** for the other. (Think of the pulley as bending the x-axis.) Label $+y$ perpendicular to x: along each normal.
- Make the positive sense of rotation match the motion: counterclockwise.
- Write out the sums: $\sum F_{1x} = m_1 a_x$, $\sum F_{1y} = m_1 a_y$, $\sum F_{2x} = m_2 a_x$, and $\sum F_{2y} = m_2 a_y$.

$m_1 g \sin 30° - T_L = m_1 a_x$, $N_1 - m_1 g \cos 30° = 0$, $T_R = m_2 a_x$, $N_2 - m_{2g} = 0$

- The example on page 135 explains which sums $m_1 g \sin 30°$ and $m_1 g \cos 30°$ go in.
- Also sum the torques acting on the pulley: $\sum \tau = I\alpha$.

$$R_p T_L \sin 90° - R_p T_R \sin 90° = \frac{1}{2} m_p R_p^2 \alpha$$

- The angles are 90° because tension is tangential to the pulley (perpendicular to the radius). Find the moment of inertia of a solid disc in Chapter 32: $I_p = \frac{1}{2} m_p R_p^2$.
- Make the substitution $R_p \alpha = a_x$ (since the angular acceleration of the pulley must match the acceleration of the masses and cord in order for the cord to rotate with the pulley without slipping). Divide both sides of the torque equation by R_p and simplify. You should get:

$$T_L - T_R = \frac{m_p a_x}{2}$$

- Add the x-equations together in order to cancel the tension forces.

$$m_1 g \sin 30° = m_1 a_x + m_2 a_x + \frac{m_p a_x}{2}$$

- Plug in numbers and solve for acceleration.
- The answer is $a_x = \frac{5}{2}$ m/s^2.

Chapter 34: Rolling without Slipping

201. Put i where the sphere starts and f just before the bottom. Put RH at the bottom.

- Study the **example** on page 365. This problem is similar to that example.
- Apply the law of conservation of energy.
- $PE_{g0} = mgh_0$ since i is **not** at the same height as RH.
- $PE_{s0} = 0$ since at i there isn't a spring compressed or stretched from equilibrium.
- $PE_0 = PE_{g0} + PE_{s0} = mgh_0$.
- $KE_0 = 0$ since the hollow sphere is at **rest** at i.
- $W_{nc} = 0$ even though there is friction because friction doesn't subtract mechanical energy from the system in the case of **rolling without slipping** (see page 364).
- $PE_g = 0$ since f is at the same height as RH.
- $PE_s = 0$ since at f there isn't a spring compressed or stretched from equilibrium.
- $PE = PE_g + PE_s = 0$.
- $KE = \frac{1}{2} mv^2 + \frac{1}{2} I\omega^2$ since the hollow sphere is **rolling** at f. (It has $\frac{1}{2} mv^2$ because its center of mass is moving and it has $\frac{1}{2} I\omega^2$ because it is rotating.)
- Write out conservation of energy: $PE_0 + KE_0 + W_{nc} = PE + KE$.
- Substitute the previous expressions into the conservation of energy equation.

$$mgh_0 = \frac{1}{2}mv^2 + \frac{1}{2}I\omega^2$$

- Plug in the expression for the moment of inertia of a hollow sphere (Chapter 32).
- The moment of inertia of a hollow sphere is $I = \frac{2}{3}mR^2$.
- Use the substitution $R^2\omega^2 = v^2$ to eliminate both R and ω.
- Mass cancels. Solve for the final speed.

$$v = \sqrt{\frac{6gh_0}{5}}$$

- Note that $\frac{v^2}{2} + \frac{v^2}{3} = \left(\frac{1}{2} + \frac{1}{3}\right)v^2 = \left(\frac{3}{6} + \frac{2}{6}\right)v^2 = \frac{5}{6}v^2$.
- Plug numbers into the previous equation for v.
- The final speed is $v = 30$ m/s.

---

202. Put i at the bottom and f when it reaches its highest point. Put RH at the bottom.

- Contrast this problem with the previous problem. This donut rolls up the incline, whereas the sphere in the previous problem rolled down an incline.
- Apply the law of conservation of energy.
- $PE_{g0} = 0$ since i is at the same height as RH.
- $PE_{s0} = 0$ since at i there isn't a spring compressed or stretched from equilibrium.
- $PE_0 = PE_{g0} + PE_{s0} = 0$.
- $KE_0 = \frac{1}{2}mv_0^2 + \frac{1}{2}I\omega_0^2$ since the donut is **rolling** at i. (It has $\frac{1}{2}mv_0^2$ because its center of mass is moving and it has $\frac{1}{2}I\omega_0^2$ because it is rotating.).
- $W_{nc} = 0$ even though there is friction because friction doesn't subtract mechanical energy from the system in the case of **rolling without slipping** (see page 365).
- $PE_g = mgh$ since f is **not** at the same height as RH.
- $PE_s = 0$ since at f there isn't a spring compressed or stretched from equilibrium.
- $PE = PE_g + PE_s = mgh$.
- $KE = 0$ since the donut runs out of speed at f (otherwise it would rise higher).
- Write out conservation of energy: $PE_0 + KE_0 + W_{nc} = PE + KE$.
- Substitute the previous expressions into the conservation of energy equation.

$$\frac{1}{2}mv_0^2 + \frac{1}{2}I\omega_0^2 = mgh$$

- Plug in the expression for the moment of inertia of the donut given in the problem.
- Use the substitution $R^2\omega^2 = v^2$ to eliminate both R and ω.
- Mass cancels. Solve for the final height.

$$h = \frac{7v_0^2}{8g}$$

- Note that $\frac{v_0^2}{2} + \frac{3v_0^2}{8} = \left(\frac{1}{2} + \frac{3}{8}\right) v_0^2 = \left(\frac{4}{8} + \frac{3}{8}\right) v_0^2 = \frac{7}{8} v_0^2$.
- The maximum height is $h \approx 140$ m. (It's 143 m if you don't round gravity.)
- However, height is **not** the final answer. The problem asks for the distance traveled.
- Draw a right triangle and apply trig to solve for the distance traveled (s) along the incline in terms of the final height (h).

$$\sin 30° = \frac{h}{s}$$

- The distance traveled up the incline is $s = 280$ m (or 285 m if you don't round).

Chapter 35: Conservation of Angular Momentum

203. Apply the law of conservation of angular momentum.

$$I_{d0}\vec{\boldsymbol{\omega}}_{d0} + I_{m0}\vec{\boldsymbol{\omega}}_{m0} = I_d\vec{\boldsymbol{\omega}}_d + I_m\vec{\boldsymbol{\omega}}_m$$

- One object is the merry-go-round (disc) and the other object is the monkey.
- Treat the merry-go-round as a solid disc and the monkey as a pointlike object.
- Look up the formulas for the moment of inertia of a solid disc (it's equivalent to "rolling mode") and a pointlike object (Chapter 32).

$$I_{d0} = I_d = \frac{m_d R_d^2}{2} \quad , \quad I_{m0} = 0 \quad , \quad I_m = m_m R_m^2$$

- $I_{20} = 0$ because $R_{m0} = 0$ (initially the monkey is standing at the center).
- The other distances are $R_{d0} = R_d = R_m = 8.0$ m (one-half the diameter).
- The initial angular speed of the merry-go-round (disc) is $\omega_{d0} = \frac{1}{4}$ rev/s.
- The merry-go-round and monkey have the **same** final angular speed: $\omega_d = \omega_m = \omega$.
- Replace $\vec{\boldsymbol{\omega}}_d$ and $\vec{\boldsymbol{\omega}}_m$ with $\vec{\boldsymbol{\omega}}$ in the conservation of angular momentum equation.
- Solve for the final angular speed.

$$\omega = \frac{\frac{m_d}{2}}{\frac{m_d}{2} + m_m} \omega_{d0}$$

- The final angular speed is $\omega = \frac{1}{6}$ rev/s.

204. Apply the law of conservation of angular momentum.

$$I_0\vec{\boldsymbol{\omega}}_0 = I\vec{\boldsymbol{\omega}}$$

- (There is just one object in the system: the earth.)
- Treat the earth as a solid sphere (Chapter 32): $I_0 = \frac{2}{5} m R_0^2$ and $I = \frac{2}{5} m R^2$.
- Plug these formulas into the conservation of angular momentum equation.
- Earth's mass will cancel out. Solve for the final angular speed.

$$\omega = \left(\frac{R_0}{R}\right)^2 \omega_0$$

- Note that $R = \frac{R_0}{3}$. Therefore $\frac{R_0}{R} = 3$ and $\left(\frac{R_0}{R}\right)^2 = 9$. Solve for ω.
- The final angular speed is 9 times greater than the initial angular speed: $\omega = 9\omega_0$.
- However, the problem **didn't** ask for angular speed.
- The problem asked for **period**. Use the equations $\omega = \frac{2\pi}{T}$ and $\omega_0 = \frac{2\pi}{T_0}$ (Chapter 16).
- Plug these formulas into the previous equation. Solve for the final period.

$$T = \frac{T_0}{9}$$

- What is the initial period of the earth?
- You should know that it normally takes $T_0 = 24$ hr for the earth to complete one revolution about its axis. (There is no need to convert.)
- The final period is $T = \frac{8}{3}$ hr (or 2.67 hr).

---

205. Apply the law of conservation of angular momentum.

$$I_0\vec{\boldsymbol{\omega}}_0 = I\vec{\boldsymbol{\omega}}$$

- (There is just one object in the system: the 200-g pointlike object.)
- Use the formula for the moment of inertia of a pointlike object (Chapter 32):

$$I_0 = mR_0^2 \quad , \quad I = mR^2$$

- Use the formulas $v = R\omega$ and $v_0 = R_0\omega_0$. Substitute all of these formulas into the equation for the conservation of angular momentum.

$$mv_0R_0 = mvR$$

- Mass cancels. Solve for the final speed.

$$v = \frac{R_0}{R}v_0$$

- The final speed is $v = \frac{25}{2}$ m/s (or 12.5 m/s).

WAS THIS BOOK HELPFUL?

A great deal of effort and thought was put into this book, such as:

- Breaking down the solutions to help make physics easier to understand.
- Careful selection of examples and problems for their instructional value.
- Multiple stages of proofreading, editing, and formatting.
- Two physics instructors worked out the solution to every problem to help check all of the final answers.
- Dozens of actual physics students provided valuable feedback.

If you appreciate the effort that went into making this book possible, there is a simple way that you could show it:

<u>Please take a moment to post an honest review.</u>

For example, you can review this book at Amazon.com or BN.com (for Barnes & Noble).

Even a short review can be helpful and will be much appreciated. If you're not sure what to write, following are a few ideas, though it's best to describe what's important to you.

- Were you able to understand the explanations?
- Did you appreciate the list of symbols and units?
- Was it easy to find the equations you needed?
- How much did you learn from reading through the examples?
- Did the hints and intermediate answers section help you solve the problems?
- Would you recommend this book to others? If so, why?

<u>Are you an international student?</u>

If so, please leave a review at Amazon.co.uk (United Kingdom), Amazon.ca (Canada), Amazon.in (India), Amazon.com.au (Australia), or the Amazon website for your country.

The physics curriculum in the United States is somewhat different from the physics curriculum in other countries. International students who are considering this book may like to know how well this book may fit their needs.

THE SOLUTIONS MANUAL

The solution to every problem in this workbook can be found in the following book:

100 Instructive Calculus-based Physics Examples

Fully Solved Problems with Explanations

Volume 1: The Laws of Motion

Chris McMullen, Ph.D.

ISBN: 978-1-941691-17-5

If you would prefer to see every problem worked out completely, along with explanations, you can find such solutions in the book shown below. (The workbook you are currently reading has hints, intermediate answers, and explanations. The book described above contains full step-by-step solutions.)

VOLUMES 2 AND 3

If you want to learn more physics, volumes 2 and 3 cover additional topics.

Volume 2: Electricity & Magnetism

- Coulomb's law
- Electric field and potential
- Electrostatic equilibrium
- Gauss's law
- Circuits
- Kirchhoff's rules
- Magnetic field
- The law of Biot-Savart
- Ampère's Law
- Right-hand rules
- Magnetic flux
- Faraday's law and Lenz's law
- and more

Volume 3: Waves, Fluids, Sound, Heat, and Light

- Sine waves
- Oscillating spring or pendulum
- Sound waves
- The Doppler effect
- Standing waves
- The decibel system
- Archimedes' principle
- Heat and temperature
- Thermal expansion
- Ideal gases
- Reflection and refraction
- Thin lenses
- Spherical mirrors
- Diffraction and interference
- and more

ABOUT THE AUTHOR

Chris McMullen is a physics instructor at Northwestern State University of Louisiana and also an author of academic books. Whether in the classroom or as a writer, Dr. McMullen loves sharing knowledge and the art of motivating and engaging students.

He earned his Ph.D. in phenomenological high-energy physics (particle physics) from Oklahoma State University in 2002. Originally from California, Dr. McMullen earned his Master's degree from California State University, Northridge, where his thesis was in the field of electron spin resonance.

As a physics teacher, Dr. McMullen observed that many students lack fluency in fundamental math skills. In an effort to help students of all ages and levels master basic math skills, he published a series of math workbooks on arithmetic, fractions, algebra, and trigonometry called the Improve Your Math Fluency Series. Dr. McMullen has also published a variety of science books, including introductions to basic astronomy and chemistry concepts in addition to physics textbooks.

Dr. McMullen is very passionate about teaching. Many students and observers have been impressed with the transformation that occurs when he walks into the classroom, and the interactive engaged discussions that he leads during class time. Dr. McMullen is well-known for drawing monkeys and using them in his physics examples and problems, applying his creativity to inspire students. A stressed-out student is likely to be told to throw some bananas at monkeys, smile, and think happy physics thoughts.

Author, Chris McMullen, Ph.D.

PHYSICS

The learning continues at Dr. McMullen's physics blog:

www.monkeyphysicsblog.wordpress.com

More physics books written by Chris McMullen, Ph.D.:

- An Introduction to Basic Astronomy Concepts (with Space Photos)
- The Observational Astronomy Skywatcher Notebook
- An Advanced Introduction to Calculus-based Physics
- Essential Calculus-based Physics Study Guide Workbook
- Essential Trig-based Physics Study Guide Workbook
- 100 Instructive Calculus-based Physics Examples
- 100 Instructive Trig-based Physics Examples
- Creative Physics Problems
- A Guide to Thermal Physics
- A Research Oriented Laboratory Manual for First-year Physics

SCIENCE

Dr. McMullen has published a variety of **science** books, including:

- Basic astronomy concepts
- Basic chemistry concepts
- Balancing chemical reactions
- Creative physics problems
- Calculus-based physics textbook
- Calculus-based physics workbooks
- Trig-based physics workbooks

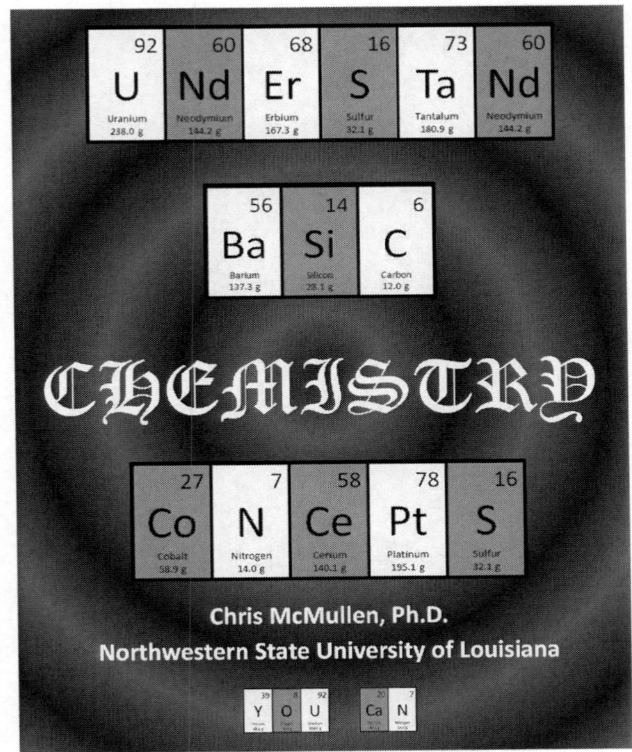

MATH

This series of math workbooks is geared toward practicing essential math skills:

- Algebra and trigonometry
- Fractions, decimals, and percents
- Long division
- Multiplication and division
- Addition and subtraction

www.improveyourmathfluency.com